Phylogenies and the Comparative Method in Animal Behavior

Phylogenies and the Comparative Method in Animal Behavior

Edited by Emília P. Martins

New York Oxford
OXFORD UNIVERSITY PRESS
1996

Oxford University Press

Oxford New York
Athens Auckland Bangkok Bombay
Calcutta Cape Town Dar es Salaam Delhi
Florence Hong Kong Istanbul Karachi
Kuala Lumpur Madras Madrid Melbourne
Mexico City Nairobi Paris Singapore
Taipei Tokyo Toronto

and associated companies in
Berlin Ibadan

Published by Oxford University Press, Inc.,
198 Madison Avenue, New York, New York 10016

Library of Congress Cataloging-in-Publication Data
Phylogenies and the comparative method
in animal behavior / edited by Emília P. Martins.
p. cm. Based on papers from a symposium organized
for the Animal Behavior Society in Seattle, July 1994.
Includes bibliographical references and index.
ISBN 0-19-509210-4
1. Animal behavior—Congresses.
2. Phylogeny—Congresses.
3. Psychology, Comparative—Congresses.
I. Martins, Emília P.
QL785.P48 1996 591.51—dc20 95-47942

9 8 7 6 5 4 3 2 1

Printed in the United States of America
on acid-free paper

Preface

The main purpose of this volume is to discuss the past, present and future impact of phylogenetic comparative methods on the field of animal behavior. The comparative method has been an integral component of the study of animal behavior since its inception. As behavior patterns only rarely leave traces in fossils, behavioral biologists, beginning with the founders of classical ethology, have compared the behavior of extant species to infer the processes and patterns of behavioral evolution. The field of comparative psychology has also relied heavily on comparative studies in the search for underlying similarities or "universals" in behavior across species with the belief that information about the behavior of one species can be used to infer the behavior of others. In more recent fields of behavioral ecology, neuroethology, and sociobiology, the comparative approach remains an important technique for understanding animal behavior.

Phylogenetic comparative studies can be used to infer the ancestral states of behavior patterns and to suggest the pattern of evolutionary changes that the behavior has undergone. They can be used to develop hypotheses about the processes of evolutionary change underlying behavioral characters, including the magnitude and direction of forces such as mutation, selection and random genetic drift. Phylogenetic comparisons of species phenotypes can also be used to test for links between the evolution of two or more characters that might suggest coevolutionary processes or other types of evolutionary constraints. Finally, phylogenetic studies can be used to correct the statistical problems of interspecific analyses (e.g., the nonindependence of species data) while answering questions that are not explicitly evolutionary in nature. Much of the modern literature in comparative studies discusses different ways of incorporating phylogenetic information into comparative analyses.

Although the phylogenetic comparative method raises issues that are of concern to scientists in all fields of biology, they are of particular interest to the study of animal behavior because of the traditional reliance of behavioral researchers on comparative studies. However, relatively few behavioral biologists have begun to consider the importance of phylogenies in their own research. The main purpose of this volume is to promote discussion on the impact of modern phylogenetic comparative methods on the study of animal behavior. In particular, we would like to I) synthesize and review some of the main issues that need to be considered in the analysis of interspecific behavioral data, II) consider some new advances in comparative method research and discuss how they might be particularly useful in the study of animal behavior, and III) review the main questions of interest in a few sub-disciplines of animal behavior in which the comparative method has been particularly important, discussing the advantages and disadvantages of this approach and suggesting new directions for future research in these areas.

This book originated in a symposium organized for the Animal Behavior Society in July 1994 (Seattle). We have added some chapters, and eliminated others. Funds for the symposium were provided by the Animal Behavior Society, the program on Comparative Studies in Brain and Behavior of the National Institutes of Mental Health (R13 MH52515), and the Animal Behavior program of the National Science Foundation (IBN-9320443). We would like to thank L. Drickamer, J. Ha, J. Felsenstein, I. Lederhendler, M. Lynch, F. Stollnitz, 33 reviewers, and the audience of our ABS symposium for constructive criticism, positive support, and generous gifts of time. Thanks are also due to each of the book authors and symposium speakers, and to L. Trimble.

Emília P. Martins
Eugene, Oregon
August 1995

Contents

Contributors

C. Gregory Anderson, Department of Ecology and Evolutionary Biology, University of Tennessee, Knoxville, Tennessee 37996–1610, USA

Sydney A. Cameron, Department of Biological Sciences, University of Arkansas, Fayetteville, Arkansas 72701, USA

Leslie K. W. Chan, Department of Anthropology, Scarborough College, University of Toronto, Scarborough, Ontario, M1C 1A4, Canada; chan@macpost.scar.utoronto.ca

Bernard J. Crespi, Department of Biosciences, Simon Fraser University, Burnaby B.C., V5A 1S6 Canada; crespi@sfu.ca

Alan de Queiroz, Department of Ecology and Evolutionary Biology, University of Arizona, Tucson, Arizona 85721, USA. *Current address*: University Museum and Department of Environmental, Population, and Organismic Biology, Campus Box 334, University of Colorado, Boulder, Colorado 80309-0334 USA.

Susan A. Foster, Department of Biological Sciences, University of Arkansas, Fayetteville, Arkansas 72701, USA. *Current address*:: Department of Biology, Clark University, 950 Main Street, Worcester, MA 01610-1477, USA; sfoster@vax.clarku.edu

John L. Gittleman, Department of Zoology, University of Tennessee, Knoxville, Tennessee 37996–1610, USA; gittlema@utkvx.utk.edu

Alan Grafen, Department of Plant Sciences, University of Oxford, South Parks Road, Oxford OX1 3PS UK.

Thomas F. Hansen, Division of Zoology, Department of Biology, University of Oslo, P. O. Box 1050, Blindern, N-0316, Oslo 3, Norway; thomas.hansen@bio.uio.no

Paul H. Harvey, Department of Zoology, University of Oxford, South Parks Road, Oxford OX1 3PS UK; harvey@vax.ox.ac.uk

Rebecca E. Irwin, Department of Biological Sciences, University of Tennessee at Martin, Martin, TN 38238, USA; rirwin@utm.edu

Mark Kot, Department of Applied Mathematics, University of Washington, Seattle WA 98195, USA. *Current address*: Department of Mathematics, University of Tennessee, Knoxville, Tennessee, USA; kot@buteo.math.utk.edu.

GEORGE V. LAUDER, School of Biological Sciences, University of California, Irvine, California 92717, USA; glauder@uci.edu

HANG-KWANG LUH, Departments of Mathematics and Ecology and Evolutionary Biology, University of Tennessee, Knoxville, Tennessee 37996–1610, USA

EMÍLIA P. MARTINS, Department of Biology, University of Oregon, Eugene, Oregon 97403, USA; emartins@work.uoregon.edu

SEAN NEE, Department of Zoology, University of Oxford, South Parks Road, Oxford OX1 3PS UK; snee@vax.ox.ac.uk

SUE TAYLOR PARKER, Anthropology Department, Sonoma State University, Rohnert Park, California, 94928, USA; parker@sonoma.edu

DANIEL E. L. PROMISLOW, Department of Biology, Queen's University, Kingston, Ontario, K7L 3N6 Canada. *Current address*: Department of Genetics, University of Georgia, Athens, Georgia 30602-7223, USA; promislo@uga.ucc.uga.edu

ANDREW F. READ, Institute of Cell, Animal and Population Biology, University of Edinburgh, West Mains Road, Edinburgh EH9 3JT, UK; aread@festival.ed.ac.uk

STEPHEN M. REILLY, Department of Biological Sciences, Ohio University, Athens, Ohio 45701, USA; reilly@mail.oucom.ohiou.edu

MARK RIDLEY, Departments of Anthropology and Biology, Emory University, Atlanta, Georgia, 30322, USA. *Current address*: Department of Zoology, University of Oxford, South Parks Road, Oxford OX1 3PS UK.

MICHAEL J. RYAN, Department of Zoology, University of Texas, Austin, Texas 78712, USA; mryan@mail.utexas.edu

PETER H. WIMBERGER, Department of Biology; University of Puget Sound; Tacoma, Washington 98416, USA; pwimberger@ups.edu

Phylogenies and the Comparative Method in Animal Behavior

CHAPTER 1

Phylogenetics in Behavior: Some Cautions and Expectations

Michael J. Ryan

Phylogenetics has had an interesting history in studies of animal behavior (Burghardt & Gittleman 1990; Brooks & McLennan 1991). For some of the early ethologists with backgrounds in comparative anatomy, species comparisons were the essence of their science. Coincident with the formulation of kin selection theory (Hamilton 1964), however, emphases in animal behavior shifted toward issues in population genetics and population biology, and sociobiology seemed quickly to forget its ethological (both its historical and mechanistic) roots. But another science was being revolutionized at the time (Hull 1988). Hennig's (1966) cladistic methodologies were being argued among systematists with a vigor and style that were sometimes more characteristic of theological debates (Hull 1988). Once the acrimony subsided and the smoke cleared, some scientists outside of systematics became convinced that this field offered tools critical for understanding the evolution of behavior — once again behaviorists realized that evolution is something that happened in the past and that we must resort to techniques used in historical analysis to study it fully.

This volume shows that there are now many behaviorists enthusiastic about utilizing historical approaches to address a variety of behavioral problems. Although far from universal — e.g., Reeve and Sherman (1993) refer to phylogenetics as one of the "recent challenges to adaptationism" — there appears to be some consensus that historical investigations are a necessary supplement to the field of animal behavior.

I was asked to discuss briefly some of my expectations for the future roles of phylogenetics in animal behavior. Before doing so, however, I would like to raise some cautionary notes about what we are doing now. There are some issues that need resolution, and until they are resolved they at least need our attention. After reviewing these issues, I then offer suggestions as to some more imaginative approaches in which phylogenetics can be used to study behavior.

Some cautions

A. Phylogenies

> "We will first look for a taxonomy in the literature. Then, when we have found one, we will assume it is true."
>
> (Ridley 1983, p 19)

It is obvious to at least some that a phylogenetic interpretation of behavior is only as good as the phylogeny used as the foundation for that interpretation. Thus behaviorists need to consider seriously the field of phylogenetics and the phylogenetic problems they wish to analyze. Ridley (1983) gives a rather glib methodology for discriminating among competing phylogenies. Alternatively, others behaviorists have approached this issue in a more serious manner (e.g., Lanyon 1992). But some of animal behavior is in its current state because there has been scant attention paid to historical pattern, especially by those who attest to be more interested in current processes.

Felsenstein (1985) and Harvey (e.g., Harvey & Pagel 1991) performed an important service for animal behavior in emphasizing the problem of evolutionary independence in statistical tests of adaptation. The comparative approach typically involves testing predictions of correlations of variables, these predictions being derived from hypotheses of how natural selection causes the evolution of adaptation. These predictions are tested by contrasting the variables of interest among different taxa. Since the hypotheses predict evolutionary origin rather than maintenance (but see below), one cannot necessarily use each taxon as an independent datum, since taxa might be similar in the variables of interest due to shared ancestry rather than independent evolution. It appears to be generally accepted that a comparative approach must have a phylogenetic component to help mitigate this problem of independence. This puts behaviorists in a bind, however, if one wants to utilize the comparative approach to test hypotheses of adaptation but there is no phylogeny available for the group under study. What is the solution? Unfortunately, to employ the comparative approach, to compare species statistically, one must know whether the traits exhibited by these species are statistically independent. No phylogeny, no test.

When there is a phylogeny available, it should never be written in stone. A frustration of behaviorists, myself included, is that phylogenies are constantly being reevaluated, whatever the technique. A case in point is the studies that my colleagues and I have been conducting on the evolution of mate recognition in the túngara frog (*Physalaemus pustulosus*) and its close relatives. This study incorporates aspects of calling behavior, auditory neurophysiology, female phonotaxis, vocal morphology and, of course, phylogeny. The original description of the *Physalaemus pustulosus* species group was provided by Cannatella and Duellman (1984; Fig. 1*a*). This hypothesis suggested two clades within the species group, those on the western side of the Andes (*P. coloradorum*, and *P. pustulatus*) constituting one clade, and the Amazonian species (*P. petersi*) and the primarily Central American species (*P. pustulosus*) constituting the other clade. This hypothesis was derived from consideration of several morphological characters. This hypothesis also led us to interpret certain neurophysiological data as supporting the notion of sensory exploitation (Ryan 1990) — that is, that males evolve traits that exploit preexisting female preferences (Ryan et al. 1990). *Physalaemus pustulosus* and *P. coloradorum* both share the neural tuning that we suggested is responsible for female *P. pustulosus* preferring calls that have been enhanced by lower-frequency chucks. Since the chucks appeared after the divergence of the clades containing *P. pustulosus* and *P. coloradorum*, this argues for the shared neural tuning preceding the evolution of the chuck, thus supporting the hypothesis of sensory exploitation.

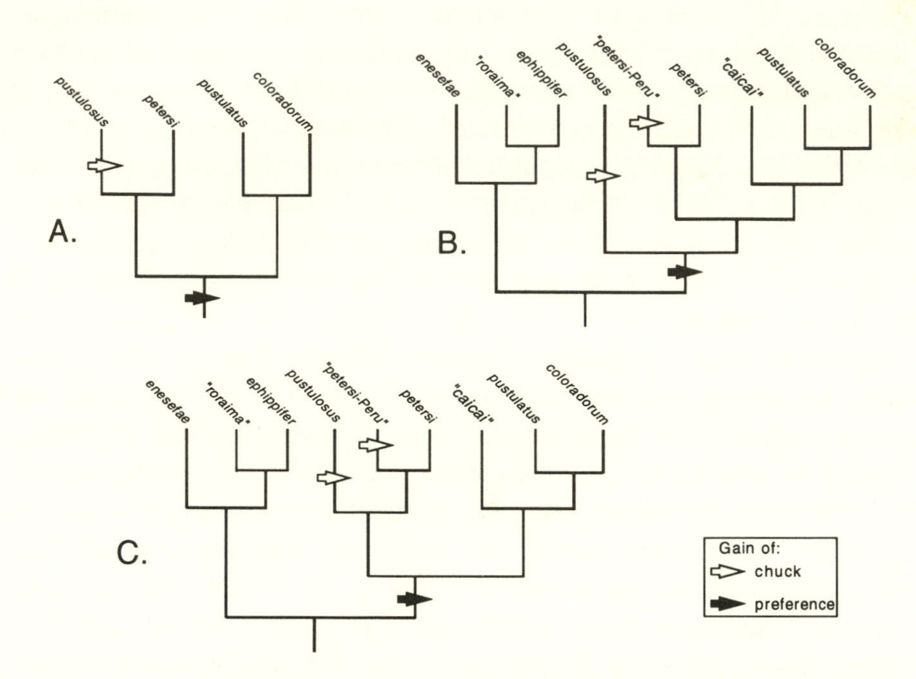

Figure 1. Various phylogenetic hypotheses and interpretations of trait-preference evolution in the *Physalaemus pustulosus* group, based on two observations: (1) Only *P. pustulosus* adds a chuck to its call; the chuck is also absent from the outgroup species and, to the best of our knowledge, in the rest of the approximately 30 species in the genus (but see *P. "petersi-Peru"*, below). (2) Females of both *P. pustulosus* and *P. coloradorum* prefer calls with chucks. (A) Phylogeny proposed by Cannatella and Duellman (1984), and the interpretation of when the chuck and the (most recent possible) preference for the chuck evolved. This phylogeny suggests two monophyletic groups within the species group. (B) Preliminary molecular analysis suggested a different tree and the discovery of a new species, *P. "petersi-Peru"* (quotes indicate undescribed species), with a suffix to its call. It is not known if this suffix is homologous to the chuck. Despite the change in phylogeny, however, parsimony still suggests that chuck preference evolved prior to the chuck (Ryan & Rand 1993). Outgroup species are also analyzed. (C) The most recent phylogenetic hypothesis is in agreement with the original hypothesis of Cannatella and Duellman (1984; A) in suggesting two monophyletic groups within the species group. It is equally parsimonious to suggest that the chuck was gained in the common ancestor of *P. pustulosus* and *P. "petersi Peru"* and lost in *P. petersi*. Nevertheless, parsimony still suggests preference for chucks evolved prior to the chuck.

Realizing that our interpretation depended on our phylogenetic hypothesis, we collaborated with several systematists (D. Cannatella, D. Hillis, and P. Chippendale) in deriving additional characters for testing the available phylogenetic hypothesis. Our preliminary analysis, which assayed 27 polymorphic allozymes and a portion (400 base pairs) of the mitochondrial 12S gene, resulted in a different tree with some additional species (Ryan & Rand 1993). This tree was pectinate with *P. pustulosus* as sister to all other taxa in the species group (Fig. 1*b*). These data led Pomiankowski (1994) to suggest that the revised phylogeny, although not rejecting the hypothesis of sensory exploitation, made such an interpretation less compelling. Upon completion of the DNA sequencing, which included 1200 bases of the 12S gene and flanking regions, however, our "final" phylogeny coincides with that originally proposed by Duellman and Cannatella, with strong statistical support from bootstrap estimates (Fig. 1*c*). But this phylogeny will be reevaluated once more when we complete our analysis of the relationships of all the approximately 40 species within the genus. If additional outgroups and data sets continue to suggest a phylogeny that supports our sensory exploitation hypothesis, we gain more confidence in our interpretation, if not, the hypothesis will need to be rejected. At no point should we be tempted by the arrogance of certainty that characterized the early debates in cladistics.

When new techniques are introduced into a field, they often assume an air of primacy. There is no question that the comparative approach, if it includes rigorous phylogenetic techniques, is a powerful method for testing hypotheses of adaptation. But it is not the only approach (Reeve & Sherman 1993). Some questions might not allow a comparative test. Consider the mammalian four-chambered heart. Few would argue against the proposition that the presence of this structure in the more than 4,000 species of mammals is due to common descent rather than 4,000 independent evolutionary events. Even if one were to have a completely resolved phylogeny of mammals, this would not aid in testing hypotheses for the adaptive significance of this structure, since there is only one evolutionary event. But all is not lost. Physiological considerations make it clear that the four-chambered heart offers advantages for circulation and correlated physiological and behavioral processes. From first principles of physiology, one could argue that the evolution of this structure would be favored by selection. Of course, by this argument alone we could never be totally confident this assertion was correct, but the same can be said of any hypothesis supported through a comparative study. It is unfortunate, but the passage of time results in serious constraints for those interested in historical sciences. The comparative method should be added to the arsenal of techniques used to study animal behavior, not substitute for them.

A related issue concerns the use of maintenance as evidence for adaptations (Reeve & Sherman 1993). Knowledge that selection is maintaining a trait in a population does indicate that the trait is adaptive, if not an adaptation (e.g., Greene 1986). This assertion, however, cannot be made based on the mere presence of the trait. It also needs to be remembered that demonstrating maintenance is not equivalent to demonstrating the role of selection in the evolution of the adaptation.

B. Reliance on parsimony

In both phylogenetic reconstruction and in optimizing character states on reconstructed phylogenies, one makes assumptions about patterns that are most likely to occur. For many, parsimony is used as a guiding principle, for others, maximum likelihood, etc. But for all, there should be sole reliance on these methods only in the absence of other information. We cannot rely on an uncritical acceptance of parsimony, maximum likelihood, or any other single method to give us the truth or even a good estimate of it. I offer some of my own work as an example.

We had suggested that the chuck of *P. pustulosus* evolved after that lineage diverged from the lineage containing *P. coloradorum*. This conclusion was based on optimizing the character on the phylogeny using the principle of parsimony. Gardner (1990) criticized this interpretation and suggested that it was more likely that the chuck evolved at the base of the species group and was subsequently lost an additional two times. Although there is no argument that a gain and two subsequent losses is less parsimonious than a single gain, it still might be true. For example, we know that when males add chucks to their calls, they not only increase their attractiveness to females but also are more at risk from predation by the frog-eating bat (Ryan 1985). If *P. pustulosus* were found in areas with much less predation pressure than that of the closely related species, these data should justify adopting the less parsimonious hypothesis, or at least entertaining it seriously. That happens not to be the case. In this example, *P. coloradorum* is not sympatric with the frog-eating bat, and patterns of vocal morphology (Ryan & Drewes 1990) are also consistent with our hypothesis of chuck evolution.

Carl Sagan, when discussing reports of alien beings piloting flying saucers, said that extraordinary claims require extraordinary evidence. Homoplasy (evolutionary convergence and reversal) is not an extraordinary claim. Almost all phylogenetic analyses show evidence of it. Parsimony and other assumptions are helpful guiding principles; they should not be accepted uncritically, however, but should be combined with our knowledge of the general biology of the animals we study.

A related issue regards the model of evolution we adopt. Do we have different expectations about rates of evolution if we are investigating foraging behavior on the one hand versus mate recognition signals on the other? We might expect the mode of evolution to be different in these cases, and we also know that phylogenetic reconstruction techniques are more or less sensitive to unequal rates of evolution (e.g., Hillis et al. 1994). Should that change how we interpret historical patterns? Another assumption we make is how new taxa arise. Some models of speciation might suggest divergence of two daughter species from an ancestral one. But in centripetal evolution, isolate populations achieve species status. Can we use the same methods to estimate ancestral characters in both cases?

C. Total evidence

There is a controversy in systematics in how to evaluate different data sets (e.g., morphology, behavior, DNA sequences) in phylogenetic reconstruction (Bull et al. 1993). A related issue involves using phylogenetic techniques to explore behavioral evolution; should behavioral characters under study contribute to the phylogenetic reconstruction that is then used to evaluate the patterns of behavioral evolution?

Systematists and behaviorists can have quite different goals in phylogenetic reconstruction. Systematists strive to present the most likely hypothesis of evolutionary relationships. A behaviorist, alternatively, wants to interpret the patterns by which behavior evolves, but the success of this task is dependent on the accuracy of the phylogeny. Thus there is validity to the proposition that the behaviorist should use all the available information to reconstruct the phylogeny. But one cannot ignore the fact that inclusion of behavioral data will influence our interpretations of how that behavior evolves. de Queiroz and Wimberger (1993) demonstrated that, for the studies they evaluated, behavioral characters were as phylogenetically informative as other characters. In our studies of call evolution in *P. pustulosus*, however, morphology, allozymes and DNA sequence suggest the same major branching patterns in their independent estimates of phylogeny, but a separate analysis of call variation results in a quite different tree. In our study we have decided to use the phylogeny reconstructed without call data to interpret patterns of call evolution. We realize this approach will have its detractors and that there are strong arguments for the alternative approach. This issue needs serious debate.

D. Homology

This has always been a slippery concept, and especially so for behavior. Atz's (1970) discussion of the difficulty in homologizing behavior might have been responsible for many researchers temporarily abandoning behavior as a useful phylogenetic character. Although determining behavioral homologies is still sometimes seen as a difficult task, it is not longer viewed as offering a serious deterrent (Wenzel 1992; Greene 1994), but perhaps it should.

The most widespread use of phylogenetics in behavior is to recover independent evolutionary events. Quite often social structure is one of the "behaviors" being analyzed (see examples in Harvey & Pagel 1991). But to what degree is social structure, such as mating systems, a single behavior, a multivariate or metabehavior, or an emergent property of behavioral interactions? Consider different forces that can influence lek evolution: hotshots, hotspots, exploding leks (Emlen & Oring 1977; Wiley 1991). If these hypotheses for lek evolution are not mutually exclusive but are more or less important in different taxa, we are probably not talking about homologous "behaviors" and thus need to exercise caution when using phylogenetics to test for evolutionary effects of leks, as an example (Höglund 1989; Oakes 1992). Homology is an old problem but that does not mean it has gone away.

Expectations

I would agree with many if not most behaviorists, and certainly with the contributors to this volume, that phylogenetics has made important contributions to our understanding of animal behavior. I do have some concerns, as just mentioned. But now what to do? Certainly, we should keep doing things that we are doing right and when possible improve on them. But we might also want to make some leaps, take some chances, and try some more imaginative but not necessarily less rigorous approaches that might strain the limits of using phylogenetics in behavior. When we exceed the limit, it should be clear — if not to us, then to our critics.

A. Reevaluate standard examples of behavioral evolution

Many of the earlier examples of patterns of behavioral evolution predate the cladistics revolution and rely on intuitions of how behavior patterns are most likely to evolve. Some of these examples have become embedded in our instructional heritage, being repeated to several generations of undergraduates in animal behavior courses. It would be instructive to know how many of these examples hold up to a phylogenetic analysis.

A classic example of a behavioral transition series deals with the bizarre courtship behavior of a group of empidids known as balloon flies. The initial observations reported on groups of males gathered in leklike mating arenas holding large, empty balloons of silk, and females choosing males as if they were judging male quality by the size of their balloons. This phenomenon begged for an explanation, and Kessel (1955) offered a very appealing one. He reported the behavior of a number of species of empidid flies that he classified into eight behavioral groups, which he arranged into a phylogenetic series (cited in Alcock 1975): (a) some species hunt small insects; (b) others use their insect prey as nuptial gifts; (c) in others, males with nuptial gifts gather in leks; (d) nuptial gifts are wrapped in a single strand of silk; (e) nuptial gifts are totally wrapped with silk; (f) nuptial gifts are wrapped in silk but the males suck the juices out of the prey prior to presenting it to the female; (g) some species feed only on nectar but use a dead insect around which the balloon if formed; and finally, (h) some species present females with a large, empty balloon.

It is interesting to know why balloon flies act as they do, and surely Kessel's hypothesis of an evolutionary elaboration of nuptial gifts deserves serious consideration. But to my knowledge, no one has examined the phylogeny of empidid flies with the goal of testing Kessel's hypothesis. That might not be surprising given that there are well over 1,000 species of empidid flies and their phylogenetic relationships are poorly known. But the little work that has been done in this area suggests that balloons have evolved several times independently (Powell 1964, Marden 1989, Cumming 1994). Until we have some better evidence to support Kessel's hypothesis of behavioral transitions, we probably want to be a bit more cautious in using this example in undergraduate courses. (Alcock seems to have done just that. This example was given in the first four editions of Alcock's widely used text *Animal Behavior* [1975–1989] but is omitted from the fifth edition [1993]).

B. Coevolution

In ecology, phylogenetic patterns have become quite useful for testing hypotheses of coevolution resulting from such effects as plant-insect interactions (Maddison & Maddison 1992). A related but slightly different approach in behavior is asking how the existence of one trait biases the evolution of a second trait. Höglund (1989) and Oakes (1992) used this approach in testing the hypothesis that lek breeding in birds promotes the evolution of male sexual dimorphism.

This approach has also proven useful in sexual selection studies. Although there has been long controversy about the efficacy of runaway sexual selection versus good genes in explaining the evolution of female mating preferences, both models predict that the preference evolves as a correlated response to selection on the male trait with which it is genetically correlated (Kirkpatrick & Ryan 1991). Ryan (1990) suggested that hypotheses based on genetic correlations could be tested by reconstructing historical patterns of trait and preference evolution (cf. Fig 1). Several studies have now shown that female preferences and male traits do not coevolve but instead that preferences exist prior to male traits; thus males evolve traits to exploit preexisting preferences (Basolo 1990; Ryan et al. 1990; Proctor 1993; Ryan & Rand 1993; Basolo 1995). Although previous researchers had predicted preexisting preferences (e.g., West Eberhard 1983), the phylogenetic approach proposed by Ryan (1990) was the first explicit test of this hypothesis.

C. Behavioral transitions.

A major issue for evolutionary biology is explaining the evolution of complex adaptations. What were the transitions and how were they favored by selection? The ritualization of behavior was an important contribution of the early ethologists (e.g., Hinde and Tinbergen 1958) that would be worth reexamining with more recently available techniques. Irwin (chap. 8, this volume) suggests the importance of such an approach in her study of bird song.

A more general consideration of behavioral transitions addresses the evolution of genetic and phenotypic correlations. Many behavior patterns are multivariate (see Arnold [1994b] for an excellent review of multivariate evolution) in that they depend upon the coordination of numerous motor patterns, certain relationships between morphology and behavior, sensory feedback, and the physiological and social context in which the behavior patterns are exhibited.

Arnold (1994a) discusses multivariate characters, including behavior, in the context of evolutionary constraints. He takes a rather broad view of constraint by defining it as any factor that biases the distribution of phenotypes exhibited by a taxa. Thus selection joins the ranks of constraints along with developmental and functional influences. He suggests that the degree to which traits are constrained can be estimated by the phenotypic and genetic correlations among them. In this paradigm, behavioral innovations can occur by breaking these constraints. If we could reconstruct phenotypic or genetic correlations at ancestral nodes, we could then determine when these constraints are broken by determining when there are changes in the character correlation matrices. Furthermore, with sufficient understanding of the animal's biology, we can gain some insights into the mechanisms involved in this reorganization.

D. Reconstructing ancestral behavior

In the novel and film *Jurassic Park*, the geologist observes dinosaur behavior that corroborates hypotheses about homeothermy and complex parental care. Although it is clearly fictional, is there anyone interested in the evolutionary history of behavior who did not feel an acute pang of jealously? Most researchers who study history would like to have more direct access to the phenomena they study.

A. S. Rand and I have tried to partially circumvent this problem imposed by the passage of time in our studies of communication evolution. For members of the *P. pustulosus* species group and their close relatives, we have estimated the calls at ancestral nodes (Fig. 2),

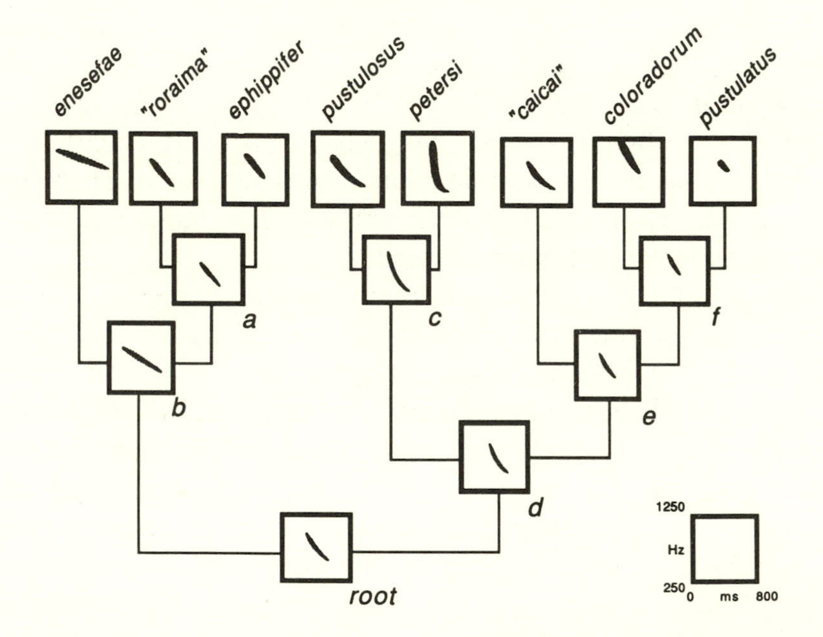

Figure 2. Sonograms of the *Physalaemus pustulosus* species group and three close relatives. Also depicted are sonograms of calls reconstructed at the ancestral nodes (Ryan & Rand, 1995). Not all species indicated in Fig. 1 are used in this analysis.

synthesized these calls, and determined the degree to which these calls elicit phonotaxis from female *P. pustulosus*. We have used both local squared-change parsimony (Felsenstein 1985) and squared-change parsimony (Huey & Bennett 1987) to estimate the nodal values for each of eight individual call characters that we use to synthesize calls (see Maddison [1991] for a comparison of these techniques, and Martins and Hansen in chap. 2 of this volume as to the appropriateness of each for estimating ancestral behavior). The estimates from these techniques usually were quite similar (within 5% of one another), and future studies will determine if these differences are relevant to female call preferences.

Our results allow some insights into how call recognition evolves in this group of frogs. There is not much evidence of finely tuned coevolution. Females do not discriminate between the call of their most recent ancestor and their conspecifics, even though these calls are significantly different from one another. When females are given heterospecific calls in the absence of conspecific calls, those of both extant and ancestral species elicit phonotaxis, i.e., they are (perhaps mistakenly) recognized as signaling an appropriate mate (Fig. 3). Alternatively, the degree to which a female discriminated between a conspecific and heterospecific call is predicted by phylogenetic distance; the more distant the nodal call, and thus the longer since the signal and receiver are thought to have diverged, the weaker the female's response. Thus, in this group, signals and receivers diverge in concert over evolutionary time, but that divergence is not tightly correlated and calls of both extant and hypothesized ancestral species contain key stimuli for eliciting female responses.

It could be argued that the reconstructed ancestral nodes in this study are not likely to be accurate estimates of species past. If that were true, this approach would still offer valuable insights into evolutionary interaction of signals and receivers. We now realize that in a variety of animals, a female's preference function is broader than the distribution of conspecific–male sexually–selected traits (Ryan 1994). One approach to defining this preference function is to determine how females respond to traits absent in their own but present in other (extant) species. Thus we find that *P. coloradorum* females prefer their own calls with chucks of *P. pustulosus*; *P. pustulosus* females prefer their own calls, to which

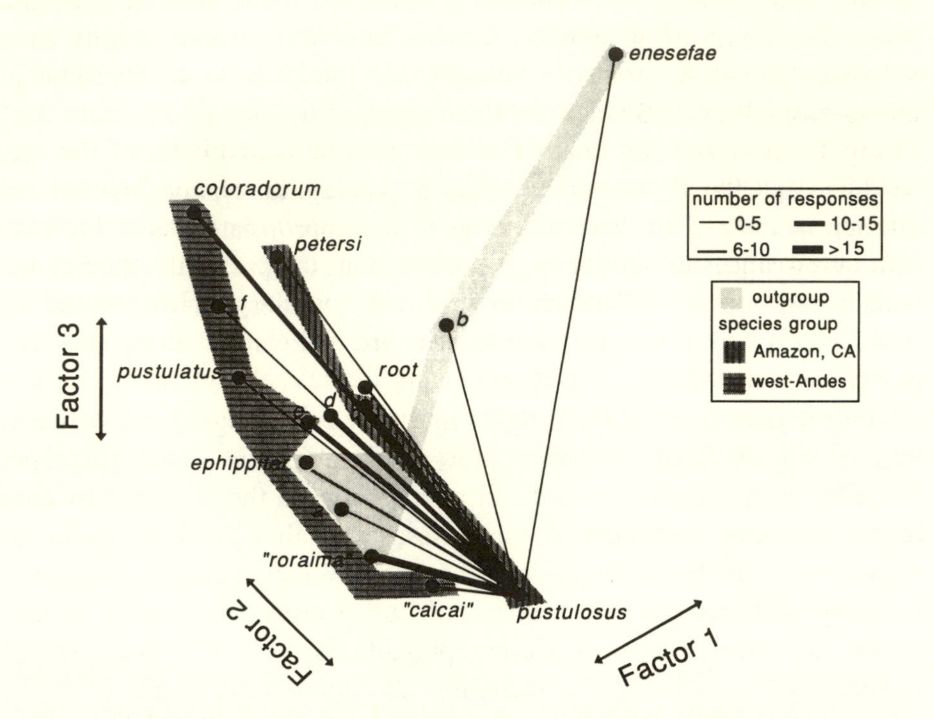

Figure 3. A plot of the first three factors of a principle component analysis of the calls shown in Fig. 2. These factors explain 87% of the variation among calls. Also depicted is the number of responses female *Physalaemus pustulosus* exhibited to each call when it was paired with a white noise stimulus (total N = 20), and the phylogenetic groupings of the taxa and nodes being tested.

the prefix of the *P. pustulatus* call has been added (Ryan & Rand 1993), and in other taxa, female platyfish prefer males with swords over their own unsworded males (Basolo 1990), female birds prefer complex repertoires although their males produce only simple song types (Searcy 1992), and female fiddler crabs prefer males that construct pillars even though this trait is only present in closely related species (Christy & Salmon 1991).

Another axis of variation along which the preference function can be explored is historical. What were the pathways along which complex signals were likely to diverge and how does the female preference interact with such signal variation? Thus, even if the calls at ancestral nodes are not good estimates of what ancestral species might have sounded like, these estimates still provide guidance as to the type of stimuli that can be used to probe the breadth of the female's preferences. Figure 3 shows that the axis of variation along which calls of the two clades within the *P. pustulosus* species group evolved are disjunct but parallel, and along this axis responses of *P. pustulosus* females decrease with decreasing call similarity. It appears that the outgroup species are evolving along an orthogonal axis of call evolution. How would *P. pustulosus* females respond to more fine-grained change along this axis as well as other axes of variation? Using digital call synthesis, we can determine female responses to an almost infinite array of acoustic stimuli. Phylogenetic reconstruction can guide us to those forms of stimulus variation that might be most informative in understanding female preference evolution.

E. Studying history as it happens

In systematics, there has been considerable debate as to the accuracy with which various techniques reconstruct phylogenetic history. Most investigations into this problem utilize numerical simulations; Hillis et al. (1992) took a more direct approach. They used a bacteriophage, T7, that was propagated in the presence of a mutagen to construct what might be the first completely known phylogeny. Restriction-site maps of the terminal taxa were used in phylogenetic reconstruction, and these reconstructions were compared to the true phylogeny. Encouragingly, most of the commonly used phylogenetic techniques were good predictors, and parsimony also accurately predicted ancestral restriction maps.

The ability to create a known phylogeny clearly offers important potential for studies of phylogenetics and microbial evolution. But can it be applied to any central problems in animal behavior? Perhaps. One of the galvanizing issues in sociobiology is the evolution of cooperation (Wilson 1975; Dawkins 1976; Maynard Smith 1982). Bull and Molineux (1992) addressed this fundamental issue using another phage, f1, in their study of "molecular sociobiology. " (Some might argue that viral benevolence is hardly what most of us mean by behavior;

nevertheless, the example should be instructive.) This phage and a bacterial host were serially propagated so that the phage could increase its numbers only by increasing the growth rate of the host; infection of new hosts were not permitted, thus there was selection for benevolent or cooperating phages. Bull and Molineux then described the genetic changes that occurred under selection for cooperation. From this research, it seems that microbial selection studies can be used to address some important issues in the evolution of adaptation, and if combined with the generation of phylogenetic diversification, as described above, could provide a system for identifying the genetic changes that accompany the evolution of adaptations. Whether there are any systems in which this approach could be replicated for studies of more conventional animal behavior is not clear.

F. How behavior influences phylogenetic diversification

A focus of phylogenetic studies of behavior, especially with the availability of MacClade (Maddison & Maddison 1992), has involved examining how behavior changes over a phylogenetic tree. Another issue is how behavior itself influences phylogenetic diversification. The notion that key adaptations or innovations can influence rates of diversification is an old one (e.g., Liem 1973) that generates continued interest (e.g., Craig et al. 1994). Theoreticians have suggested that the potential for divergence of mate-recognition systems influences the rate of speciation (Lande 1982; West Eberhard 1983). There is some empirical support for this notion, suggesting that accelerated speciation rates accompany increasing complexity of the frog's inner ear (Ryan 1986), evolution of syringeal structures and song learning in birds (Fitzpatrick 1988; Brenowitz 1992, but see Baptista and Trail 1992), and elaboration of complex androconia (pheromone dispersing organs) in nystaleine moths (Weller 1989).

One method of testing the relationship between behavioral evolution and speciation rates is to examine the shape of the phylogenetic trees. If all species are equally likely to speciate, then highly asymmetrical trees are unlikely. Guyer and Slowinski (1993) and Kirkpatrick and Slatkin (1994) have recently developed the statistical techniques necessary for testing such hypotheses.

G. Phylogenies and conservation

Brooks et al. (1992) discuss the contribution that phylogenetics can make to conservation. The most intriguing possibility for behaviorists is using phylogenetic information to infer the behavior and ecology of unstudied and inaccessible species. For example, it is critical to define the nesting habitat and habits of species for sensible attempts at rescue or conservation. If the species under question is nested well within a group that shares a common nesting syndrome, then one can evaluate the relative costs and benefits of relying on phylogenetic inference to assume this critical knowledge versus the alternative of disturbing the species at risk.

Summary

In an influential paper in comparative psychology, Frank Beach (1950) warns against basing the foundations of a "comparative" field on studies of the white rat conducted under unnatural laboratory conditions. Animal behaviorists were sometimes traveling down a different but similarly hazardous path by doing "comparative" biology without considering evolutionary history. This volume suggests that we now understand that history can no longer be ignored; to incorporate it we need to rely on methods developed for historical analyses. Phylogenetics offers some methods that have been utilized fruitfully to incorporate estimates of the past into evaluating how behavior evolves. It might be that these methods will need to be refined and reevaluated, but they presently offer possibilities for study not available twenty years ago. But if there is a discipline in the life sciences that needs to be multidisciplinary, animal behavior is it. Phylogenetics should be added to our methodological repertoire but it cannot substitute for all other analyses and approaches. It appears that the best behavioral studies are still those that address the general biology of behavior.

ACKNOWLEDGMENTS

I thank L. Dries, J. Ellingson, G. Rosenthal, and P. Warren for comments on the manuscript and H. Greene for helpful discussion of some of these topics.

References

Alcock, J. 1975. *Animal Behavior, An Evolutionary Approach*, 1st ed. Sunderland, Massachusetts: Sinauer and Associates.

Arnold, S. J. 1994a. Constraints on phenotypic evolution. In: *Behavioral Mechanisms in Evolution Ecology* (Ed. by L. Real), pp. 258–278. Chicago: University of Chicago Press.

Arnold, S. J. 1994b. Multivariate inheritance and evolution: A review of concepts. In: *Quantitative Genetic Studies of Behavioral Evolution* (Ed. by C. R. B. Boake), pp. 17–48. Chicago: University of Chicago Press.

Atz, J. W. 1970. The application of the idea of homology to behavior. In: *Development and Evolution of Behavior* (Ed. by L. R. Aronson, E. Tobach, D. S. Lehrman, & J. S. Rosenblatt), pp. 53–64. San Francisco: Freeman.

Baptista, L. F., & P. W. Trail. 1992. The role of song in the evolution of passerine diversity. *Syst. Biol.*, 41, 242–247.

Basolo, A. L. 1990. Female preference predates the evolution of the sword in swordtails. *Science*, 250, 808–810.

Basolo, A. L. 1995. Phylogenetic evidence for the role of preexisting biases in sexual selection. *Proc. R. Acad. Sci.*, 259, 307–312.

Beach, F. A. 1950. The snark was a boojum. *Am. Psychol.*, 5, 115–124.

Brenowitz, E. A. 1991. Evolution of the vocal control system in the avian brain. *Semin. Neurosci.*, 3, 399-407.

Brooks, D. R., & D. A. McLennan. 1991. *Phylogeny, Ecology, and Behavior*. Chicago: University of Chicago Press.

Brooks, D. R., R. L. Mayden, & D. A. McLennan. 1992. Phylogeny and biodiversity: Conserving our evolutionary legacy. *Trends Ecol. Evol.*, 7, 55–59.

Bull, J. J., J. P. Huelsenbeck, C. W. Cunningham, D. L. Swofford, & P. J. Waddell. 1993. Partitioning and combining data in phylogenetic analyses, *Syst. Biol.*, 42, 384–397.

Bull, J. J., & I. J. Molineux. 1992. Molecular genetics of adaptation in an experimental model of cooperation. *Evolution*, 46, 882–895.

Burghardt, G. M., & J. J. Gittleman. 1990. Comparative behavior and phylogenetic analyses: New wine, old bottles. In: *Interpretation and Explanation in the Study of Animal Behavior*. (Ed. by M. Bekoff & D. Jamieson) pp. 192–225, Boulder, Colorado: Westview Press.

Cannatella, D. C., & W. E. Duellman. 1984. Leptodactylid frogs of the *Physalaemus pustulosus* group. *Copeia*, 1984, 902–921.

Christy, J. H., & M. Salmon. 1991. Comparative studies of reproductive behavior in mantis shrimp and fiddler crabs. *Am. Zool.*, 31, 329–337.

Craig, C. L., G. D. Bernard, & J. A. Coddington. 1994. Evolutionary shifts in the spectral properties of spider silks. *Evolution*, 48, 287–296.

Cumming, J. M. 1994. Sexual selection and the evolution of dance fly mating systems (Diptera: Empididae: Empidinae). *Can. Entomol.*, 126, 907-920.

Dawkins, R. 1976. *The Selfish Gene*. Oxford, England: Oxford University Press.

de Queiroz, A., & Wimberger, P. H. 1993. The usefulness of behavior for phylogeny estimation: Levels of homoplasy in behavioral and morphological characters. *Evolution*, 47, 46–60.

Emlen, S. T., & L. W. Oring. 1977. Ecology, sexual selection, and the evolution of mating strategies. *Science*, 197, 215–223.

Felsenstein, J. 1985. Phylogenies and the comparative method. *Am. Nat.*, 125, 1–15.

Fitzpatrick, J. W. 1988. Why so many passerine birds? A response to Raikow. *Syst. Zool.*, 37, 71–78.

Gardner, R. 1990. Mating calls. *Nature*, 344, 495.

Greene, H. W. 1986. Diet and arboreality in the emerald monitor, *Varanus prasinus*, with comments on the study of adaptation. *Fieldiana Zool.*, 31, publ. 1370.

Greene, H. W. 1994. Homology and behavioral repertoires. In: *Homology: The Hierarchical Basis of Comparative Biology* (Ed. by B. K. Hall), pp. 369–391. San Diego, California: Academic Press.

Guyer, C., & J. B. Slowinski. 1993. Adaptive radiation and the topology of large phylogenies. *Evolution*, 47, 253–263.

Hamilton, W. D. 1964. The genetical evolution of social behaviour (I and II). *J. Theor Biol.*, 7, 1–16, 17–52.

Harvey, P. H., & M. D. Pagel. 1991. *The Comparative Method in Evolutionary Biology*. Oxford, England: Oxford University Press.

Hennig, W. 1966. *Phylogenetic Systematics*. Urbana, Illinois: University of Illinois Press.

Hillis, D. M., J. J. Bull, M. E. White, M. R. Badgett, & I. J. Molineaux. 1992. Experimental phylogenetics: Generation of a known phylogeny. *Science*, 255, 589–592.

Hillis, D. M., J. P. Huelsenbeck, & C. W. Cunningham. 1994. Application and accuracy of molecular phylogenies. *Science*, 264, 671–677.

Hinde, R. A., & N. Tinbergen. 1958. The comparative study of species-specific behavior. In: *Behavior and Evolution* (Ed. by A. Rue & G. G. Simpson) pp. 251–268, New Haven, Connecticut: Yale University Press.

Höglund, J. 1989. Size and plumage dimorphism in lekking birds: A comparative analysis. *Am. Nat.*, 134, 72–87.

Huey, R. B., & A. F. Bennett. 1987. Phylogenetic studies of coadaptation: preferred temperatures versus optimal performance temperatures of lizards. *Evolution*, 41, 1098–1115.

Hull, D. L. 1988. *Science as a Process.* Chicago: University of Chicago Press.

Kessel, E. L. 1955. The mating activities of balloon flies. *Syst. Zool.*, 4, 97–104.

Kirkpatrick, M., & M. J. Ryan. 1991. The paradox of the lek and the evolution of mating preferences. *Nature*, 350, 33–38.

Kirkpatrick, M., & M. Slatkin. 1994. Searching for evolutionary pattern in the shape of a phylogenetic tree. *Evolution*, 47, 1171–1181.

Lande, R. 1982. Rapid origin of sexual isolation and character divergence in a cline. *Evolution*, 36, 213–223.

Lanyon, S. M. 1992. Interspecific brood parasitism in blackbirds (Icterinae): A phylogenetic perspective. *Science*, 255, 77–79.

Liem, K. F. 1973. Evolutionary strategies and morphological innovations: Cichlid pharyngeal jaws. *Syst. Zool.*, 22, 425–441.

Maddison, W. P. 1991. Squared-change parsimony reconstructions of ancestral states for continuous-valued characters on a phylogenetic tree. *Sys. Zool.*, 40, 304–314.

Maddison, W. P., & D. R. Maddison. 1992. *MacClade. Analysis of Phylogeny and Character Evolution.* Version 3. Sunderland, Massachusetts: Sinauer and Associates.

Marden. J. H. 1989. Effects of load-lifting constraints on the mating system of a dancefly. *Ecology,* 70, 496–502.

Maynard Smith, J. 1982. *Evolution and the Theory of Games.* Cambridge, England: Cambridge University Press.

Oakes, E. J. 1992. Lekking and the evolution of sexual dimorphism in birds: Comparative approaches. *Am. Nat.*, 140, 665–684.

Pomiankowski, A. 1994. News and views. *Nature*, 368, 494–495.

Powell, J. A. 1964. Observations on flight behavior and prey of the dance fly *Rhamphomyia curvipes* Coquillet (Diptera: Empedidae). *Wassman J. Biol.*, 22, 311–321.

Proctor, H. C. 1993. Sensory exploitation and the evolution of male mating behaviour: A cladistic test using water mites (Acari: Parasitengona). *Anim. Behav.*, 44, 745–752.

Reeve, H. K., & P. W. Sherman. 1993. Adaptation and the goals of evolutionary research. *Q. Rev. Biol.,* 68, 1–32.

Ridley, M. 1983. *The Explanation of Organic Diversity. The Comparative Method and Adaptations for Mating.* Oxford, England: Clarendon Press.

Ryan, M. J. 1985. *The Túngara Frog, A Study in Sexual Selection and Communication.* Chicago: University of Chicago Press.

Ryan, M. J. 1986. Neuroanatomy influences speciation rates among anurans. *Proc. Natl. Acad. Sci. USA*, 83, 1379–1382.

Ryan, M. J. 1990. Sensory systems, sexual selection, and sensory exploitation. *Oxford Surv. Evol. Biol.*, 7, 157–195.

Ryan, M. J. 1994. Mechanisms underlying sexual selection. In: *Behavioral Mechanisms in Evolution Ecology* (Ed. by L. Real), pp. 190–215. Chicago: University of Chicago Press.

Ryan, M. J., & R. C. Drewes. 1990. Vocal morphology of the *Physalaemus pustulosus* species group (Family Leptodactylidae): Morphological response to sexual selection for complex calls. *Biol. J. Linn. Soc.*, 40, 37–52.

Ryan, M. J., J. H. Fox, W. Wilczynski, & A. S. Rand. 1990. Sexual selection for sensory exploitation in the frog *Physalaemus pustulosus*. *Nature*, 343, 66–67.

Ryan, M. J., & A. S. Rand. 1993. Sexual selection and signal evolution: the ghost of biases past. *Phil. Trans. R. Soc. ser. B.*, 340, 187–195.

Ryan, M. J., & A. S. Rand. 1995. Female responses to ancestral advertisement calls in túngara frogs. *Science*, 269, 390–392.

Searcy, W. A. 1992. Song repertoire and mate choice in birds. *Am. Zool.*, 32, 71–80.

Weller, S. J. 1989. Phylogeny of the Nystaleini (Lepidoptera: Noctuoidea: Notodontidae). Ph.D. dissertation. Austin, Texas: University of Texas, Austin.

Wenzel, J. W. 1992. Behavioral homology and phylogeny. *Annu. Rev. Ecol. Syst.*, 23, 361–381.

West Eberhard, M. J. 1983. Sexual selection, social competition and speciation. *Q. Rev. Biol.*, 58, 155–183.

Wiley, R. H. 1991. Lekking in birds and mammals: Behavioral and evolutionary issues. *Adv. Study Behav.*, 20, 201–291.

Wilson, E. O. 1975. *Sociobiology*. Cambridge, Massachusetts: Belknap Press.

CHAPTER 2

The Statistical Analysis of Interspecific Data: A Review and Evaluation of Phylogenetic Comparative Methods

Emília P. Martins and Thomas F. Hansen

In the last few decades, the comparative method has undergone a virtual renaissance in evolutionary biology as researchers develop new ways to incorporate taxonomic and phylogenetic information into the design and analysis of interspecific data (see Brooks & McLennan 1991; Harvey & Pagel 1991; McKitrick 1993; Miles & Dunham 1993; and Maddison 1994 for recent reviews). Although phylogenies were an integral part of many classical ethological studies, their use has been largely neglected by other fields in the study of animal behavior. Recent advances in systematics and evolutionary biology have shown the critical importance of incorporating phylogenies into comparative studies in all areas of biology. From a statistical perspective, phylogenies are needed to transform comparative data so that they do not violate the assumptions of standard statistical analyses (e.g., regression, ANOVA, chi-squared tests). From a more biological perspective, phylogenies and new phylogenetic comparative methods (PCMs) allow researchers to infer the patterns and processes of character evolution from the patterns observed in data measured from extant species.

The first part of this chapter is a discussion of some general issues involved in comparative analysis. The second part reviews and critically evaluates most of the PCMs available today, and is intended as a reference manual. We discuss the ideas, assumptions and interpretations of the methods and evaluate them from a statistical point of view. We do not give the kind of technical description necessary for use of the methods. Although there are many excellent reviews of comparative methods, there is no substitute for reading the original papers when it comes to application. In the Discussion, we explore the role and importance of statistical models in any comparative analysis.

Why phylogenetic comparative methods are used

A. To solve the statistical problem of dependent data.

One reason for incorporating phylogenetic information into a comparative study is to address the problem of statistical dependence. This is the "degrees of freedom" or "effective sample size" problem that has been mentioned frequently in the recent evolutionary and systematics literature. In statistics, the accuracy of a parameter estimate or hypothesis test depends on the number of degrees of freedom available, which in turn depends on the effective sample size of measured data. This effective sample size depends on the number of *independent* data points in the study rather than on the number of samples taken. For example, if 15 data points are measured from one animal and 15 more are measured from a second, the effective sample size is somewhat less than 30, because individual animals are often consistently different from one another (e.g., Martin & Kraemer 1987; Boake 1989), and only two individuals were measured. This "pooling fallacy" (Machlis et al. 1985) occurs when statistics are conducted as if the effective sample size were the number of data points measured (in this case, 30) and the dependence due to individual differences in behavior has been ignored. Standard errors can be greatly underestimated by this problem, such that spurious patterns may be found. Similar sorts of pooling fallacies can occur whenever important factors (e.g., individual identity, sex, preferred habitat, body size) are left out of the models used in statistical analyses. Whenever these missing variables are correlated with the variables of interest, excluding them from the analysis can lead us to judge a pattern unimportant even

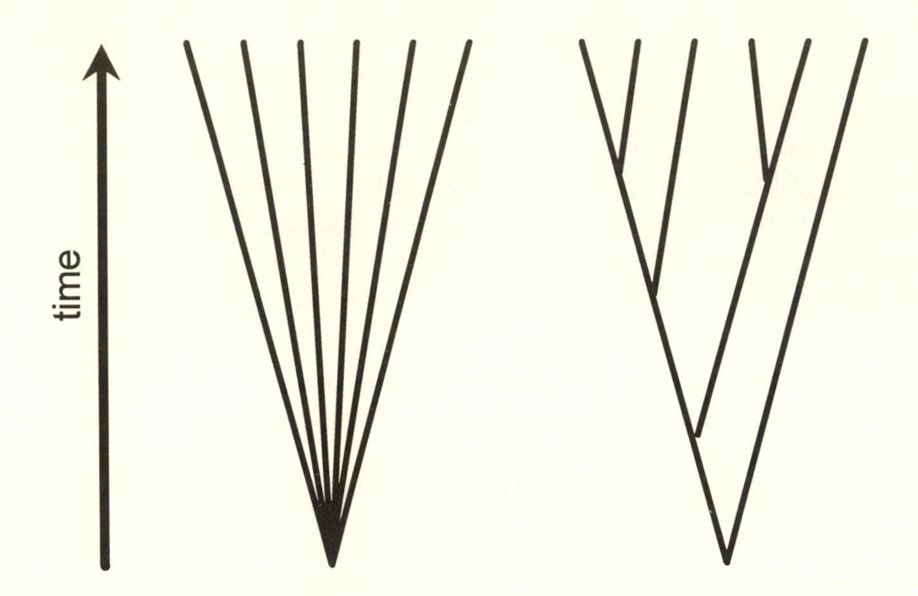

Figure 1. The phylogeny on the left depicts the situation assumed by most statistical analyses of comparative data. Species arise instantaneously from a single common ancestor in a "star" radiation. The phylogeny on the right depicts a more realistic situation as would be allowed by any one of a number of available phylogenetic comparative methods. Although there are six species in this group, the effective sample size of a study in which the data consist of measurements of each of these species will be somewhat less than six.

though it is important. This is particularly aggravated if the pattern of character variation differs among different phylogenetic or taxonomic groups and phylogeny is not taken into account.

Comparative data serve as an illustration of a particularly common and frequently ignored pooling fallacy that can occur whenever multiple data points are collected from the same genus, family, or phylogenetic clade. Closely related species are often more similar to each other than they are to distantly related species because of their shared evolutionary histories. In many studies, however, data are pooled as if there were no dependence among species due to phylogenetic history. Unless the data are transformed in some way to take phylogenetic information into account, the number of species measured in a set of interspecific data is likely to be an inflated estimate of the number of independent data points or effective sample size for the study (Fig. 1).

It has been shown both analytically and empirically that dependence of this sort in comparative studies can lead to serious statistical problems (e.g., Felsenstein 1985; Grafen 1989; Martins & Garland 1991b; Gittleman & Luh 1992, 1993; Martins in review). Accuracy of estimation is usually decreased and hypothesis tests are inadequate. For example, a Pearson correlation coefficient calculated between two traits measured in several different species is likely to exaggerate the absolute magnitude of the relationship between those two traits such that an association is found much more often than expected by chance alone when the two traits are actually independent of one another. Phylogenetic comparative methods allow incorporation of available taxonomic or phylogenetic information into the calculation of correlation coefficients and other statistics, thereby taking the degree of statistical dependence into account and greatly improving the reliability of parameter estimates and hypothesis tests.

B. To answer evolutionary questions

Dependence due to phylogenetic relationships among species in comparative data contains information about the evolutionary process. In general, we expect there to be an underlying correspondence between the phenotypes of existing species and the historical patterns of speciation. We will refer to this correspondence as *phylogenetic correlation*. Phylogenetic correlation is what allows us to reconstruct phylogenies from morphological or molecular data. When the phylogeny is known, we can also use any phylogenetic correlation that might exist in other traits (e.g., behavior) to infer the form and direction of phenotypic evolutionary change along that phylogeny. To do this adequately, we must first consider why phylogenetic correlation arises.

Phylogenetic correlation, as discussed by Hansen & Martins (in press; Martins & Hansen 1995), arises through common ancestry. Phenotypes of extant species are expected to be correlated with those of their ancestors. Because all species share some common ancestors, the phenotypes of extant species are also expected to be correlated with each other. Usually, the similarity between the phenotype of a species and that of its ancestor decreases with time as mutations appear and species phenotypes respond to natural selection in a changing environment. In general, we expect distantly related species to be less similar to each other than are closely related species, and thus exhibit less phylogenetic correlation. Phylogenetic correlation is expected for complex behavior patterns and behavior patterns governed by complex developmental rules (e.g., communicative displays), because such traits are often slowly evolving and constrained in the changes that can occur. Similarly, behavior patterns that are of extreme importance in a wide variety of environmental conditions (e.g., antipredator strategies, parent-offspring interactions) may be resistant to evolutionary change, and may therefore retain more phylogenetic correlation.

On the other hand, many behavioral and life history traits show high levels of within species variation and are capable of rapid adaptation and smooth tracking of changes in the environment. Such traits may be more influenced by recent selective history than by ancient conditions and are therefore not likely to show phylogenetic correlation directly. This is illustrated by the success of optimality models in behavioral ecology (e. g., Krebs & Davies 1987; Parker & Maynard Smith 1990) which predict species behavior based on present day conditions (without regard to phylogeny). However, the constraint functions that are an integral part of the optimality models may themselves be subject to phylogenetic correlation, and even adaptive traits may be phylogenetically correlated due to indirect mechanisms. Phylogenetic correlation may also be caused by forces other than direct evolution when animals seek out and live in environments similar to those of their ancestors. In these cases, selective constraints are inherited along with the environment, and similarities among species due to their shared evolutionary history results. In conclusion, we expect some degree of phylogenetic correlation in most comparative data sets, and the burden of proof is on those who wish to argue for its absence.

Even when the goals of the comparative study are not explicitly evolutionary, it is important to consider the sources of phylogenetic correlation. Phylogenetic comparative methods are usually based on either statistical or evolutionary models that describe how the phylogenetic correlation of comparative data came to be. Some methods (e.g., phylogenetic autocorrelation, nested ANOVA; see below) are designed to find and incorporate whatever phylogenetic "effect" is present in the data, regardless of how it arose. Other methods (e.g., independent contrasts, microevolutionary model-based methods; see below) are based on explicit models of phenotypic evolution which assume that the actual causes of phylogenetic correlation are known and are of a particular sort. Which sort of method is more appropriate depends on the details of the biological question of interest. Phylogenetic comparative studies can be used to answer questions such as: (a) Is a behavior pattern phylogenetically constrained? If so, by how much? (b) What was the ancestral state of the behavior? (c) How quickly did the behavior evolve? (d) In what order did a set of behavior patterns appear? Did they appear before or after some other trait? (e) Did two behavior patterns evolve in a correlated fashion, or did a behavior evolve in association with ecological factors? (f) Did a behavior pattern evolve neutrally via random genetic drift alone, or was it subjected to selection? Our ability to answer these questions will often depend on the development of a specific method for addressing that question. When the sources of phylogenetic correlation are not explicitly stated in the description of a specific method, problems of interpretation may arise (Frumhoff & Reeve 1994; Leroi et al. 1994; Martins & Hansen 1995; Nee et al. Chap. 13, in this volume).

Not taking phylogenetic information into account in standard statistical analyses is equivalent to assuming that the phylogenetic relationships among species are irrelevant and that any phylogenetic correlation is insignificant. The differences between a mouse and a dove due to one being a mammal and the other a bird may be unimportant, for example, if all species in the study arose simultaneously from a single common ancestor in a "star" radiation (Fig. 1). In this case, phylogenetic history does not lead to the sorts of dependence described above because all of the measured species are related to each other to the same degree. The assumption that phylogenetic relationships are irrelevant may also be reasonable if the response to selection is fast in comparison to the rate of speciation, such that historical constraints are quickly erased from species phenotypes. In any of these latter situations, our ability to estimate parameters of the evolutionary process and to answer the questions listed above will also be quite limited.

Getting comparative data

The first step in conducting a phylogenetic comparative study is to obtain a set of comparative or interspecific data. The richest source of comparative data is the library. With animal welfare, conservation, and funding concerns limiting the use of animals in behavioral research, comparative or metaanalyses have become increasingly important as ways of generating hypotheses and gathering preliminary evidence. The study of animal behavior has existed for long enough now that many important data have already been collected and can be reused through a comparative approach to provide fresh insights into old problems in animal behavior. In other cases, data must be collected specifically for the comparative project from laboratory or field experiments.

Whether collecting from the field, laboratory or library, there are a number of issues researchers need to be concerned with in choosing and collecting comparative data. Several of these problems occur with any "metaanalysis" when data are combined from separate studies done by different researchers. First, data collected in different studies are likely to vary in their statistical and biological reliability due to differences in data collection methods and numbers of individuals sampled. This variability among data sets should be taken into account whenever possible, possibly by weighing the data from different studies by their reliability. Second, most experiments are designed to address only a few specific questions, and often make small, hidden assumptions that may be inconsequential to the aims of the original study but are crucial to the comparative study. Each data point in a metaanalysis must be carefully researched to ensure that the data are appropriate to answer the question at hand. Gaillard et al. (1994) provide an illustration of such problems in an excellent discussion of this issue. Finally, a sample of results gleaned from the behavioral and ecological literature is almost certainly biased because of the unfortunate tendency to think that results that do not show statistically significant patterns are not interesting and therefore not publishable. When these results are compiled in a metaanalysis, the final conclusions are likely to be an exaggerated view of the real world. This problem is particularly aggravated in metaanalyses based on studies that report results only as the outcome of statistical tests or as p values. This is only one of many reasons that we support efforts to publish raw data whenever possible and highly recommend that comparative biologists restrict their statistical analyses of combined data sets to situations in which raw data or at least parameter estimates are available.

Even when all of the data for a comparative study are collected by a single researcher, other issues need to be considered. First, when data are collected from several species, it is not always a trivial problem to determine that the same behavior pattern is being measured in all the different species. This is the classical problem of homology, which has been discussed at length in the evolutionary systematics literature (Hall 1994; in this volume, Lauder & Reilly, Chap. 4; de Queiroz & Wimberger, Chap. 7; Irwin, Chap. 8). Ontogenetic, physiological, and morphological information about the mechanistic bases of the behavior pattern may be useful in ensuring that the behavior of interest is essentially the same pattern with variations for different species. Second, most phylogenetic comparative methods still assume that there is no variation among individuals or populations within a species and that a single number (i.e., the species mean phenotype) can represent the entire species (in this volume, see Foster & Cameron, Chap. 5 for further discussion). If there are considerable sex or age differences in the behavior, it may be important to analyze data for the different sexes or age groups separately. Often, it will also take substantial effort to find the right measures (e.g., mean, median, coefficient of variation) to describe each species effectively. Finally, as in any other behavioral study, it is important to remember that body size or any other confounding variables should, as always, be measured and taken into account in the comparative study. Phylogenetic correlation in these confounding variables should also be incorporated into the analysis.

Throughout this chapter, we refer to data measured in different species, but much of the discussion is also true whether the data are measured from different populations, genera, or even genetically related individuals within populations. Evolutionary biology suggests that all taxa are related to one another. Thus, data measured from any living organisms are likely to be dependent to some extent or another. The relative importance of this dependence to statistical or biological concerns will depend on the details of the questions being considered and the underlying microevolutionary processes. Although the evolutionary processes leading to the relationships among species will differ from those that cause relationships among different populations or individuals, there will often be a direct extension of interspecific methods that can be used at the population or individual level. When doing analyses at levels other than the species level, though, special care must be taken in the interpretation of results.

Getting phylogenetic information

Phylogenies can be obtained from a number of different sources, and means of doing so have been discussed at length (see Hillis & Moritz 1990; Felsenstein 1988 for reviews). At best, direct fossil evidence will exist showing the patterns and timing of speciation events in the clade. More commonly, phylogenetic hypotheses derived from morphological or molecular information will be available, and branch lengths on the tree will be given in units of genetic sequence divergence or the minimum number of morphological evolutionary changes required. In some cases, estimates of the time since divergence of various species on the phylogeny (branch lengths) will also be available. At worst, taxonomic information can be used to infer hierarchical topological relationships (assuming that all large taxonomic groups are monophyletic and that species within genera, genera within families, families within orders, and so on have evolved independently of one another).

All statistical analyses of interspecific data require that some phylogenetic information be either available or assumed. In most cases, it is assumed that both the phylogenetic relationships among species (i.e., a phylogenetic topology) and the process of phenotypic evolution underlying the particular character(s) that have been measured (usually described as branch lengths on the tree in units of expected variance or amount of phenotypic change) are known. For example, most standard statistical methods (without making any attempt to incorporate phylogenetic information) assume that: a) the similarities among species expected due to phylogenetic relationships can be described as a "star" phylogeny (e.g., Fig. 1*a*) and b) the character has evolved at the same rate and for the same length of time along each branch of this topology. Other methods (see below) allow for more flexibility of assumptions and for information from molecular phylogenies or quantitative genetic experiments to be incorporated into the analysis. The accuracy of a phylogenetic comparative study depends on the accuracy of the phylogenetic information provided, so it is well worth the effort to find a reasonable tree.

Once a phylogenetic topology is obtained, branch lengths must be estimated, inferred, or assumed. In most systematics studies, branch lengths are reported in units of time. For phylogenetic comparative methods, however, branch lengths are usually needed in units of the relative amount of change expected in the characters being analyzed (these units are often referred to as "expected variance of character change"). This expected amount of change will be a function of the evolutionary process (e.g., random genetic drift, selection) underlying the characters of interest as well as the phylogeny, and usually branch lengths must be transformed from units of time, sequence divergence, or the minimum number of evolutionary changes into these character-specific units. For example, if phenotypic evolution occurs as a gradual, "clocklike" process (e.g., if the character is evolving only by random genetic drift), branch lengths in units of expected variance of change will be directly proportional to branch lengths in units of time. Alternatively, if phenotypic evolution is thought of as a punctuational or "burstlike" process in which the character undergoes bursts of change at speciation events followed by long intervals of stasis, then branch lengths in units of expected variance of change will be proportional to the number of speciation events occurring along each branch. If a phylogeny based on molecular or morphological information is available, we might also assume that branch lengths in units of expected variance of change are proportional to branch lengths in units of phenetic similarity (usually measured as sequence divergence or number of evolutionary changes). This assumes that the characters used to reconstruct the phylogeny evolved at a rate that is linearly proportional to the rate at which the characters in the comparative study evolved.

There are a number of statistical procedures that infer phylogenetic topologies or branch lengths in units of expected variance of change using the information available in the comparative data themselves. The resulting branch lengths have intrinsic interest, as they are essentially a description of the microevolutionary process underlying phenotypic change. Because most of these procedures are specific to one type of phylogenetic comparative method or another, we will discuss them individually below. They should be used cautiously because they are generally conservative. There is only so much variability present in a set of comparative data. These procedures choose to use that variation first to determine phylogenetic relationships and only afterwards to infer adaptive evolution. With convergent evolution, for example, these procedures first try to explain the similarity between species as being due to shared evolutionary history and only afterwards as the result of a response to similar selective pressures. The resulting estimates of evolutionary parameters may be conservative or even biased. For the same reason, phylogenetic information should be derived independently of the characters used in the comparative study whenever possible.

Often, several competing phylogenetic hypotheses and/or sets of branch lengths exist. If no independent phylogenetic information at all is available, it is also possible to generate a large set of possible phylogenies using computer simulation techniques (Losos 1995, Martins in press). Most phylogenetic comparative methods can incorporate most sorts of phylogenetic information, with the accuracy of the results depending primarily on how well the phylogenetic information actually resembles evolutionary "truth." When several possible "truths" exist, we recommend that the comparative analyses be conducted on each of the possible trees and that the variance in results be incorporated explicitly into the estimation and hypothesis testing of any resulting phylogenetic statistics (Martins in press). This provides results that are not specific to any one of the possible phylogenies. Phylogenetic methods (reviewed below) also differ in terms of their robustness to different types of phylogenies and branch length transformation procedures as well as in terms of their robustness to inaccurate phylogenetic information. Thus, accuracy of phylogenetic information and robustness to inaccurate information should be an important consideration in the choice of method.

Choosing a phylogenetic comparative method

Several types of phylogenetic comparative methods have been proposed recently and many of the questions mentioned above can now be answered given the right sort of data. Methods differ in their theoretical perspective, biological assumptions and statistical properties, the types of variables for which they were designed, the types of phylogenetic information they require, and the availability and accessibility of computer programs needed to apply them. Most of the methods have been proposed as ways of estimating the evolutionary relationship between two traits (answering, for example these questions: Has the evolution of one trait constrained evolution of a second? Did two traits evolve in a correlated fashion?), but they can also be used to infer the actual sequence and pattern of evolutionary changes acting in single traits. A few methods can shed light on the question of evolutionary process (e.g., whether the trait has been evolving neutrally or was subjected to selection). In general, phylogenetic comparative methods involve making some assumptions about the phylogeny and process of phenotypic evolution in the clade and then fitting a statistical model or using a computer algorithm to transform the mean phenotypes of extant species into phylogenetically relevant units. For continuously varying characters, these new phylogenetic statistics can then be used in correlations, regressions, or other analyses instead of the raw species data. For categorical or "state" traits, techniques have been devised to answer specific questions (e.g., Have two traits evolved in a correlated fashion? What is the relationship between gains and losses in a single trait?). Herein, we review seven general classes of alternative techniques that can be used to analyze comparative data phylogenetically.

Before choosing a method, we highly recommend doing some exploratory data analysis. A certain familiarity with the data eases the choice of method, helps to generate hypotheses and guards against making unreasonable assumptions. In the case of comparative data, it is a good idea to get some feeling for the pattern of phylogenetic correlations before applying complex statistical procedures. Several of the methods described below include or can be used as exploratory tools. In particular we recommend *MacClade* (Maddison & Maddison 1992), a computer program, which friendly graphical interface invites you to play with your data. *MacClade* is, as mentioned by Felsenstein, "positively addictive".

A. Inferred changes methods

1. The methods

a. General description — Inferred changes methods are the most popular type of phylogenetic comparative method in use today (see Brooks & McLennan 1991; Harvey & Pagel 1991; Maddison & Maddison 1992; Miles & Dunham 1993; and Maddison 1994 for reviews). These methods begin by using prescribed algorithms (usually parsimony) to infer the character states of hypothetical ancestors on a known phylogeny and the corresponding magnitude and direction of evolutionary changes occurring along each branch. In many cases, this inference is the final goal of the study and only qualitative conclusions are made. Alternatively, once the magnitude and direction of evolutionary changes in the trait along the tree are known, these changes can be used instead of the raw species data in explicit statistical procedures (e.g., regressions). This requires some further manipulation, as inferred evolutionary changes on a phylogeny are not statistically independent and violate the assumptions of most statistical procedures.

Note that once the character reconstruction algorithm has been applied, the data points for the study are evolutionary changes rather than species phenotypes. Thus, the effective sample size of any comparative study conducted using these methods can be no larger than the number of evolutionary changes that have occurred, regardless of the number of species measured. If only a single change has occurred in each of two behavior patterns during the entire history of a clade, trying to determine whether the two traits have evolved in a correlated fashion is similar to trying to estimate a correlation after having measured only a single data point. In these situations, any errors in the phylogeny or the data can have extreme effects. For example, Basolo (1990a, b) showed that both female swordtail (*Xiphophorus helleri*) and platy (*Xiphophorus maculatus*) fish prefer males with longer swords (an extension of the lower caudal fin), even though male *X. maculatus* fish do not normally have swords on their tails. Since a phylogenetic reconstruction suggested that swordlessness was the ancestral state for the genus *Xiphophorus*, Basolo (1990b) concluded that female preference for long swordtails evolved before the existence of the male trait and that the "preexisting bias" hypothesis of sexual selection was supported. Unfortunately, Basolo's (1990b) interpretation relies on the location of a single evolutionary change (from nonsworded fish to sworded fish; Fig. 2), and Basolo's analysis is comparable to estimating a correlation (between the existence of male swords and female preference for sword tails) with a single data point. Thus, it is not surprising that when a phylogenetic reconstruction was done on a more recent phylogeny (Meyer et al. 1994), this new phylogenetic information suggested that the common ancestor of all *Xiphophorus* species was probably sworded, thereby overturning Basolo's original conclusion. Future analyses using several sworded and unsworded species from the genus and a more quantitative and powerful method (e.g., Maddison 1990) may provide a more conclusive answer to whether the "preexisting bias" hypothesis is a reasonable model for sexual selection.

b. Continuous traits — Although there are many ways in which the ancestral states of a continuous character might be estimated from

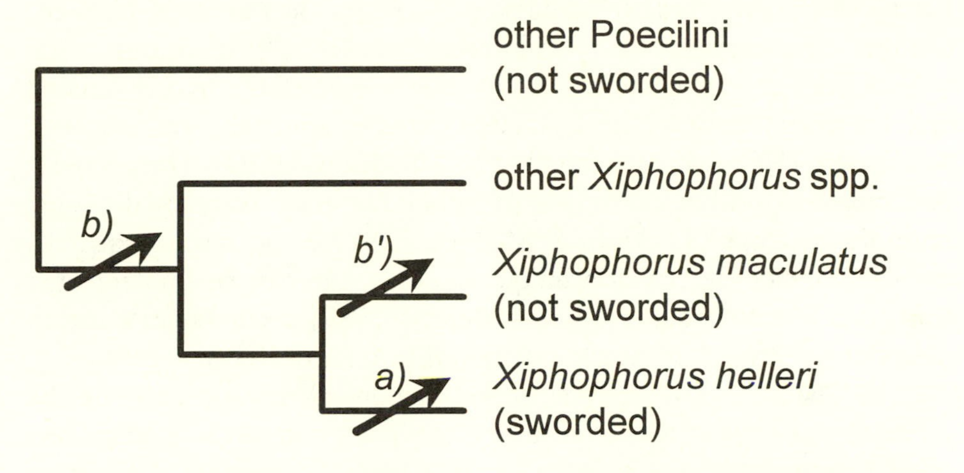

Figure 2. An example from Basolo (1990a, b) and Meyer et al. (1994) of how even minor changes in a phylogenetic hypothesis can have a serious impact on the conclusions of a study when a small number of evolutionary changes (in this case one versus two) are being considered. Using a parsimony reconstruction, Basolo (1990b) suggested that the common ancestor of all *Xiphophorus* species was swordless, with a single evolutionary change at (a) leading to the presence of swords in present-day *X. helleri*. Using empirical manipulations of sword length, Basolo (1990a, b) also showed a female preference for long swords on males in both *Xiphophorus maculatus* and *X. helleri*. Given only this information, it seems likely that the common ancestor of the two species also had a female preference for long male swords. A phylogenetic reconstruction in Basolo (1990b) concluded that there was evidence for the preexisting bias hypothesis of sexual selection because female preference for male swords seems to predate the presence of male swords in this genus. In contrast, using a more recent phylogeny, Meyer et al. (1994) found that swordtails have appeared repeatedly in the evolution of *Xiphophorus*, and that there were probably two evolutionary changes of interest: b) an evolutionary change towards swordedness before the ancestor of all *Xiphophorus*, and b') a loss of swordedness in *X maculatus*. If this second phylogenetic reconstruction is true, Basolo's (1990b) study does not provide evidence for the preexisting bias hypothesis.

comparative data, two algorithms have been in common use. Both are parsimonylike algorithms designed to minimize the total amount of phenotypic evolution occurring along a phylogeny. "Wagner" or "linear" parsimony is based on the absolute value of the evolutionary change occurring along each branch of the tree and minimizing the sum of those changes across the entire tree (Edwards & Cavalli-Sforza 1967; Farris 1970; Swofford & Maddison 1987). In the "sum of squared-changes parsimony" or "minimum evolution" algorithm, the evolutionary change occurring along each branch of the tree is squared and the overall sum of these squared changes along the entire tree is minimized (Huey & Bennett 1987, Maddison 1991). The Wagner parsimony algorithm is simple to apply but may occasionally give multiple solutions (Swofford & Maddison 1987). The sum of squared-changes method results in a unique solution and can be implemented using iterative procedures (Huey & Bennett 1987), recursive calculation (Maddison 1991) or direct computation (McArdle & Rodrigo 1994).

The above algorithms are applied independently to each of the characters of interest. Once the ancestral phenotypes of those characters have been estimated, the evolutionary change occurring along each branch of the phylogeny can be computed. The resulting evolutionary changes can then be used in standard statistical procedures (e.g., regression, correlation, principal components analysis) to estimate possible relationships among characters or between characters and environments. As there will be $2(N\text{-}1)$ evolutionary changes available for each set of N species, it is not immediately obvious what sample size should be used for computing confidence intervals. Huey and Bennett (1987), for example, suggested that although they had inferred $2(N\text{-}1)$ evolutionary changes, it was more reasonable to assume that the effective sample size was closer to N. Martins and Garland (1991) showed how phylogenetic randomization tests can be used to estimate the true effective sample size and to conduct hypothesis tests using the sum of squared-changes method.

c. Categorical traits — Ridley's (1983) book was the first modern account of comparative methods from an inferred changes perspective. He suggested that character transitions reconstructed by parsimony methods could be regarded as independent evolutionary changes and proposed that these changes be used as independent trials in testing for adaptation. For example, the hypothesis that a specific character state (e.g., male precopulatory behavior) is an adaptation to a specific "environment" (e.g., predictable receptivity in females) could be tested by comparing the number of times the character state had been attained in this as opposed to alternative environments. There are many algorithms available to reconstruct the ancestral states of categorical or "state" characters evolving under parsimony. As these have been reviewed frequently and recently (e.g., Maddison & Maddison 1992; Maddison 1994), we will not do so herein. Instead we concentrate on methods that address what to do once the ancestral states are obtained.

Sillén-Tullberg (1988) used Ridley's approach to study the evolution of warning coloration (aposematism) of butterfly larvae. She found that out of 23 evolutionary changes toward gregariousness, 15 to 18 occurred in areas of the phylogeny in which the taxa were aposematic. Without using statistics, she rejected the popular hypothesis that aposematism evolved by kin-selection among gregarious and unpalatable caterpillars, concluding instead that "unpalatability is an important predisposing factor for the evolution of egg clustering and larval gregariousness." Maddison (1990) criticized this approach because it did not take the distribution of aposematism on the phylogeny into account. Aposematism was much more common than crypsis in Sillén-Tullberg's (1988) data set. Thus, we expect most evolutionary changes towards gregariousness to be associated with aposematism even without any causal or functional relationship between the two characters. Maddison (1990) developed his concentrated-changes test to correct this sort of problem. Under the null hypothesis that gregariousness is uniformly distributed on the phylogeny, Maddison's method calculates the expected distribution of changes in one first variable (e.g., gregariousness) given the state of a second causal variable (e.g., coloration) and given the actual number of changes (gains and losses of gregariousness) that occurred on the phylogeny as a whole. This distribution can then be used to evaluate whether the observed number of gains in gregariousness is associated with aposematism. As an illustration, Maddison reanalyzed Sillén-Tullberg's data and found that 15 to 18 evolutionary gains in gregariousness in aposematic taxa, given 23 gains and six losses on the phylogeny as a whole, are well within the range expected from the null hypothesis of no association between the traits (Fig. 3).

Use of Maddison's method assumes that the data are either a random or a complete sample of the clade under study. As discussed in Maddison (1990), any change in that phylogeny produced by adding or subtracting taxa will affect the outcome of his method. Sillén-Tullberg (1993) responded to Maddison's reanalysis of her data by pointing out that the taxa included in her original study (Sillén-Tullberg 1988) were not a random sample of butterfly taxa. Aposematic taxa were, in fact, strongly overrepresented. Thus, the results of Maddison's reanalysis were due more to the dependence of his method on which taxa were sampled than to butterfly biology. Sillén-Tullberg (1993) proposed a variant of Maddison's method, the contingent states test, which uses a contingency table to test whether gains of gregariousness are dependent on aposematism. In this method, Sillén-Tullberg considers only those branches along which gains of gregariousness are possible (i.e. branches in which the solitary state is ancestral). Thus, the results of the contingent states test depend somewhat less on which species are sampled in the clade. Using this new method and a more representative data set, Sillén-Tullberg again found a strong relationship between aposematism and gregariousness.

If the phylogeny is known and all relevant species in the clade have been measured, Maddison's approach gives a reasonable test of relationship between two characters in that clade. If the study is being used to make inferences about a large clade from a smaller sample of measured species, then the way in which those species were sampled becomes important, and Sillén-Tullberg's approach may be preferred. Both Sillén-Tullberg's and Maddison's methods are directional or causal in their approach as they only consider evolutionary changes that can be unambiguously associated with the state of the second causal, variable (i.e., they disregard branches along which the causal variable has changed). These methods can also be used to ask more general evolutionary questions such as whether evolutionary changes in a trait are concentrated in defined parts of the phylogeny.

In his general discussion of the inferred changes approach, Ridley (1983) proposed using contingency tables to test whether two characters have evolved independently of each other or whether a single trait has evolved independently of the environment. This method uses only those branches along which at least one of the characters has changed. Grafen and Ridley (in press, and, in this volume, Ridley & Grafen, Chap. 3) formalized this into the "independent character evolutions" or ICE test and developed a second method, the ICDE (independent character-differences evolution method), which also tests a null hypothesis of independent evolution but avoids some of the ancestral reconstructions and thereby some of the problems of the ICE.

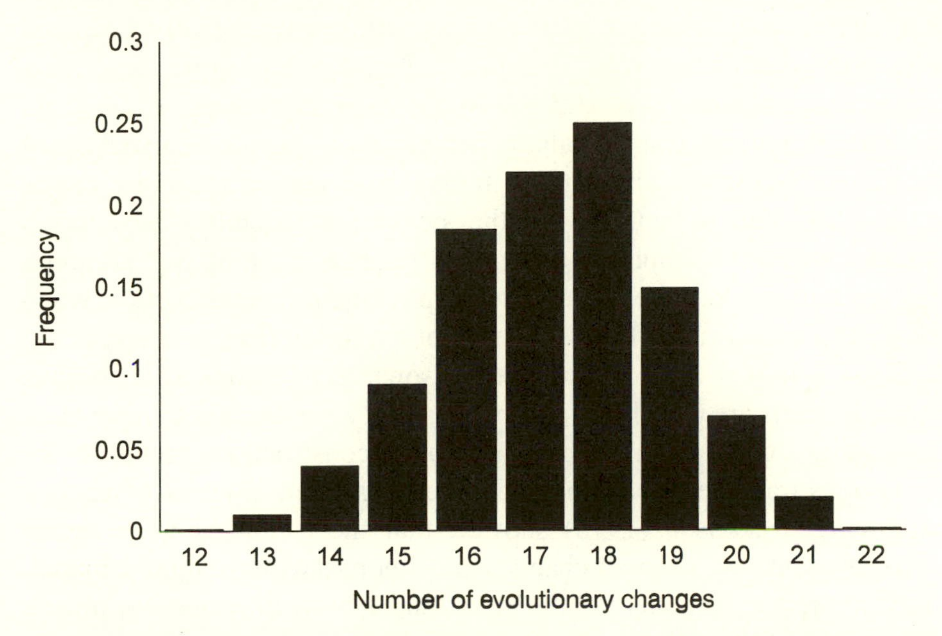

Figure 3. A frequency distribution from Maddison (1990) illustrating the use of his method to analyze data from Sillén-Tullberg (1988). Computer simulation was used to determine the number of times gregariousness in butterfly larvae is expected to evolve in taxa that also exhibit warning coloration, given 1) that there have been 23 gains and 6 losses in gregariousness during the evolution of the entire clade of 136 taxa (determined by using parsimony to reconstruct the ancestral states of gregariousness), and 2) that gains and losses are distributed randomly along the phylogeny. The histogram describes the number of computer simulation trials in which a particular number of evolutionary gains in gregariousness were observed along branches of the phylogeny in which warning coloration has evolved. Sillén-Tullberg (1988) observed 15-18 evolutionary gains in gregariousness. Results in this range occur in more than 90% of the trials and are thus very compatible with a null hypothesis of no association between gregariousness and warning coloration (but see Sillén-Tullberg 1993).

2. Assumptions

As is true for most phylogenetic methods, all of the above methods assume that within-species variation does not exist or is negligible in comparison to the level of among-species variability and that the phylogeny is known. The use of parsimony to reconstruct the ancestral states implies a few additional assumptions. Notably, parsimony implies that evolutionary changes in the character(s) are relatively rare and uniformly distributed over branches. Exactly how rare and how uniformly distributed is determined by the details of the algorithm (e.g., the sum of squared changes algorithm assumes that the distribution of evolutionary changes which gives the smallest sum of squared changes is the most likely to be true). Whether or not the parsimony assumption is reasonable is an empirical question. In particular, note that although this assumption may be reasonable in phylogeny reconstruction when the traits are chosen for their slow rate of evolution (but see Felsenstein & Sober 1986), it may often be unreasonable in comparative studies when the traits are usually chosen because they are thought to have been responding to selection. The ancestor reconstruction methods for continuous traits described above can be interpreted in terms other than parsimony. Maddison (1990) showed that the estimates of ancestral states obtained by squared-changes parsimony (with weighted branch lengths) are the same as would be obtained by a likelihood approach assuming that the traits evolved as if by Brownian motion (a common model of evolution discussed under independent contrasts below).

The inferred changes methods also assume that the ancestral states are known and not, as is almost always the case, estimated. Although evolutionary changes on different branches of the phylogeny may be independent, changes estimated using a parsimony algorithm will probably not be statistically independent. Thus, the above methods do not completely solve the problem of statistical dependence. Grafen and Ridley's (in press) ICDE methods are less vulnerable to this problem as most ancestor reconstruction in this method is either avoided or replaced with randomization. Using computer simulation, Maddison (1990) argued that his method is relatively robust to inaccurate parsimony reconstruction of ancestral states and is unlikely to give consistently misleading results if conservative significance levels are used.

Many of the above methods also assume that all branches are of equal length (i.e., that the same amount of phenotypic evolution is expected to occur along each branch of the phylogeny). This is sometimes associated with a model of punctuated evolution in which all changes occur during speciation. A model of punctuationed evolution, however, is only reasonable in such analyses if all speciation events (including those leading to extinct taxa) are known and included in the phylogeny. Thus, when alternative branch lengths are available, they can and should be incorporated into the algorithm for estimating evolutionary changes. For example, the iterative sum of squared changes algorithm (e.g., Huey & Bennett 1987) proceeds by estimating the phenotype of each ancestor as the average of its three closest relatives on the phylogeny. Variable branch lengths can be incorporated by estimating ancestral phenotypes as weighted rather than simple averages, using the branch lengths as the weights (Maddison 1991; Martins & Garland 1991b). Similarly, Sanderson (1991) proposed a variant of Maddison's (1990) method which allows the probability of character change to vary on different branches.

3. Variations for dealing with unknown phylogenetic information

None.

4. Strengths and weaknesses

Although relatively few computer simulation studies have included tests of the inferred changes methods, the results have been quite encouraging. Again, for categorical characters, Maddison (1990) found that his method is relatively robust to inaccurate parsimony reconstruction of ancestral states. For continuous characters, the sum of squared changes method with appropriately weighted branch lengths (Maddison 1991, Martins & Garland 1991b) has been found to result in consistently good estimates of evolutionary relationships that are sometimes superior to those produced by other methods for continuous characters (Martins & Garland 1991b; Martins in review).

For categorical variables, the greatest weakness of the inferred changes methods is the lack of development of parameter estimation as opposed to testing of null hypotheses. With the exception of Sanderson's (1991) method, these methods are further restricted in the assumption that all branches on the tree are of equal length. For continuous variables, the greatest weakness is the lack of explicit, analytical tools for computing standard errors and conducting hypothesis tests. The randomization procedure proposed by Martins and Garland (1991) is adequate but time-consuming and requires making a number of explicit assumptions about the microevolutionary processes guiding change in the character of interest. Finally, for both continuous and categorical variables, no extensions of the inferred changes methods have been developed to address the problem of inaccurate or unknown phylogenetic information.

5. Computational difficulties and computer programs

The popularity of inferred changes methods is almost certainly due to their relative computational simplicity and intuitive appeal. There are several user-friendly computer programs available to infer the ancestral states of characters along a known phylogeny under most of the available parsimony algorithms (e.g., Maddison & Maddison 1992; Felsenstein 1993; Swofford 1993; and Martins & Garland 1991a). We particularly recommend Maddison & Maddison's (1992) *MacClade* as an exceptional heuristic tool for any biologist interested in working with categorical comparative data. Once the ancestral states are known and evolutionary changes inferred, changes in continuous characters can be analyzed by hand, or using any standard statistical package. For categorical characters, Ridley's method can be applied by hand or using any statistical package that perform chi-squared tests; and probabilities from Maddison's (1990) method are calculated by the "concentrated changes" test of *MacClade* (Maddison & Maddison 1992). Sanderson (1991) and Sillén-Tullberg (1993) offer computer programs to calculate probabilities using their variants of Maddison's method.

B. Independent Contrasts and related methods

1. The methods

<u>a. General description</u> — There are several methods that can be grouped under the broad category of "independent contrasts," and which originated with the method proposed by Felsenstein (1985). The

independent contrasts method was originally devised as a way of addressing the problem of statistical dependence of comparative data within a population genetics framework. As described by Felsenstein (1985), this method assumes that the evolution of the measured trait can be described using a Brownian motion model of phenotypic evolution. Brownian motion is a stochastic process in which the evolutionary change in phenotype occurring during any interval of time is normally distributed, has variance proportional to the time interval, and is independent of the state of the phenotype at the beginning of the interval. It may be used to describe the evolution of a continuous character undergoing random genetic drift, a character responding to selection when the direction of selection is constantly shifting back and forth at random, or a character under stabilizing selection around an optimum that itself evolves according to Brownian motion (Felsenstein 1988, Hansen & Martins in press). Under such a model, although the species data are not independent, the differences or "contrasts" between certain pairs of species on the phylogeny are independent of one another (Fig. 4). If the phylogenetic relationships among species and the amount of expected change along each branch of the phylogeny are known, a simple algorithm can be used to transform the measured species data into a set of N - 1 standardized and independent "contrasts" (where N is the number of species that were measured).

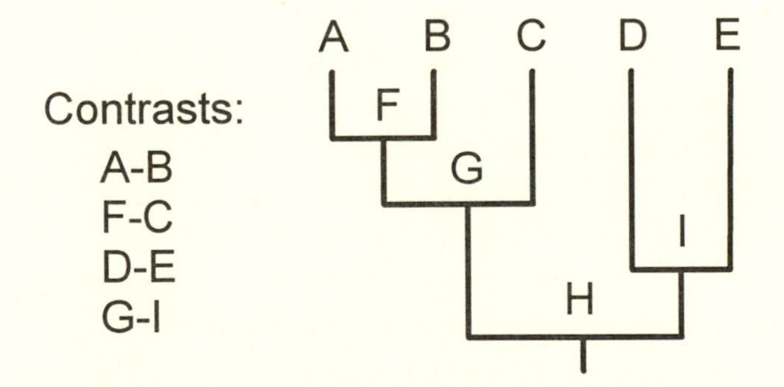

Figure 4. Illustration of Felsenstein's (1985) method of independent contrasts. Under a Brownian motion model of phenotypic evolution, although the measured species (A-E) are not statistically independent of one another, certain "contrasts" or differences between pairs of taxa are. Felsenstein's (1985) method consists of a numerical algorithm for calculating N-1 of these contrasts (where N is the number of measured species) and then standardizing them appropriately, such that they can be used in any standard statistics.

Under the Brownian motion assumption, contrasts are normally distributed. If the phylogeny is known, contrasts are also independent of one another. Standardization ensures that the final contrasts have the same variance (i.e., be homoscedastic) and a mean of zero. Thus, Felsenstein's contrasts are independent, normally distributed, and homoscedastic variables and fit the assumptions of most statistical tests. Once independent contrasts are formed and standardized, they can be used instead of the raw species data in any standard statistical procedure (e.g., regressions and ANOVA) as a way of conducting the analysis while taking phylogenetic information into account. Note that because the expectation of the contrasts is zero, any statistical model involving these contrasts should not include a grand mean. For example, in a regression of contrasts on other variables, the regression should be forced through the origin without estimating a nonzero y intercept (see Garland et al. 1992 for further explanation).

A number of "pairwise" contrast methods have arisen from Felsenstein's (1985) suggestion that we consider data collected from "two seals, two whales, two bats, two deer, etc." (e.g., Burt, 1989, Møller & Birkhead 1992, and in this volume, Nee et al., Chap. 13). If the pairs are formed using only closely related pairs of extant species (e.g., two bats, two whales) they are independent under less stringent assumptions than a Brownian motion (see below). On the other hand, as they use less of the information contained in the data, they lead to less precise and less powerful estimates of evolutionary parameters.

A method related to independent contrasts is the generalized least squares technique proposed by Grafen (1989, 1992). Grafen describes a set of comparative data as the sum of interspecific values of many other measured and unmeasured traits in a multiple regression format. He then uses generalized least squares and a "contrasts" approach to incorporate phylogenetic information into a multiple regression model describing the importance of each trait in predicting variation in the single response trait (e.g., using regression slopes, correlation coefficients, coefficients of determination). Grafen's method is based on statistical rather than population genetic assumptions, and requires that the error term in the multiple regression model consist of normally distributed variables with known variances and covariances. Grafen's "standard regression" calculates these variances and covariances of the error term using a known phylogeny and an assumption of character evolution (e.g., branch lengths in units of expected character change as in Felsenstein's method).

Felsenstein's (1985) contrasts method can be viewed as a model based on restricted maximum likelihood (REML), whereas Grafen (1989) relies on generalized least squares (GLS). When the phylogeny and branch lengths are known, Felsenstein's and Grafen's methods yield identical standardized contrasts and measures of the relationships among traits. The main practical differences between the two methods occur (a) when the phylogeny and branch lengths are uncertain (see below), (b) when categorical variables are included in the analysis (see below), and (c) in the assumptions and interpretations of the results. The results of statistics conducted on Felsenstein contrasts can be interpreted directly in historical terms. For example, a correlation between Felsenstein contrasts in two behavioral characters is an estimate of the correlation between the evolutionary changes occurring in the two traits at each generation. A regression slope describing the regression of one set of Felsenstein contrasts on another is an estimate of the amount of variation in one behavior pattern explained by the second at each generation through the evolutionary history of the clade. Grafen (1989, 1992) argues that most phylogenetic correlation is lost through evolutionary time due to the action of selection. His method partitions variation in one trait (the response or Y variable) into several components. Some of these components are other measured traits (predictor or X variables). The regression model estimates the relationship between the response and predictor variables and interprets this as due to recent adaptation (i.e., occurring at the tips of the phylogeny). The last component is a residual error term that describes the phylogenetic correlation due to evolutionary change in the response variable along the rest of the phylogeny.

Note that although one step of Felsenstein's algorithm is described as estimating "ancestral" states, the method was not intended as a means of estimating the ancestral states or magnitude of evolutionary changes occurring along the phylogeny. The "ancestral states" calculated as an intermediate step of the contrasts algorithm are the weighted averages of the two descendant species resulting from each node. This is a *local* estimate of the ancestral states (i.e., a reasonable estimate if the entire tree consists only of the ancestor and its descendants). A better estimate of the ancestral states would be a weighted average of all the extant species phenotypes (Maddison 1991).

b. Continuous vs. categorical traits — Felsenstein's version of independent contrasts is designed for use only with continuously varying characters (e.g., body size, display rate). Grafen's (1989, 1992) version allows for the use of categorical predictor ("independent" or "*X*") variables described as sets of dichotomous "dummy" variables (0/1; see Draper & Smith 1981 for an excellent description of how this is done). When these are incorporated into a multiple regression, the relationship between continuous and categorical traits can be estimated. This difference between the two methods is due to the fact that Felsenstein derived his method from an assumption that the traits evolve as if by Brownian motion, whereas Grafen assumes that the error terms in his multiple regression models are normally distributed with means of zero and known variance and covariance. Although categorical variables cannot evolve by Brownian motion nor be normally distributed, the residual error terms can.

2. Assumptions

Felsenstein's (1985) original method assumes that: (a) within-species variation does not exist or is negligible, (b) phylogenetic relationships are known and can be described as a standard binary tree structure (i.e., the phylogenetic topology is known), (c) the process of phenotypic evolution for the characters being considered can be described as a Brownian motion process, and (d) the relative rate of this process along each branch of the tree (i.e., branch lengths in units of expected amount of character change) is also known (or that branch lengths in units of time are available and that the rate of phenotypic evolution is constant throughout the clade).

Grafen's (1989) regressions assume that (a) within-species variation does not exist or is negligible, (b) that the predictor or *X* variables are independent of each other and the error term, and (c) that the variance-covariance of the error term has been taken into account by the GLS procedure. When the variance-covariance of the error term is compatible with what would be obtained from a Brownian motion, the last two assumptions of Grafen's method are comparable to the last two assumptions of Felsenstein's method, but Grafen's method may also apply to other situations. Grafen (and others, see below) also relax the assumptions that the phylogeny and branch lengths are known by using algorithms and estimation procedures to infer some of this information from the comparative data themselves.

The pairwise comparison methods relax the Brownian motion assumption of Felsenstein's (1985) method but still assume that each pair of species has been diverging for the same length of time and at the same rate as every other pair (such that each contrasts has the same variance) and that the differences are normally distributed. If nonparametric statistics are used, these assumptions are also relaxed, but interpretations are much less clear.

3. Variations for dealing with unknown phylogenetic information

Several variants of the independent contrasts method have been proposed to address the problems of obtaining accurate phylogenies and branch lengths for the analysis. In the original description of his method, for example, Grafen (1989) proposed "the phylogenetic regression," which includes a means of reducing unresolved polytomies into binary topologies. Beginning with an initial "working" phylogeny and Ridley's (1983) "radiation principle," Grafen suggests that a single data point be "extracted" from each unresolved polytomy on the tree by setting the nodal value equal to the weighted average of the trait values of its descendants. Grafen (1989) then incorporates the error generated from this procedure into the variance-covariance structure of the generalized least squares model. Pagel (1992) presents a similar approach, which differs mostly in suggestions that some phenetic or other information be used to reduce multiple nodes initially before applying his "expansion" procedure. The "pairwise" methods can be seen as an extreme form of these topology reductions designed for situations in which the available phylogenetic information only allows the identification of closely related pairs of extant species (e.g., two bats, two whales). All of these topology reduction methods are conservative and may be useful when polytomies are due to uncertainties in the available phylogenetic information rather than to a belief that several species really did radiate simultaneously from a single point. In the latter case, Felsenstein's (1985) suggestion of resolving the polytomy into a set of bifurcations by inserting branches of zero length is more appropriate (which species are paired with which does not matter).

Other methods consider ways of transforming branch lengths on a rough "working" phylogeny into units of expected amount of phenotypic change for particular characters in the study. For example, Grafen's (1989) phylogenetic regression also includes a statistical estimator, ρ, for inferring the relative amount of phenotypic evolution expected along each branch of the topology (i.e., branch lengths in units of expected variance of character change). He suggests that each branch length (h) from the working phylogeny (scaled such that the tree has a total length of one) be raised to a power ρ, and then that $1 - h^{\rho}$ be considered as a measure of the expected amount of similarity due to phylogenetic relationships (i.e., the variance of the trait value or contrast). Grafen (1989) then shows how ρ can be estimated along with the other parameters of his regression model using maximum likelihood techniques. Alternatively, using the framework of Felsenstein's method, Martins (1994) proposed an iterative least squares procedure in which branch lengths in units of time can be transformed into units of expected variance of character change using a Brownian motion model underlying each character rather than the full regression model used by Grafen. Martins' method can also transform branch lengths based on an Ornstein-Uhlenbeck model of phenotypic evolution under random genetic drift with stabilizing selection. Thus, although Martins' (1994) procedure was described originally as a means of estimating the rate of phenotypic evolution of a single character from comparative data, it is also useful as a way of testing whether the Brownian motion model underlying Felsenstein's (1985) method adequately describes the interspecific variation observed. Finally, from a more subjective approach, Garland et al. (1991) suggest that the absolute value of each standardized independent contrast be plotted versus its standard deviation (the square root of the branch lengths in units of expected variance of change) and that a qualitative, visual analysis be used to determine whether there are any general patterns which would suggest that the branch lengths are in the wrong units. They then suggest that statistical transformations of the data or branch lengths be used to correct this problem. One potential problem with this approach is that any statistical transformations of the data or branch lengths are, in effect, changing the microevolutionary assumptions underlying the comparative analysis. Purely statistical solutions to the problem of unknown or incomplete phylogenetic information can lead to difficulties in interpreting results.

4. Strengths and weaknesses

If the assumptions and the phylogenetic information provided are correct, the independent contrasts method is guaranteed analytically to correct the problem of statistical dependence (Felsenstein 1985, Grafen 1989). Since Felsenstein's (1985) method may be derived from population genetic principles, it also has the advantage of yielding results that can be interpreted from an evolutionary perspective. Simple extensions of the method can be used to estimate related parameters such as the rate of phenotypic evolution (e.g., Garland 1992; Martins 1994). Both Felsenstein's (1985) and Grafen's (1989) methods have also been shown to provide quite good estimates of evolutionary relationships between traits, often much better estimates than provided by other methods, including inferred changes or spatial autocorrelation techniques (Martins and Garland 1991b; Grafen & Ridley in press; Martins in review). Even when phylogenetic information is incorrect, Felsenstein's contrasts method usually matches or exceeds the statistical performance of other comparative methods proposed for continuous traits.

Most of the criticisms and discussion of independent contrasts have centered on the obvious difficulties of obtaining accurate and complete information about both the phylogenetic relationships (the tree structure) and the process of phenotypic evolution (branch lengths in units of expected character change). The topology reduction algorithms and branch length transformation procedures proposed to address the problem of unavailable phylogenetic information may not always perform as well as the original method and should be used with caution (Gittleman & Luh 1992, 1993; Purvis et al. 1994; Martins in review). These methods are statistically conservative in the sense that the resulting confidence intervals are too wide, and the power of statistical tests to detect real evolutionary patterns is low. When these methods are used, any variation in the comparative data is explained first by reference to the phylogeny and will only later be considered as the possible result of correlated evolution, convergence, or other responses to selection. Thus, relationships between traits can be obscured, and the absolute values of estimates may be downwardly biased. Again, if the Brownian motion model is appropriate and the phylogenetic information is known, most of the above variations of the basic independent contrast method yield identical results, and Felsenstein's (1985) original method or Grafen's (1989) "standard regression" should be applied. Use of the pairwise methods, topology reduction algorithms, and branch length estimation procedures should yield similar results but may be less accurate or even biased as the methods use less of the available information. If Brownian motion is not appropriate or if the estimated phylogenetic information is inaccurate, none of these variants of the independent contrasts approach is guaranteed to correct the problem of statistical dependence.

5. Computational difficulties and computer programs

For small phylogenies, Felsenstein's method can be computed by hand. For larger phylogenies, his method has been implemented in *PHYLIP* which is menu-driven, user-friendly, and available for DOS/Windows, Macintosh and UNIX platforms (Felsenstein 1993). Grafen's method can be implemented using a program that is available from the author. This program is in *GLIM*, and requires substantial knowledge of that programming language for implementation. Purvis and Rambaut's (in press) *CAIC* offers a user-friendly Macintosh program to calculate independent contrasts using Pagel's (1992) topology reduction procedure. Martins and Garland's (1991a) *CMAP* implements several variants of Felsenstein's independent contrast method for DOS machines. Martins' (1995) *COMPARE* is available for DOS/Windows, Macintosh, and UNIX platforms; it calculates Felsenstein contrasts, estimate the rate of phenotypic evolution using Martins' (1994) method, and implement several other comparative methods.

C. Explicit model based methods for categorical traits

1. The methods

There are a number of methods which, like Felsenstein's (1985) independent contrasts, are based on explicit models of phenotypic evolution (e.g., Harvey & Pagel 1991, p101; Janson 1992; Sanderson 1993; and Pagel 1994). These methods differ from the contrasts methods, however, in that they were explicitly designed for use with dichotomous or categorical (i.e., yes/no or "state") variables and therefore cannot be used to form "contrasts." The methods begin by building a probability model of the possible transitions between character states. Most then estimate these transition probabilities between states using maximum likelihood techniques.

These probability model based methods were designed to address a rather wide variety of evolutionary questions. Janson's (1992) method asks whether a single trait has evolved neutrally along a phylogeny or whether it was subjected to evolutionary constraints of various defined sorts. Sanderson's (1993) method considers whether there have been significantly more gains than losses in a trait along a tree (Fig. 5). Pagel's method (Pagel 1994; Harvey & Pagel 1991, p. 101) is concerned with determining whether two traits have evolved in a correlated fashion, by estimating the probability that evolutionary changes in one trait have been associated with evolutionary changes in a second. Similar probability models have been used by authors considering the use of DNA sequences to reconstruct phylogenetic trees, and questions in molecular evolution (e.g., Barry & Hartigan 1987; Ritland & Clegg 1987; Hendy 1989; Hendy & Penny 1989; Reeves 1992). Although the models proposed in these latter studies could be applied to the question of phenotypic evolution, as of yet they are primarily concerned with tree reconstruction and are beyond the scope of this chapter.

2. Assumptions

All of the methods described in this section are based on specified probability models. The methods usually require that each character exhibit only a few (usually two) states, that the probability of change between those states be constant throughout evolution in the clade, that within-species variability be negligible, and that the phylogeny be known without error. Each method also makes a number of other simplifying assumptions that differ from method to method. All except Pagel (1994) assume that the ancestral character states are known. In practice, when applying Janson's (1992) and Sanderson's (1993) methods, for example, the ancestral states are estimated from the extant species data using some sort of parsimony algorithm, and then used as if they were known without error. In Pagel's (1994) method, the likelihood is summed over all possible assignments of ancestral states.

3. Variations for dealing with unknown phylogenetic information

None.

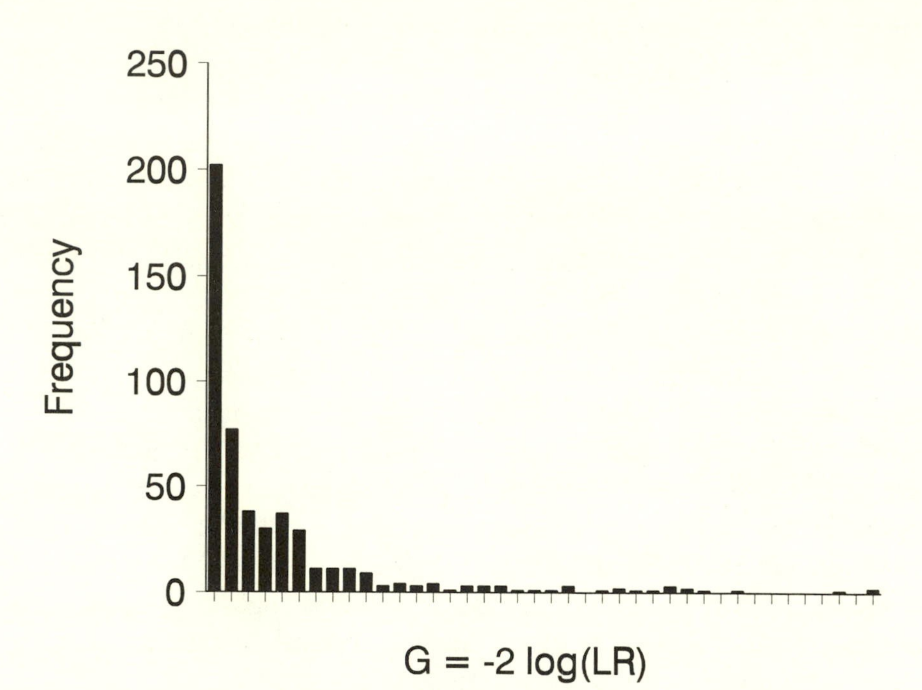

Figure 5. A frequency distribution modified from Sanderson (1993), illustrating the use of his method. Beginning with reconstructions of the ancestral states of a phenotypic character, Sanderson's (1993) method can be used (among other things) to test whether the rate at which a dichotomous (i.e., yes/no) trait is gained equals the rate at which the same character is lost along a known phylogeny. Use of the method yields a quantity, $G = -2 \log (LR)$, where LR is a likelihood ratio calculated from the measured data. This figure depicts the frequency distribution of G obtained by computer simulation for a phylogeny of 128 taxa. Such distributions can be used to conduct simple hypothesis tests about biases in evolutionary gains and losses. For example, an empirical measure of G calculated for a specific trait evolving along this phylogeny of 128 taxa is said to be significantly different from that expected under the null hypothesis of equal rates of gains and losses if it falls in the rightmost 5% of the distribution depicted in this figure.

4. Strengths and weaknesses

The main advantage of the methods discussed in this section lies in their explicit reliance on probabilistic models. As with Felsenstein's (1985) contrasts method, these probabilistic models make it much easier to relate both the assumptions of the methods and the interpretations of resulting parameter estimates to the underlying microevolutionary processes (Hansen & Martins in press). Although such comparisons have not been pursued with most of the methods reviewed in this section, the possibility of doing so remains one of the strengths of the existing methods.

As with many of the inferred changes methods, the primary weakness of most of the probability model methods is the requirement that ancestral character states be known without error. As discussed above, this can lead to inflated sample sizes and estimated parameters may reflect more about the method of reconstructing the ancestral states than about the evolutionary process being inferred. Pagel (1994) solves this problem by utilizing as likelihood the marginal distribution of the extant species which does not depend on the ancestral character states. This leaves only the state of the root to be specified

5. Computational difficulties and computer programs

Pagel's method is computationally intensive, requiring several onerous computations (especially with more than a few possible character states). Janson's and Sanderson's methods are less computer-intensive. Programs to implement these procedures are available from each author on request.

D. Spatial autocorrelation

1. The method

Cheverud et al. (1985, Cheverud & Dow 1985a) suggested the use of spatial or network autocorrelational techniques (as adapted from Cliff & Ord 1981) to incorporate phylogenetic information into comparative analyses. Their method uses a spatial autoregressive model to partition the variation in each measured species phenotype into a factor describing the predicted phenotype of the species given the phylogeny and the measured phenotypes of all the other species in the clade (i.e., the "phylogenetic component") and a factor unique to that species (the "residual component"). In mathematical terms, the model is $\mathbf{y} = \rho\mathbf{W}\mathbf{y} + \varepsilon$ where $\mathbf{y}$ is a vector of the observed species phenotypes, $\rho\mathbf{W}\mathbf{y}$ is the phylogenetic component, and ε is the residual component. In the phylogenetic component, $\mathbf{W}$ is a N x N "phylogenetic connectivity matrix" where N is the number of species. The off-diagonal elements of $\mathbf{W}$ describe the expected similarity among species due to phylogenetic relationships, whereas the diagonal elements are set to zero such that for each species the phylogenetic component is the phenotype for that species predicted by the phenotypes of all the other species not including itself. In practice, the degree of expected phenotypic similarity (i.e., each element of $\mathbf{W}$) is a combination of the known phylogeny and a particular hypothesis regarding how much the character is expected to change as it evolves along the branches of the phylogeny. It is loosely comparable to the "branch lengths in units of expected variance of change" required for Felsenstein's (1985) contrasts method.

The autoregressive model is fit by estimating the autocorrelation coefficient, ρ, (which is completely unrelated to Grafen's ρ above) using maximum likelihood techniques. This coefficient describes the amount of variation in the character explained by the particular phylogenetic hypothesis (represented by $\mathbf{W}$). Positive values of ρ imply that closely related species are phenotypically similar to one another in the way specified by the $\mathbf{W}$ matrix, whereas negative values suggest that character displacement or other forces have led to closely related species being more phenotypically different than distantly related species.

As pointed out by Cheverud et al.(1985), a measure of the overall fit of the model [e.g., $r^2 = 1 - \Sigma(\varepsilon_i^2)/\Sigma(y_i^2)$, where the summations are over all species i] can be used as an estimate of the amount of variation in the character explained by the phylogenetic hypothesis (i.e., a measure of the phylogenetic correlation discussed above). A low value of r^2 implies either that the species phenotypes are not well predicted by the phenotypes of the rest of the clade or that the phylogenetic hypothesis and/or autoregressive model is incorrect. If the model describes the data reasonably well, the residual components (obtained by subtraction; $\varepsilon = \mathbf{y} - \rho\mathbf{W}\mathbf{y}$) will be statistically independent variables on which further statistical analyses (e.g., regression, ANOVA, multivariate statistics) can be performed. Thus, fitting this model to measured comparative data is equivalent to conducting a linear transformation or filtering of those data that takes phylogenetic information into account.

In an extension of the basic method, Gittleman and Kot (1990) developed the use of Moran's *I* and spatial autocorrelograms as diagnostic tools to determine whether or not the model fits comparative data well (see also Gittleman & Luh 1992, 1993, Purvis et al. 1994, and in this volume, Gittleman et al., Chap. 6). An autocorrelogram is a histogram plot of Moran's *I* by historical distance. Moran's *I* is a relative measure of the expected similarity among species phenotypes due to a specific phylogenetic connectivity matrix. High values of Moran's *I* indicate that the proposed phylogenetic or taxonomic hypothesis does indeed explain some of the variation present in the data. If *I* is small or negative, the spatial autoregressive model may not provide a reasonable fit to the comparative data, and/or incorporation of historical information may be unnecessary. An autocorrelogram can be obtained by slicing a taxonomy or phylogeny into several vertical segments and calculating Moran's *I* to describe the extent of phylogenetic correlation within each segment. For example, if only taxonomic information is available, we might calculate Moran's *I* at the species, genus, family and order levels (e.g., Fig. 6). When phylogenetic information is available, we might slice the phylogeny into vertical segments to describe the phylogenetic correlation in data from species which have a common ancestor 0–10 million years ago, 10–20 my ago, and so on. In the latter case, it is not clear what the best interval of time (or genetic distance, or number of evolutionary changes, or other unit of branch length) is most useful for calculation of Moran's *I* (see Purvis et al. 1994 for a suggestion). Once the autocorrelogram has been obtained, it can be used to determine whether an autoregressive model is appropriate for a particular set of data and relationship matrix (**W**). If an autoregressive model is appropriate, we expect Moran's *I* to be positive at several levels (indicating that phylogenetic correlation can be detected) and to decrease with increasing taxonomic or phylogenetic distance. If particular values of Moran's *I* are small or negative, we might consider altering the relationship matrix (e.g., by using a cut-off value) to reduce the effects of those parts of the phylogeny which do not fit the autoregressive model. Thus, visual inspection of a correlogram can serve as a diagnostic tool in applying the spatial autoregressive model.

The spatial autoregressive model proposed by Cheverud et al. (1985) is based on linear regression techniques and is designed primarily for the analysis of continuous characters. As illustrated by the authors, dichotomous (yes/no) characters can also be analyzed using this model, and by extension, categorical or "state" characters can be

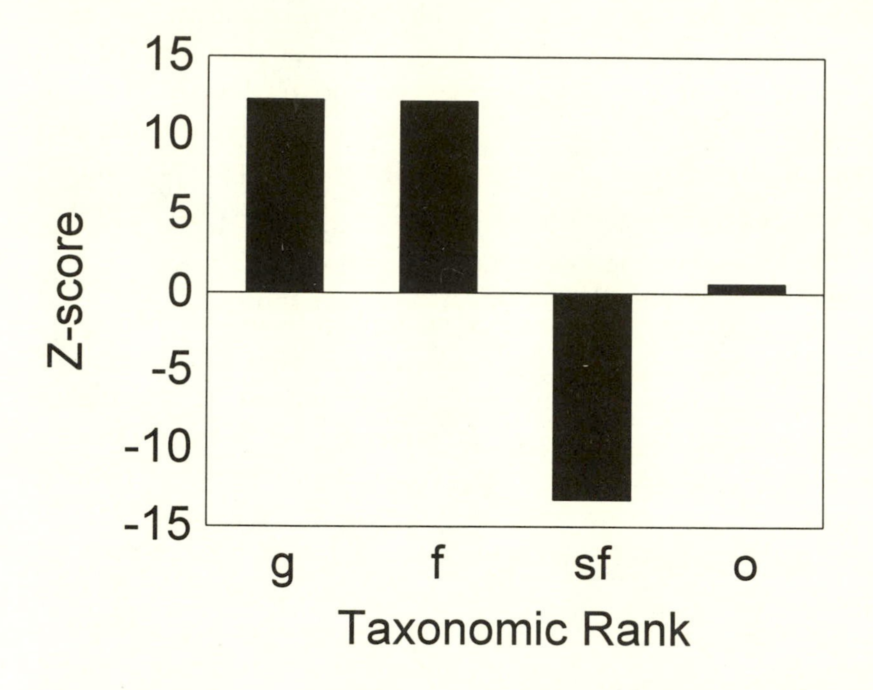

Figure 6. Phylogenetic correlogram modified from Gittleman & Kot's (1990) analysis of carnivore body weight (using Nowak & Paradiso's 1983 classification of 123 species). By varying the phylogenetic relationship matrix, normalized values of Moran's I and their respective z-scores were calculated at the genus, family, superfamily and order levels. These values describe the amount of phylogenetic correlation (i.e., a relationship between character variation and the phylogeny) present at each taxonomic rank, and can be used to suggest the best taxonomic level for further analysis, and to determine whether phylogenetic information need be taken into account for such analyses. Z-scores with absolute values greater than 1.96 are considered to be significantly different from zero. In this case, the correlogram shows that phylogenetic relationships among species would have a significant effect on statistical analyses conducted at the genus, family, or superfamily levels.

considered if they are first converted into sets of dichotomous "dummy" variables (e.g., see Draper & Smith 1981).

2. Assumptions

The biological assumptions of the spatial autoregressive method have not been explicitly stated but include at least the following partially overlapping points: (a) within-species variation is negligible or incorporated into the analysis, (b) the phylogeny is known, (c) each species phenotype can be described as a linear function of the phenotypes of all of the other species on the phylogeny (i.e., an autoregressive model), (d) a species phenotype can be represented as a simple sum of phylogenetic and a nonphylogenetic components, and (e) only the non-phylogenetic component (ε) is of interest in the further study of the trait.

The spatial autoregressive method is a purely statistical approach which was developed in the absence of an explicit model of phenotypic evolution. Although this gives the method substantial robustness and flexibility to variation in the microevolutionary process, it also makes evolutionary interpretations difficult. In principle, the **W** matrix can be specified so as to correspond to a given phylogenetic correlation pattern, but the relationship is quite complex (Martins & Hansen, 1995). In computer simulation tests of the method, the **W** matrix has not been set in this way to correspond to the model actually used to generate the data (such as Brownian motion). Thus, it is not surprising that computer simulation results show that although this method provides reasonable estimates of the evolutionary relationship between two continuous characters, these estimates are almost always somewhat less accurate than those provided by the inferred changes and independent contrasts methods (Martins in review). The residual components of the autoregressive model are particularly difficult to interpret in an evolutionary framework. One possibility is that the residual components represent recent adaptations to a changing environment that occurred after each species split away from its most recent common ancestor. If species can respond that quickly to selective forces, however, it is not clear why their phenotypes would retain any phylogenetic component at all or why we would want to consider possible adaptation to the environment along the most recent branches of the phylogeny but ignore it elsewhere.

3. Variations for dealing with unknown phylogenetic information

In another extension of the spatial autoregressive model, Gittleman and Kot (1990) proposed the use of a statistical estimator that allows flexibility in the phylogenetic connectivity matrix (**W**) when phylogenetic information is uncertain or unavailable. Elements of the **W** matrix describe the expected degree of phenotypic similarity between species. For example, the element of **W** corresponding to species i and j (w_{ij}) is the inverse of the expected phenotypic divergence between those two species ($w_{ij} = d_{ij}^{-1}$, usually given as the inverse of the time that the two species have been evolving independently of one another). Gittleman and Kot (1990) proposed that $d_{ij}^{-\alpha}$ (or $\exp[-\alpha d_{ij}]$) be used, instead of d_{ij}^{-1} as the elements of **W**, and showed how α can be estimated using maximum likelihood techniques. The parameter α stretches and shrinks the phylogeny (in much the same way as Grafen's ρ parameter above) to vary the phylogenetic connectivity matrix **W** when the phylogeny or expected amount of phenotypic change along that phylogeny are not known. Alpha also serves as a rough estimate of the rate or tempo of phenotypic evolution in a particular character, given a particular phylogeny and value of ρ.

4. Strengths and weaknesses

The spatial autoregressive method has great heuristic appeal because it offers not only a solution to the problem of statistical dependence but also an estimate of the magnitude of this problem (i.e., the amount of variation in the trait that is due to phylogenetic history). Gittleman and Kot (1990) have contributed substantially to the method by providing useful exploratory and diagnostic tools to test the assumptions of the model and a statistical estimator (α) that allows for flexibility when phylogenetic information is unknown or uncertain. Representation of the phylogeny and model of phenotypic evolution as a connectivity matrix also gives this method somewhat more flexibility than the independent contrasts approach in that phylogenetic relationships need not be represented as a binary phylogeny. This may be of substantial importance for studies involving (a) species hybridization or horizontal gene transfer, (b) ecological factors which may not evolve by the usual microevolutionary models or along a phylogeny, and (c) population or individual level comparisons (e.g., Edwards & Kot in press; Foster & Cameron, Chap. 5 in this volume).

Possibly the most important weakness of this method is that it can have extremely poor statistical performance when the model does not fit the data well. Computer simulation results suggest that if the comparative data really have little or no phylogenetic correlation or if the sample size is too small for any phylogenetic correlation to be detected (fewer than about 40 species), the spatial autoregressive method can give far more inaccurate parameter estimates and hypothesis tests than would not taking phylogeny into account at all (Martins in review; but see Gittleman & Luh 1992, 1993). Fortunately, Gittleman and Kot's (1990) diagnostic techniques can and should be used to identify such situations and avoid these potential problems (see also Gittleman & Luh 1992, 1993; Purvis et al. 1994). Martins' (in review) results also suggest that Gittleman and Kot's α parameter should be used cautiously. When a strong correlation between two characters exists, use of the α parameter can lead to biased results (strong correlations will be diminished). Thus, we recommend that analyses always be conducted both with and without the α parameter and that differences between the results of the two analyses be interpreted cautiously.

A further difficulty is that the method is purely univariate and must be applied to each trait separately. Correlations between traits are ignored in the estimation and correction of phylogenetic effects. Thus, when the method is used to estimate the correlation between two traits, we first assume that the relationship between characters is insignificant in comparison to the effect of phylogenetic history and fit spatial autoregressive models to each of the two traits independently. We then calculate a correlation coefficient between the residuals (i.e., the nonphylogenetic components) from the two models to determine whether there is a relationship between the two traits. If a substantial relationship between the traits exists, then the original assumption that the relationship between characters is insignificant in comparison to the phylogenetic effect has been violated. A better approach would include both traits in a more complex transformation model (e.g., as suggested by Ely & Kurland 1989).

5. Computational difficulties and computer programs

The spatial autoregressive method is computationally intensive and requires a computer to do the analyses. Programs to conduct the analyses are available from several researchers. Cheverud and Dow's (1985b) MINRHO3 is a simple BASIC program which requires calls to a separate set of commercial matrix manipulation software and fits the original autoregressive model. Luh et al. (1994) offer a Macintosh program that will fit the spatial autoregressive model to a set of comparative data with the modifications proposed by Gittleman and Kot (1990). Martins' (1995) COMPARE is available for DOS/Windows, Macintosh and UNIX platforms, and will fit the spatial autoregressive model with and without the modifications by Gittleman and Kot (1990).

E. Comparing relationship matrices

1. The methods

As in the methods described above, given an appropriate model of phenotypic evolution, variation in any set of comparative data can be described as either a branching tree structure (a phenogram) or a phylogenetic relationship matrix. Thus, there have been several studies in which researchers constructed a phenogram or distance matrix describing the patterns of similarity observed in a set of interspecific data and compared this to a known phylogeny or phylogenetic distance matrix. The comparison can be made qualitatively or statistically, and relationships between the two types of trees or matrices can be interpreted as a measure of the amount of phylogenetic correlation or constraint in the data.

If interspecific measurements of a behavior pattern are described as a phenogram, this branching tree can then be compared to other trees (e.g., a phylogenetic hypothesis based on morphological or molecular data) to determine how similar the evolution of the behavior has been to the patterns of speciation in the clade. Phenograms based on behavioral trees can also be compared to trees based on other sorts of characters to test whether the evolution of behavior has been qualitatively different from the evolution of other traits (e.g., morphology). Such comparisons can be conducted using standard techniques such as consistency or retention indices borrowed from traditional systematics (e.g., in this volume, de Queiroz & Wimberger, Chap. 7; Irwin, Chap. 8).

Alternatively, many authors have described a set of comparative data as an N x N phylogenetic relationship matrix where N is the number of taxa, and the elements of the matrix are some measure of the similarity or differences between each pair of taxa. Given an appropriate evolutionary or statistical model of phenotypic evolution, these matrices can be developed for individual characters or groups of characters and then compared, usually using some sort of permutation test. Matrix comparisons or "matrix correlation analysis" has been particularly popular in anthropology and population genetics when the taxa being considered are populations rather than species and where comparisons are generally conducted using Mantel's test (a permutation test; see Smouse and Long 1992 for review). At the species level, Legendre et al. (in press) proposed an alternative way of using multiple regression to compare the variation in a set of interspecific data (described as a phenetic distance matrix) with a phylogenetic relationship matrix based on independent information. In this method, a correlation coefficient is used to describe the relationship between the two matrices and a permutation test is used to correct for the non-independence within each matrix. The correlation coefficient provides a measure of phylogenetic correlation.

This general approach can be used with either continuous or categorical characters. Whether a specific method can be used with categorical or continuous characters depends primarily on the algorithms used to describe interspecific variation as a branching tree structure or as a relationship matrix.

2. Assumptions

There are several assumptions implicitly buried in (a) the creation of the branching trees or matrices from real data, and (b) the permutation tests or other statistics used to compare matrices. Algorithms for constructing phenograms or relationship matrices from comparative data usually make several assumptions about the process of phenotypic evolution underlying the character that has been measured (e.g., Martins & Hansen 1995). Permutation tests or statistics that summarize the shape of a branching tree necessarily reduce the variation present in a set of comparative data by concentrating on a few important aspects of the matrix or tree. The choice of aspects to consider implicitly assumes that only these aspects are important in the evolution of the characters.

3. Variations for dealing with unknown phylogenetic information

None.

4. Strengths and weaknesses

The basic matrix comparison approach is intuitively simple and easy to perform. As mentioned below, calculations can be conducted using a wide variety of available computer programs.

There are many different ways of comparing matrices or branching structures and each makes its own set of limiting assumptions. Little effort has been made to view these methods in an explicitly evolutionary framework, so it is quite difficult to understand the meaning and effect of the exact assumptions made by different methods. Both matrix and tree comparison methods can also be extremely vulnerable to various scaling effects and may consider two sets of comparative data that are simply scaled versions of each other to be qualitatively different phenomena. Given the difficulties regarding both assumptions and interpretation of the results, we do not recommend use of these methods without a deep understanding of the evolutionary implications of the models underlying the analyses of the particular method being used.

5. Computational difficulties and computer programs

Most computer packages offering ways of reconstructing phylogenies also offer ways of building phenetic treelike structures or relationship matrices from sets of comparative data (e.g., Maddison & Maddison 1992; Felsenstein 1993; Swofford 1993). These programs also usually calculate summary statistics such as consistency indices that can be used to compare branching trees to each another. Permutation tests for comparing relationship matrices are becoming increasingly common in standard statistical packages (e.g., SPSS, SAS).

F. Lynch's method

1. The method

Lynch (1991) developed another linear model to incorporate degree of relatedness into comparative analyses. He drew an analogy between individuals within populations and species within a phylogeny and expanded an existing quantitative-genetic model to partition variation in the observed species mean phenotypes ($\bar{z}$) into a phylogeny wide mean phenotype (μ), a phylogenetically heritable component (a) and a residual component (e) due to nonadditive phylogenetic effects, environmental effects and measurement error ($\bar{z} = \mu + a + e$). Maximum likelihood techniques can be used to estimate each of these components which might then be used in further analyses to examine the evolutionary histories of individual traits or of relationships among traits. For example, ancestral values of a character might be estimated from the mean phenotype (μ) and the heritable component (a) without the use of the nonheritable, residual component (e). The full model is multivariate, such that relationships between two characters can be estimated directly from the model as the correlation between the heritable components of the two traits.

The method is based on linear regression techniques and is therefore designed for use with continuously varying characters. As with Grafen's (1989) and Cheverud et al.'s (1985) methods (described above), it may also be reasonable to consider categorical or "state" characters as sets of dichotomous "dummy" variables with this method.

2. Assumptions

Lynch's (1991) method assumes that (a) phylogenetic information is available and can be represented as a matrix (**G**) of expected similarities among species and (b) interspecific variation in a trait resulting from shared evolution along a phylogeny is well described by a linear model with normally distributed, additive, phylogenetically heritable components and phylogenetically uncorrelated residual components that have a mean of zero. The first assumption is similar to that used in the spatial autocorrelation methods discussed above. The two methods differ substantially in interpretation, however, as the spatial autocorrelation methods conduct further analyses using the residual component of the model, whereas Lynch suggests that the phylogenetically heritable component should be used. Essentially, Lynch's method views character evolution and phylogenetic correlation as one and the same process in much the same way as Felsenstein's contrasts method does. The spatial autoregression is more similar to Grafen's (1989) regression method in interpretation.

3. Variations for dealing with unknown phylogenetic information

None.

4. Strengths and weaknesses

The parameters of Lynch's model are easy to interpret via the analogy with similar parameters in quantitative genetics. As with the spatial autoregressive method, Lynch's technique also has the advantage of not being limited by the binary structure of a phylogeny, and it provides an estimate of the degree of phylogenetic correlation (his "phylogenetic heritability"). Lynch's method further improves on earlier methods by incorporating measures of within-species variability into the analysis.

The method is computationally difficult and requires an iterative procedure to converge on a final result. Some further development is required before the method can be put to practical use (Lynch, pers. comm.). The method has not yet been implemented in any distributed computer programs, and has not yet been applied to even a single set of real data. It must be regarded as promising but still in a preliminary stage of development.

5. Computational difficulties and computer programs

See above.

G. Nested ANOVA

1. The method

Nested ANOVA models (e.g., Clutton-Brock & Harvey 1977, 1979, 1984, Bell 1989) were the first to be suggested in the recent comparative method literature (see Harvey & Pagel 1991 for review). The authors of these methods suggest that the phylogenetic relationships among species can be described as a set of hierarchical factors in ANOVA models. For example, if data were measured from 10 individuals in each of the six species in Fig. 1*b*, we might consider analyzing these data with a nested ANOVA in which the species data are nested within three groups representing the three different taxonomic classes of animals (mammals, birds and insects). With more species, a researcher might nest species within genera, genera within families, families within orders and so on. Thus, the method provides estimates of the effects due to belonging to a certain genus, family, and so on.

When this is done, adaptation can be investigated by choosing a taxonomic level that is relatively free of phylogenetic correlation for the analysis (Clutton-Brock & Harvey, 1977). For example, if species are very influenced by the genus to which they belong, but the genus effect is not strongly influenced by the family to which the genus belongs, the analysis can be performed with genus means as data points. Stearns' (1983) "phylogenetic-subtraction" method performs the analysis on the species level but first subtracts all the estimated higher order effects from the species data. This differs fundamentally in interpretation from Clutton-Brock and Harvey's original suggestion as Stearns uses the information which is discarded in the Clutton-Brock & Harvey approach and visa versa. The phylogenetic subtraction method studies recent species-specific adaptation by attempting to control for phylogenetic correlation due to evolution occurring prior to the species level.

ANOVA models of this sort were designed originally for use with continuous variables. Although nested logistic regression and log-linear models (the analog of ANOVA for categorical variables) could be used in much the same way as nested ANOVA to take phylogenetic relationships into account, this application has not been developed.

2. Assumptions

Nested ANOVA methods assume that: (a) phylogenetic relationships can be described as a strictly hierarchical tree structure in which each taxonomic group within a larger group (e.g., all species within a genus, all genera within a family) have evolved independently of one another as a "star" phylogeny and (b) the process of phenotypic evolution has been constant throughout the entire clade and that each taxonomic group has evolved in the same way (at the same rate and for the same length of time) as all other taxa on the tree. These are essentially the same assumptions as those of the nonphylogenetic approach, adding only the hierarchical structure of a taxonomy and assuming that this taxonomy directly reflects phylogenetic and evolutionary relationships. The validity of the nested ANOVA approach depends primarily on the extent to which the taxonomic clustering used in the nested ANOVA is similar to a phylogeny describing the relationships among species.

3. Variations for dealing with unknown phylogenetic information

None.

4. Strengths and weaknesses

The nested ANOVA approach has been popular, as it is relatively simple to apply using most standard statistical computer packages (e. g., SPSS, SAS). It is useful as an exploratory tool to describe the broad patterns of phylogenetic correlation. Harvey & Pagel (1991) termed this method obsolete because it neither solves the problem of statistical dependence nor proposes better estimates of evolutionary relationships. Rather than assuming that species are independent, it assumes that genera, families or orders are independent. When only hierarchical taxonomic information is available, these assumptions may seem reasonable and the method may still be useful. If any alternative phylogenetic information is available, any of the methods that allows for greater flexibility in the phylogenetic structure may be preferable.

5. Computational difficulties and computer programs

Nested ANOVA is a standard parametric statistical technique that is implemented in most commercial statistical software.

Discussion

The main goal of most statistical analyses in animal behavior is to obtain an estimate of a statistical parameter and some measure of its uncertainty. For example, we might use statistics to tell us whether the home range sizes of dominant animals are larger or smaller than those of subordinate animals and whether that difference is greater than what would be expected by chance alone. A comparison of the mean home range size for each group of animal will give us an answer to the first question, and a standard error, a confidence interval or — less informatively — a *p* value from a hypothesis test can be used to answer the second. Both answers, however, will be given within the bounds of some assumed statistical model and the range of available data. For example, if the data in our study were measured from individual sparrows, we would have some reservations about any conclusions made about the behavior of lizards from those data. Similarly, any conclusions or predictions made outside the bounds of the statistical model may be subject to error.

When interspecific data and phylogenetic comparative methods are considered, the statistical model is a complex one, consisting of three parts: (a) any hypothesis of the patterns of speciation underlying the measured species (i.e., a phylogeny, often with branch lengths), (b) a model or assumption regarding how this phylogeny translates into the expected relationships among species phenotypes (e.g., parsimony, the Brownian motion assumption underlying Felsenstein's independent contrasts, the $\mathbf{y} = \rho\mathbf{W}\mathbf{y} + \varepsilon$ model of the spatial autocorrelation method); and (c) a statistical model or assumption describing how the parameter we are interested in (e.g., difference in home range size) can be estimated from the measured data (e.g., a specified regression model). Thus, any conclusion based on the statistical analysis of interspecific data will depend on the accuracy and appropriateness of the assumptions made in each of these three parts, whether or not these assumptions are explicitly specified.

For example, imagine a study in which we want to know how much of the variation in mammal home range size can be predicted by differences in body size. We have measured the home range and body sizes of several different species of mammals and estimate the slope of a simple linear regression (b_1) of home range size (the response or y variable) on body size (the predictor or x variable) without making any attempt to take phylogenetic information into account. We use the statistical model: $y = b_0 + b_1x + e$ (where b_0 gives us a y intercept and e is an error term), and make a number of assumptions. For example, if we want the regression slope to be an unbiased estimate of the "true" relationship between these two characters, we might be assuming in the above three-part assumption, that: (a) the phylogeny of mammals is a "star" phylogeny, as in Fig. 1*a*, (b) both home range and body size have evolved at the same rate along each branch of that phylogeny, and (c) home range size is linearly related to body size, no within-species variation exists, and the error term in the regression model consists of independent, normally distributed variables. To estimate a standard confidence interval on the resulting regression slope (or to conduct a hypothesis test), we must also assume that the error terms in the regression model are statistically uncorrelated with each other and have the same variance. Other sets of assumptions are also possible. For example, instead of the first two of the above assumptions, we might assume that brain and body size were responding so quickly to selective pressures in these species that no evidence of shared phylogenetic histories remains in the data. In this case, the shape of the phylogeny (assumption a), and the rate of phenotypic evolution (assumption b) need not be known. The third assumption remains unchanged.

Lack of specificity or vagueness in the underlying model does not mean that a method is more robust or general. For example, Felsenstein's independent contrast method has been viewed skeptically due to its Brownian motion assumption and its requirement of phylogenetic information. All of the existing methods, however, make comparable sorts of assumptions either implicitly or explicitly, since they must all address the three parts of the comparative method model listed above. For example, as reviewed above, Grafen's (1989) phylogenetic regression assumes that the error term in his multiple regression model consists of normally distributed variables with specified variances and covariances. As it stands, Brownian motion is the only microevolutionary model grounded in a population and quantitative genetic framework that produces the variance-covariance matrix actually used in Grafen's applications. What Grafen's method gains by allowing for the possibility of other, unknown microevolutionary models, it loses in terms of the evolutionary interpretation of the results. Similarly, any specification of the connectivity matrix in the phylogenetic autocorrelation method implicitly assumes a model on the same level of detail as Brownian motion. In fact, none of the popular, existing microevolutionary models will produce the patterns of covariation among species assumed by existing applications of the method (Martins & Hansen 1995). In our opinion, given that most of the existing methods make comparable assumptions, explicitness of those assumptions should be viewed as a strength rather than a weakness of a phylogenetic comparative method.

In fact, the specification of the above three-part assumption is a major (and usually unquantified) source of error in the statistical analysis of comparative data. The robustness of different phylogenetic methods to errors in these assumptions and the incorporation of different aspects of this complex model lie at the forefront of the development of modern phylogenetic comparative methods. Most modern comparative methods were originally designed to incorporate known phylogenetic information into the analysis of comparative data, and thereby explicitly address assumptions a and b. Several other methods have been designed to infer the needed phylogenetic information when it is not known (e.g., Grafen 1989, Gittleman & Kot 1990, Pagel 1992; reviewed above), or to incorporate uncertainty in that information into the final estimates of the uncertainties of parameter estimates (e.g., Losos 1995, Martins in press).

Computer simulation studies have shown that several of the available phylogenetic comparative methods can produce estimates of evolutionary relationships with reasonable rates of Type I error. One concern regarding these studies is that many put an unfortunate emphasis on statistical hypothesis testing over parameter estimation. The ubiquitous *p* values found in empirical studies can be very misleading in comparative analyses (or any behavioral study; see Yoccoz 1991 for documentation and discussion). The most common and serious of these misinterpretations is to confuse statistical significance with biological importance. For example, in determining whether body size is a good predictor of home range size, the question should not be whether a relationship between the two traits exists at all but rather whether the relationship is large enough to be of biological importance. Given any association between the two traits, no matter how weak or biologically unimportant, we only need a large enough sample size to make the estimated correlation significantly different from zero. Similarly, when the sample size is small, a statistical hypothesis test may fail to reject a null hypothesis of no relationship between two traits on some specified significance level, even though an important biological relationship exists and is, in fact, indicated by the estimate.

The best conclusion in this situation is not that "there is no evidence for a relationship" but rather that the best estimate of the parameter indicates an important pattern, although this estimate is too uncertain to be taken very seriously.

On an absolute scale, all of the phylogenetic comparative methods that have been tested (i.e., several versions of inferred changes, independent contrasts, and spatial autoregressive methods; all designed for use primarily with continuous characters) produce unbiased, reasonably accurate estimates that are usually far superior to those produced using a nonphylogenetic approach. Although some techniques will not perform well when the species are actually relatively independent of one another (i.e., as depicted in the phylogeny on the left of Fig. 1) or when convergence of unrelated species diminishes the impact of evolutionary history, diagnostic techniques can be used to identify these situations and to avoid poor method performance. Other methods perform reasonably well with any phylogeny and model of evolutionary change when these are known, and all of the tested methods are also relatively robust to inaccuracies in the available phylogenetic information.

More importantly, phylogenetic methods have opened up many new horizons in animal behavior, by allowing us to answer evolutionary questions that did not seem possible previously given traits that do not appear in the fossil record and which may not always be suitable for phylogenetic reconstruction (but see, in this volume, Irwin, Chap. 8, and de Queiroz & Wimberger, Chap. 7). Using phylogenetic methods, we can ask about the rate of behavioral change (e.g., in this volume, Gittleman et al., Chap. 6), the sequence of changes (e.g., in this volume, Crespi, Chap. 9), etc. We can also solve a troublesome statistical problem that has been the subject of much recent discussion. As new techniques for addressing the problems of unknown or uncertain phylogenetic information are developed, there seems to be very little to lose by applying a phylogenetic comparative method and much to gain. Thus, phylogenetic analyses of comparative studies will probably one day become as widely used as repeated-measures ANOVA or paired sample *t*-tests as a simple way of correcting a common statistical problem and a much more powerful tool for exploring questions in the evolution of behavior.

ACKNOWLEDGMENTS

This work was supported by a grant from the National Science Foundation to EPM (#DEB-9406964).

References

Barry, D., & J. A. Hartigan. 1987. Statistical analysis of hominoid molecular evolution. *Stat. Sci.*, 2, 191–210.

Basolo, A. L. 1990a. Female preference for male sword length in the green swordtail, *Xiphophorus helleri* (Pisces: Poeciliidae). *Anim. Behav.*, 40, 332–338.

Basolo, A. L. 1990b. Female preference predates the evolution of the sword in swordtail fish. *Science*, 250, 808–810.

Bell, G. 1989. A comparative method. *Am. Nat.*, 133, 553–571.

Boake, C. R. B. 1989. Repeatability: its role in evolutionary studies of mating behavior. *Evol. Ecol.*, 3, 173–182.

Brooks, D. R., & D. A. McLennan. 1991. *Phylogeny, Ecology, and Behavior*. Chicago: Chicago University Press.

Burt, A. 1989. Comparative methods using phylogenetically independent contrasts. *Oxford Surveys in Evolutionary Biology* , 6, 33–53.

Cheverud, J. M., & M. M. Dow. 1985a. An autocorrelation analysis of genetic variation due to lineal fission in social groups of rhesus macaques. *Am. J. Phys Anthropol.*, 67, 113–121.

Cheverud, J. M., & M. M. Dow. 1985b. *MINRHO3*. Distributed by the authors. Evanston, Illinois: Northwestern University.

Cheverud, J. M., M. M. Dow, & W. Leutenegger. 1985. The quantitative assessment of phylogenetic constraints in comparative analyses: sexual dimorphism in body weights among primates. *Evolution*, 39, 1335–1351.

Cliff, A. D., & J. K. Ord. 1981. *Spatial Processes: Models and Applications*. London: Pion Press.

Clutton-Brock, T. H., & P. H. Harvey. 1977. Primate ecology and social organization. *J. Zool. Lond.*, 183, 1–39.

Clutton-Brock, T. H., & P. H. Harvey. 1979. Comparison and adaptation. *Proc. R. Soc. Lond. B*, 205, 547–565.

Clutton-Brock, T. H., & P. H. Harvey. 1984. Comparative approaches to investigating adaptation. In: *Behavioral ecology: An evolutionary approach*. 2nd ed. (Ed. by J. R. Krebs & N. B. Davies), pp. 7–29. Oxford, England: Blackwell Press.

Draper, N., & Smith. 1981. *Applied Regression Analysis*. New York: John Wiley and Sons, Inc.

Edwards, A. F. W., & L. L. Cavalli-Sforza. 1964. Reconstruction of evolutionary trees In: *Phenetic and phylogenetic classification*. (Ed. by W. H. Heywood & J. McNeill), pp. 67–76. London: Systematics Association Publication No. 6.

Edwards, S. V., & M. Kot. (in press). Comparative methods at the species level: Geographic variation in morphology and group size in Gray-crowned Babblers (*Pomatostomus temporalis*). *Evolution*.

Ely, J., & J. A. Kurland. 1989. Spatial autocorrelation, phylogenetic constraints, and the causes of sexual dimorphism in primates. *Int. J. Primatol.*, 10, 151–171.

Farris, J. S. 1970. Methods for computing Wagner trees. *Syst. Zool.*, 19, 83–92.

Felsenstein, J. 1985. Phylogenies and the comparative method. *Am. Nat.*, 126, 1–25.

Felsenstein, J. 1988. Phylogenies and quantitative characters. *Annu. Rev. Ecol. Syst.*, 19, 445–471.

Felsenstein, J. 1993. *PHYLIP (Phylogeny Inference Package) version 3.5*. Distributed by the author. Department of Genetics, University of Washington, Seattle.

Felsenstein, J., & E. Sober. 1986. Parsimony and likelihood: An exchange. *Syst. Zool.*, 35, 617–626.

Frumhoff, P. C., & H. K. Reeve. 1994. "Using phylogenies to test hypotheses of adaptation: a critique of some current proposals." *Evolution* , 48, 172–180.

Gaillard, J.-M., D. Allainé, D. Pontier, N. G. Yoccoz, & D. E. L. Promislow. 1994. Senescence in natural populations of mammals: a reanalysis. *Evolution*, 48, 509–516.

Garland, T. Jr. 1992. Rate tests for phenotypic evolution using phylogenetically independent contrasts. *Am. Nat.*, 140, 509–519.

Garland, T. Jr., P. H. Harvey, & A. R. Ives. 1992. Procedures for the analysis of comparative data using phylogenetically independent contrast. *Syst. Biol.*, 41, 18–32.

Gittleman, J. L., & M. Kot. 1990. Adaptation: statistics and a null model for estimating phylogenetic effects. *Syst. Zool.,* 39, 227–241.

Gittleman, J. L., & H.-K. Luh. 1992. On comparing comparative methods. *Ann. Rev. Ecol. Syst.*, 23, 383–404.

Gittleman, J. L., & H.-K. Luh. 1993. Phylogeny, evolutionary models and comparative methods: A simulation study. In: *Phylogenetics and Ecology* (Ed. by P. Eggleton & D. Vane-Wright), pp. 103–122. London: Academic Press.

Grafen, A. 1989. The phylogenetic regression. *Phil. Trans R. Soc. B*, 326, 199–157 .

Grafen, A. 1992. The uniqueness of the phylogenetic regression. *J. theor. Biol.*, 156, 405–423.

Grafen, A., & M. Ridley. (in press). Statistical tests for discrete cross-species data. *J. Theor. Biol.*,.

Hall, B. K. (Ed.) 1994. *Homology: the Hierarchical Basis of Comparative Biology*. San Diego, California: Academic Press.

Hansen, T. F., & E. P. Martins. (in press). Translating between microevolutionary process and macroevolutionary patterns: The correlation structure of interspecific data. *Evolution*,.

Harvey, P. H., & M. D. Pagel. 1991. *The Comparative Method in Evolutionary Biology*. Oxford, England: Oxford University Press.

Hendy, M. D. 1989. The relationship between simple evolutionary tree models and observable sequence data. *Syst. Zool.*, 38, 310–321.

Hendy, M. D., & D. Penny. 1989. A framework for the quantitative study of evolutionary trees. *Syst. Zool.*, 38, 297–309.

Hillis, D. M., & C. Moritz. 1990. *Molecular Systematics*. Sunderland, Massachusetts: Sinauer and Associates.

Huey, R. B., & A. F. Bennett. 1987. Phylogenetic studies of coadaptation: Preferred temperatures versus optimal performance temperatures of lizards. *Evolution*, 41, 1098–1115.

Janson, C. H. 1992. Measuring evolutionary constraints: a Markov model for phylogenetic transitions among seed dispersal syndromes. *Evolution*, 46, 136–158.

Krebs, J. R. & N. B. Davies. 1987. *An Introduction to Behavioural Ecology*. Sec Ed. Oxford. Blackwell Scientific Publications.

Legendre, P., F.-J. Lapointe, & P. Casgrain. (in press). Modeling brain evolution from behavior: A permutational regression approach. *Evolution*.

Leroi, A. M., M. R. Rose, and G. V. Lauder. 1994. "What does the comparative method reveal about adaptation?" *Am. Nat.* 143, 381–402.

Losos, J. B. 1995. An approach to the analysis of comparative data when a phylogeny is unavailable or incomplete. *Syst. Biol.* 43, 117–123.

Luh, H-K., J. Gittleman, & M. Kot. 1994. *PA: Spatial autoregression computer program*. Distributed by the authors. Department of Zoology, University of Tennessee, Knoxville.

Lynch, M. 1991. Methods for analysis of comparative data in evolutionary biology. *Evolution*, 45, 1065–1080.

Machlis, L., P. W. D. Dodd, & J. C. Fentress. 1985. The pooling fallacy: Problems arising when individuals contribute more than one observation to the data set. *Z. Tierpsychol.*, 68, 201–214.

Maddison, D. R. 1994. Phylogenetic methods for inferring the evolutionary history and processes of change in discretely valued characters. *Annu. Rev. Entomol.*, 39, 267–292.

Maddison, W. P. 1990. A method for testing the correlated evolution of two binary characters: Are gains or losses concentrated on certain branches of a phylogenetic tree? *Evolution*, 44, 539–537.

Maddison, W. P. 1991. Squared-change parsimony reconstructions of ancestral states for continuous-valued characters on a phylogenetic tree. *Syst. Zool.*, 40, 304–314.

Maddison, W. P., & D. R. Maddison. 1992. *MacClade: Analysis of Phylogeny and Character Evolution*. Sunderland, Massachusetts: Sinauer and Associates.

Martin, P., & H. C. Kraemer. 1987. Individual differences in behaviour and their statistical consequences. *Anim. Behav.*, 35, 1366–1375.

Martins, E. P. 1994. Estimating rates of character change from comparative data. *Am. Nat.*, 144,193–209.

Martins, E. P. 1995. *COMPARE: statistical analysis of comparative data, version 1.0*. Distributed by the author. Department of Biology, University of Oregon, Eugene.

Martins, E. P. (in press). Conducting phylogenetic comparative analyses when the phylogeny is not known. *Evolution*,.

Martins, E. P. (in review). Phylogenies, spatial autoregression, and the comparative method: A computer simulation test. *Evolution*.

Martins, E. P., & T. Garland, Jr. 1991a. *CMAP: Comparative Method Analysis Package*. Distributed by the author. Department of Zoology, University of Wisconsin, Madison.

Martins, E. P., & T. Garland, Jr. 1991b. Phylogenetic analyses of the correlated evolution of continuous characters: a simulation study. *Evolution*, 45, 534–557.

Martins, E. P., & T. F. Hansen. 1995. A microevolutionary link between phylogenies and comparative data. In: *New Uses for New Phylogenies*. (Ed. by P. Harvey, J. Maynard-Smith, & A. Leigh-Brown) Oxford, England: Oxford University Press.

McArdle, B., & A. G. Rodrigo. 1994. Estimating the ancestral states of a continuous-valued character using squared-change parsimony: An analytical solution. *Syst. Zool.* 43, 573–577.

McKitrick, M. C. 1993. Phylogenetic constraint in evolutionary theory: Has it any explanatory power? *Annu. Rev. Ecol. Syst.*, 24, 307–330.

Meyer, A., J. M. Morrissey, & M. Schartl. 1994. Recurrent origin of a sexually selected trait in *Xiphophorus* fishes inferred from a molecular phylogeny. *Nature*, 368, 539–542.

Miles, D. B., & A. E. Dunham. 1993. Historical perspectives in ecology and evolutionary biology: The use of phylogenetic comparative analysis. *Annu. Rev. Ecol. Syst.,* 24, 587–619.

Møller, A. P., & T. R. Birkhead. 1992. A pairwise comparative method as illustrated by copulation frequency in birds. *Am. Nat.*, 139, 644–656.

Nowak, R. M., & J. L. Paradiso. 1983. *Walker's Mammals of the World.* Baltimore: Johns Hopkins University Press.

Pagel, M. D. 1992. A method for the analysis of comparative data. *J. Theor. Biol.*, 156, 431–442.

Pagel, M. D. 1994. Detecting correlated evolution on phylogenies: A general method for the comparative analysis of discrete characters. *Proc. R. Soc. Lond. B*, 255, 37–45.

Parker, G. A., & J. Maynard Smith. 1990. Optimality theory in evolutionary biology. *Nature*, 348, 27–33.

Purvis, A. & A. Rambaut. (in press). Comparative analysis by independent contrasts (*CAIC*): an Apple Macintosh application for analysing comparative data. *Comput. Appl. Biosci.*

Purvis, A., J. L. Gittleman, & H-K. Luh. 1994. Truth or consequences: Effects of phylogenetic accuracy on two comparative methods. *J. Theor. Biol.*, 167, 293–300.

Reeves, J. H. 1992. Heterogeneity in the substitution process of amino acid sites of proteins coded for by mitochondrial DNA. *J. Mol. Evol.*, 35, 17–31.

Ridley, M. 1983. *The Explanation of Organic Diversity: The Comparative Method and Adaptations for Mating.* Oxford, England: Clarendon press.

Ritland, K., & M. T. Clegg. 1987. Evolutionary analysis of plant DNA sequences. *Am. Nat.*, 130, S74–S100.

Sanderson, M. J. 1991. In search of homoplastic tendencies: Statistical inference of topological patterns in homoplasy. *Evolution*, 45, 351–358.

Sanderson, M. J. 1993. Reversibility in evolution: A maximum likelihood approach to character gain/loss bias in phylogenies. *Evolution*, 47, 236–252.

Sillén-Tullberg, B. 1988. Evolution of gregariousness in aposematic butterfly larvae: A phylogenetic analysis. *Evolution*, 42, 293–305.

Sillén-Tullberg, B. 1993. The effect of biased inclusion of taxa on the correlation between discrete characters in phylogenetic trees. *Evolution*, 47, 1182–1191.

Smouse, P. E., & Long, J. C. 1992. Matrix correlation analysis in anthropology and genetics. *Yearbook Phys. Anthropol.*, 35, 187–213.

Stearns, S. C. 1983. The influence of size and phylogeny on patterns of covariation among life-history traits in mammals. *Oikos*, 41, 173–187.

Swofford, D. L. 1993. *PAUP: A computer program for phylogenetic inference using maximum parsimony*. Washington D. C.: Smithsonian Institution.

Swofford, D. L., & W. P. Maddison. 1987. Reconstructing ancestral states under Wagner parsimony. *Math Biosci.*, 87, 199–229.

Yoccoz, N. G. 1991. Use, overuse, and misuse of significance tests in evolutionary biology and ecology. *Bull. Ecol. Soc. Am.*, 72, 106–111.

CHAPTER 3

How to Study Discrete Comparative Methods

Mark Ridley and Alan Grafen

This chapter has three main purposes. The first is to show how to justify a proposed statistical method for discrete comparative data. We shall be particularly concerned with associative hypotheses. However, several authors have recently recommended methods for testing directional, rather than associative, hypotheses with discrete data. Our second purpose is to argue that the conceptual advantage of directional methods has been exaggerated and that the statistics of directional methods present difficult problems that have not been solved. Our third purpose is to compare the statistical approach we have taken with some alternatives — mainly from likelihood theory, though we also comment on bootstrapping — and we shall argue that, at least in their standard forms, they are inapplicable to phylogenetically structured data. We shall therefore set out our view of how discrete data should be studied — or, at least, of how potential methods to study discrete data should themselves be studied — and criticize the main alternative approaches being advocated in the current literature.

This chapter is concerned with discrete as opposed to continuous data. Data can be discrete for either of two reasons. One is that it is really continuous in its underlying form but has been discretized for a more or less observational reason. In an extreme case, a manifestly continuous variable like time might be divided into long/short categories for a test; in a subtler case, something we perceive categorically might have an underlying quantitative genetic or embryologically continuous control. Alternatively, the data might really be discrete, as in the example suggested by Burt (1989) of modes of X/Y sex determination.

How to justify a comparative method

A. Problems that a discrete comparative method must solve

The minimal study will have two observed characters (call them *A* and *B*), each of which can have a number of states (from 1 up to n_A and n_B); in the simplest case they have only two states (A_1 or A_2 and B_1 or B_2, which we shall often write as *A/a* and *B/b*). The states of both these observable characters are known for a number of species, all at the tips of a phylogeny.

The main statistical problem is that the different species are nonindependent. Grafen and Ridley (1995) identify three sources of nonindependence. One, which we call phylogenetic overcounting, is well known (Clutton-Brock & Harvey 1977; Ridley 1983). It occurs when a group of related species, all sharing the same character state, are entered in the test as more than one trial (usually as many trials as there are species in the group). The other two had not (we believe) been identified before and are at least more of a problem with discrete than continuous data and may be exclusive to discrete data. One of them arises in Ridley's (1983) method. The method reconstructs combined character states throughout the entire tree back to the root. It then (in the formalization of Grafen & Ridley 1995) collapses each set of contiguous uniform nodes into single nodes to produce a "character change tree." Figure 1 is an example; it expresses all the changes in the joint character states in the tree; more than one such tree may be compatible with one data set. Changes in the tree suffer from statistical nonindependence because the changes away from any one node in the character change tree have to be away from the state of that node. If the node is *AB*, a change has to be to *Ab*, *aB*, or *ab*; it cannot be to *AB* because the tree has been reconstructed by parsimony, and if any neighboring nodes had the same character states, they would have been collapsed into one. It is a parsimonious impossibility for neighboring

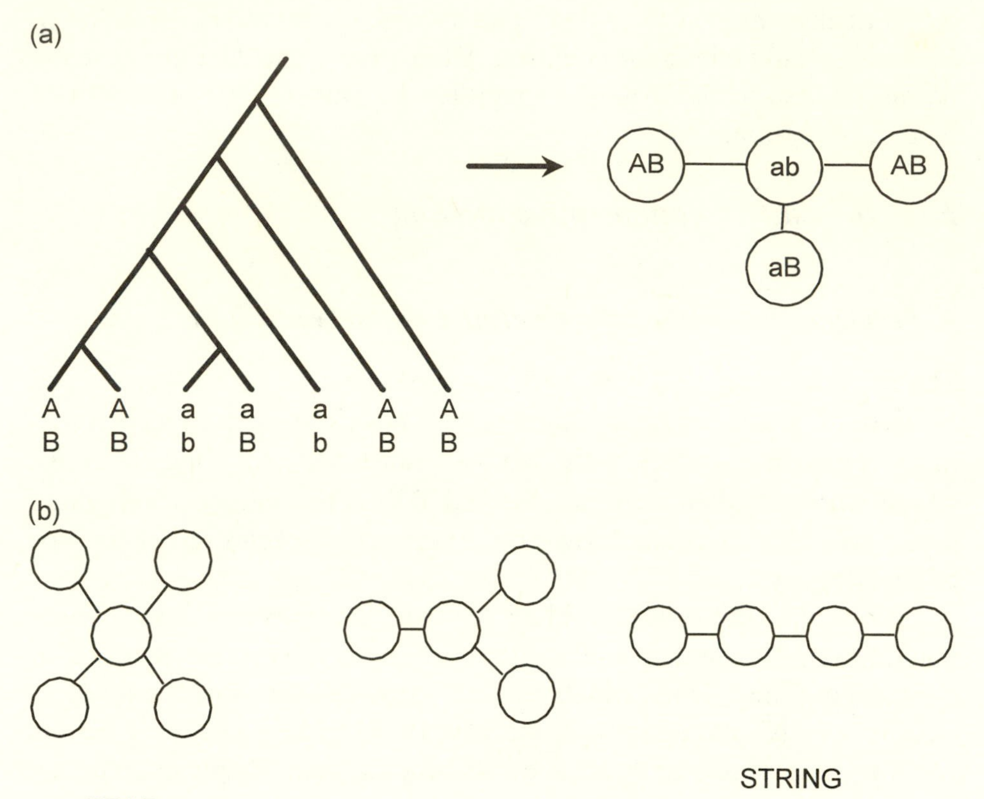

Figure 1. (a) A character change tree illustrates the character changes in a phylogeny, with all the contiguous uniform nodes collapsed into single nodes. (b) It can take on various shapes, depending on the pattern of evolutionary change.

nodes to have the same character state. This problem is likely to arise in any method that reconstructs ancestral states throughout the tree by parsimony and treats inferred changes as independent events. All the changes were separate evolutionary events, but the use of parsimony makes the resulting data points statistically nonindependent.

Other methods make more limited ancestral reconstruction and aim only to find regions of the tree — or partitions of the data — in which change is independent of other such regions. Then the pattern in all the regions is combined in a test. The independent contrasts method of Burt (1989) is one example; the randomization test described by Grafen and

Ridley (1995) is another. The pairwise comparison test of Møller and Birkhead (1992) focuses on variable regions in a tree in a similar, more restrictive, way. These tests all reduce or eliminate the amount of nonindependence due to parsimonious reconstruction because they make little or no use of it; but that is bought at the price of another kind of nonindependence, due to what Grafen and Ridley (1995) call the family problem. These methods identify separate regions in the tree in which evolutionary change must have occurred separately. In Figure 2, for instance, the *AB/ab* change in region 1 must have been a separate historical evolutionary event from the change in region 2. But although they were separate events, they can still be statistically non-independent. If a number of regions share the same local ancestral state, the transition in all those regions will be away from that ancestral state. A full analysis of the consequences of this fact is laborious (Grafen & Ridley 1995); but the net effect, provided that evolutionary change is rare, is an excess of associations within variable regions between the locally nonancestral states. There will usually only be one change in each character per variable region of the tree. If the locally ancestral state is *AB* for several regions, there will be an excess of them showing *Ab* and *aB* with null data. It is possible for there to be a double change (as in Fig. 2) to *ab*, giving *AB* and *ab* within the variable region, but this is less likely by chance than the non-ancestral association. The exact relative chance of finding *AB+ab* or *Ab+aB* depends on the rate of evolutionary change and on the way a particular method selects variable regions within the tree. However, methods that pick variable regions and do not reconstruct the higher ancestral states that are the cause of nonindependence among them are vulnerable to the bias in favor of the nonancestral associations.

These arguments illustrate the awkward fact that, with discrete data, it is not enough to identify separate events that occurred, in an evolutionary sense, independently: those events may still not be statistically independent. We suspect that almost any method for discrete data will have to face up to these problems of nonindependence; and no method that has been proposed so far has been shown, in terms of the logic of its operations, to avoid them successfully. Such are the problems: but how can we find out how much, in any given method, they matter?

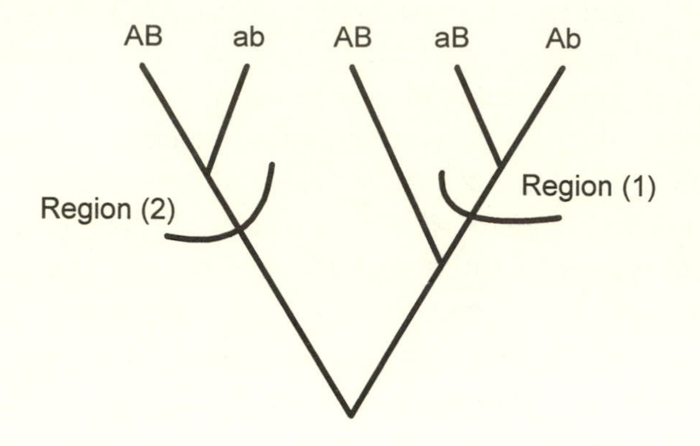

Figure 2. Identification of regions of character variation. Separate changes must have occurred in regions 1 and 2.

B. How to find out whether a proposed method solves the problems

There are two stages in the study of any statistical test. The first is to show that it has valid Type I error rates; the second is to compare the power of those tests that pass the first stage by studying Type II error rates.

Both stages require a model of evolutionary change through a phylogeny, which is a data-generation process. In the specific problem with which we are concerned, the model will be one in which the ancestral state at the root is given and the model specifies how the characters change along the branches of the tree to generate a pattern of character states at the tips. The model can be used to generate large numbers of data sets, which can then by used to scrutinize proposed tests, simply by having the tests analyze the data sets. The model is null if the states of the two characters do not influence each other's evolution, and the data sets can then be used to scrutinize Type I error rates. If the test is valid, it should find associations between the characters that are significant at, for instance, the 5% level 5% of the time, at the 1% level 1% of the time and so on. Alternatively, the state of one character may influence the evolution of the other (and vice versa) and the data sets can then scrutinize Type II error rates. It is crucially important that any method recommended for general use should be shown to be valid in this way. It is impossible to know whether a test is worth using unless its validity has been assessed against null data generated by a model; indeed, the concept that a test is valid (or, in some looser sense, "good," or "worthwhile," or "well-behaved," or "satisfactory") assumes some model of data generation. In simple cases, such as standard regression theory, the need actually to generate random data sets can be short-circuited by analytical methods; in other cases, such as nonparametric tests, it is necessary only to make some assumptions about the data-generation process and not to construct it explicitly. But the principle is the same in both cases.

In setting out this procedure for justification, we do not mean to imply that no one apart from ourselves understands how to conduct statistical research in biology. The whole or parts of the argument will be familiar to many readers, not least because research with continuous comparative data is now well penetrated by it. The argument does need to be made, however, because it shows practically no influence in any recent research with discrete data. We shall cite all the theoretical work we know of from, say, the past 5 years, and none of it has been conducted by the method, entirely standard in statistics, that we have just described and elaborate on more below. If (which we doubt) work from an earlier period did use this method, it is not having any influence on modern work, and a restatement would be timely.

An alternative to simulational scrutiny is to show, analytically or logically, that a method is valid: this has been the aim of some applications of likelihood theory, applications that we find unreliable. In the current state of statistical theory, simulations are essentially the only reliable route to justification with the complicated problems of phylogenetically structured data.

C. A model for discrete character change through a phylogeny

A statistical test that has been justified by simulation has been shown to be valid only relative to the model that was used to generate the data. It is therefore desirable that the model contain as many of the biologically interesting features as possible. Harvey and Pagel (1991) were the first to introduce an explicitly phylogenetic model of discrete character change in the comparative method; their model (and Pagel's 1994 developments) has attractive features, but we prefer a model in which the processes generating the phylogenetic structure in the data are even more explicit.

Let us return to the case of two observed characters with two states each. It is well known that if the numbers of species in the states *AB*, *Ab*, *aB*, *ab* are counted, the Type I error rates will be wrong, and too liberal; significant associations will be found in excessive frequency in null data. But why exactly is this? One version of the explanation is as follows. The chance that a character changes from one state to another will usually be influenced by the states of many other characters. We can concentrate on the null case, in which the state of *A* does not influence the evolution of *B* or vice versa. Other, unobserved characters will nevertheless be influencing the evolution of both *A* and *B* (if the other influential characters are observed, they become, for purposes of argument, like *A* and *B*). This can have consequences for the phylogenetic pattern of *A* and *B* in two ways. The simplest would be if a single "third variable" character *E* jointly influenced both *A* and *B*. This could cause phylogenetic pattern for causal reasons. Thus, one state of *E* might causally generate an association between states of *A* and *B* (for example if $E_1 \rightarrow A_1$ and B_1, and $E_2 \rightarrow A_2$ and B_2); this is a familiar kind of problem in comparative, observational, and some kinds of experimental inference (does an association of *A* and *B* mean they are causally related or they are both controlled by an unobserved variable?). However, counting independent trials will not help in this case. The relevant case is the null one, when the unobserved variable produces associations at the species level among the observed variables, but it is equally likely to do so for any of the four possible associations of character states. (In the causal case above, *E* produced particular combinations of *A* and *B*.) Suppose that *E* influenced the rate of evolution of a number of characters, including *A* and *B*; some states of *E* would be general "freezer" states that slowed down the rate of evolution of the other characters, whatever state those other characters were in. Then a block of species would end up with the same states of *A* and *B* — whatever the states were when the lineage entered the freezer *E* state — not because *A* and *B* were influencing each other's evolution but because *E* had brought all change to a stop. A test that counted species, or otherwise phylogenetically overcounted the evidence, would be liable to find spurious evidence of an association.

E is a single hidden character. An alternative is for there to be two different hidden characters (*C* and *D*) that separately interact with the observed characters *A* and *B* respectively. Again, *C* and *D* have states that can freeze the evolution of the character they interact with. But now *C* can freeze only states of *A* and *C* can freeze only states of *B*. For concreteness, suppose *C* can take on six states and the relation with states of *A* is as follows:

$$
\begin{array}{ll}
C = 1 & \mathrm{A} = 1 \\
C = 2 & A = 1 \\
C = 3 & A = 2 \\
C = 4 & A = 1 \\
C = 5 & A = 2 \\
C = 6 & A = 2
\end{array}
$$

There is a perfect deterministic relation between the states of *C* and *A*. Now suppose that transitions between some *C* states are more frequent than others: transitions between $C = 2$ and $C = 3$ and between $C = 4$ and $C = 5$ are common; but transitions out of these states, and out of $C = 1$ and $C = 6$, are rare. $C = 1$ and $C = 6$ then act as "freezer" states for *C*: if a taxon is $C = 1$, the observed state of *A* will be 1 in a large block of species, or $A = 2$ if $C = 6$. Suppose, finally, that some similar relation exists between the states of *D* and *B*.

A model of this kind will again generate null associations between states of *A* and *B* at the species level, when there is no causal relation between them (or between *C* and *D*). As *C* and *D* evolve, it will happen from time to time that a species will evolve a freezer state for both *C* and *D*. Then there will be a block of species descended from it that are likely to share the same state of *A* and *B* and a species count will find evidence of a nonexistent association. We picked the relation between *C* and *A* above such that any one of the four states (*AB*, *Ab*, *aB*, *ab*) could be frozen in this way. In practice, the evolution of the observed characters will be influenced by more than one other character, but only one character is needed to represent the essence of the problem; the effects of a realistically complicated multicharacter set of influences could be represented as some version of the "*C*" character above. At least, that is the design.

A model with explicit third variables controlling the observed variables, as well as being concrete about the evolutionary processes of interest, also makes clear why phylogenetic overcounting is indeed overcounting. A taxon with freezer states of *C* and *D* will have a number of species uniform for states of *A* and *B*. A test that counts species as independent trials will inevitably have erroneous, and liberal, Type I error rates. Maddison (1990), Harvey and Pagel (1991), and Pagel (1994) have suggested that large uniform taxa can be evidence of an association between the characters; but in our model, the safest interpretation is that the taxon has evolved a freezer state of an interacting variable.

We have used a null two-character model (with two observed characters controlled by two unobserved characters like *C* and *D*) to scrutinize the behavior of a number of proposed tests for discrete cross-species data. *C* and *D* evolve in a stochastic manner (like the observed characters in Harvey & Pagel's model). The same setup could be used to investigate Type II error rates if the model had a non-null relation between the observed characters. For example, *C* could set the state of *A* as above and then *A* and *D* could determine *B*. But we have not deployed the model in this form, nor have we investigated single general freezer characters like *E* above.

D. Type I error rates of some proposed tests for discrete characters

In Grafen and Ridley (1995) we obtained the Type I error rates for several tests. Here we only summarize some of our findings and emphasize interim practical conclusions for people working with real data. We stress, however, that our work allows no general recommendation of one particular test.

Discrete tests have been proposed by Ridley (1983), Burt (1989), Maddison (1990), Harvey and Pagel (1991), Møller and Birkhead (1992), Pagel (1994), and Grafen and Ridley (1995). We did not test Maddison's, Harvey and Pagel's, or Pagel's on the simulated data sets, mainly because those tests could be seen to suffer from phylogenetic overcounting. The effects of this kind of nonindependence are understood, and it would be easy to invent data sets in which they would behave arbitrarily badly. We did not test Møller and Birkhead's pairwise test, which is restricted to a certain kind of data set and is in important respects a refined version of Burt's test. Here we shall also exclude the randomization test proposed by Grafen and Ridley (1995) because that test is not familiar in the existing literature, it takes lengthy explanation, and we are unsure how good a test it is. See Grafen and Ridley (1995) for explanation of that test, and Type I error rates for it. We also scrutinized the phylogenetic regression (Grafen 1989), to see how it fared with discrete data, and we did naïve species counts for comparison. Table 3.1 contains a simplified summary of the results for two tree shapes; the table footnote gives some details of the simulations.

We draw attention to the following. First, Ridley's and Burt's tests are highly biased against the ancestral state in the tetratomous phylogeny. The reason is the nonindependence described above: both tests (for different reasons) find nonancestral associations more easily than ancestral ones. In the extreme cases of a "star" character change tree, it is practically impossible for either test to find a significant association on the ancestral diagonal because there is a maximum of one data point (the center of the star) in the ancestral state. The character change trees of the tetratomous null data are probably not simple stars, but they do have a large shadow of influence of the ancestral state. Both tests, however, have reasonably good Type I error rates with the more realistically shaped "Hennig" tree, though Grafen and Ridley show that the reason for the improvement in Burt's test in the Hennig tree was not that given as a rationale for the test.

The results have similar implications for the practitioner of either test. The behavior of the test is influenced by two properties. One is the shadow of influence of the ancestral state. It is not difficult to look at the character states on the phylogeny and see at a glance whether the root state permeates it all, with most evolutionary events being single changes from it. Under parsimony, the more resolved the tree, the less the influence of the ancestral state; with complete phylogenetic ignorance (that is, we can say nothing better than that all the species are equally related, in a star phylogeny) the character change tree has to be a star. The simulations show that in one realistic tree with approximately realistic rates of change, the shadow of influence of the root state has been reduced sufficiently for the bias in the tests to disappear. The condition of extensive shadow of influence of the root state is the condition for these two tests to be biased.

Table 3.1. Type I error rates for four comparative methods with discrete data:

	Tree Shape							
	Tetratomy Association on:				Hennig Association on:			
	ancestral diagonal		nonancestral diagonal		ancestral diagonal		nonancestral diagonal	
Chance of finding *p* value more extreme than:	0.05	0.01	0.05	0.01	0.05	0.01	0.05	0.01
Test:								
Species counts	0.239	0.150	0.172	0.083	0.286	0.211	0.325	0.239
Ridley (1983)	0.017	0.006	0.706	0.589	0.055	0.014	0.061	0.011
Burt (1989)	0.000	0.000	0.219	0.072	0.031	0.000	0.014	0.003
Grafen (1989)	0.055	0.008	0.072	0.011	0.044	0.006	0.039	0.014

Results of 360 null data sets, each containing 256 terminal species. The datasets were generated by the model described in the text; the parameter values were picked to make the amounts and distribution of change approximately realistic; there were about 25–30 events in the 256 species tree. The tetratomous tree contained 256 species arranged symmetrically in four member groups through four levels (4^4=256). The "Hennig" tree was abstracted from Hennig's (1981) phylogeny for the insects and aimed to have an approximately realistic amount of symmetry and frequency of polytomous and dichotomous branching. Associations on ancestral and nonancestral diagonals have the following meaning. Imagine a 2x2 contingency table for the A/a and B/b characters. If the ancestral state in the simulation was AB, the ancestral diagonal is the one with AB and ab; the nonancestral diagonal is the one with Ab and aB. See Grafen and Ridley (1995) for details and the full *p*-distributions.

When they are biased, the thing to look out for is whether the root state counts for or against the hypothesis under test. If the ancestral state supports the hypothesis, the bias (when it exists) is towards conservativeness. The test can be so conservative that it is almost unusable, and this was Proctor's (1991) reason for developing an alternative to Ridley's (1983) method; but when a significant *p* value is obtained with the ancestral state supporting the hypothesis, the evidence is more significant than the *p* value alone suggests. In practice, the method has mainly been used in cases in which the ancestral state supports the hypothesis, and this is likely to continue to be the norm;

our simulations confirm that in this case Ridley's (1983) test is conservative. However, when the ancestral state counts against the hypothesis, both tests are liberal and a significant *p* value would not be impressive evidence for a hypothesis if that ancestral state had extensive shadow of influence through the tree. Thus, if the ancestral state supports the hypothesis, the tests can be used with confidence, in the knowledge that any bias is only making it more difficult to find support; when the ancestral state counts against the hypothesis, the tests should be used with caution. We should also note that the test of Møller and Birkhead (1992) does not suffer from the same difficulties and should give valid results under mild assumptions. However, it does throw away data, and few data sets will be large enough and well enough resolved for it to be practical.

Second, the phylogenetic regression had reasonably valid Type I error rates in Grafen and Ridley's (1995) simulations. We regard these results as promising, and as expected from the theory of the phylogenetic regression (Grafen 1989); but they are not as definitive as they might appear. When running the method on the data sets, we fixed the branch lengths at their known and correct values. This is justifiable as a first stage in scrutinizing the method (if it does not work when fed the correct branch lengths, then that is the end of it); but it gives the method an advantage relative to the other tests, which were not supplied with so much information. However, an interim conclusion would be that the phylogenetic regression is probably a reasonable method to use with discrete data when we are willing to make assumptions about branch lengths.

Table 1 also reveal that the species counts are predictably awful. The 1% significance level, for example, is exceeded in about 10 to 20% of the null data sets. Naïve, prephylogenetic comparative tests should be kept at the other end of a barge pole.

Association or direction of change

The methods discussed above all test for associations between characters and leave open the direction of causation. However, a number of authors have suggested that methods that test associations are inferior to methods that test hypotheses about causal direction. Indeed, they form an orthodoxy in recent thinking about discrete comparative data.

Pagel and Harvey (1989) noticed that associative tests "conflate" changes from several evolutionary directions; Donoghue (1989, p. 1141) noticed as a "difficulty [that...] simply recording the number of times that a particular combination of states appears in a cladogram may obscure information on the sequence of character origination"; and Miles and Dunham (1993, p. 600), in the most recent review of the subject, noticed as a "weakness" of an associative test that it "ignores the sequence of transitions." Maddison (1990), whose paper inspired those remarks, more neutrally suggested that different methods are needed for different questions; but only the recent paper of Frumhoff and Reeve (1994) really stands out against this stream of argument. We agree with Frumhoff and Reeve's main point and shall incorporate it below. Ridley (1983, pp. 34–40) discussed earlier writing on the subject, including Gittleman's (1981) study of parental care in fish, using an ad hoc taxonomic method, as well as how to infer causation in some special cases.

Causal direction is less often discussed for continuous methods, perhaps because it is less clear how to infer, from inferred ancestral states, the causes of a correlation between continuous characters. Lande's (1979) paper on brain-body size allometry in mammals is one interesting example; he inferred that the correlation is more likely to have evolved from selection on body size, with brain size being dragged along in a correlated response, than vice versa. The inference follows from the heritabilities of brain and of body size, which have been measured in mice, and the observed variation in brain and body size; it is, however, more curious than convincing, because modern laboratory murine heritabilities are unlikely to be representative of the broad sweep of 60 million years or more of mammalian evolution.

Huey (1987) and Harvey and Pagel (1991, pp. 162–164), when discussing Huey, used the term "directionality" to describe another kind of inference with continuous comparative data. Huey (1987) inferred the direction of change of two continuous variables (running speed, preferred environmental temperature) from inferred ancestral to modern states. But his interest was in associations between those changes: he was not trying to infer which character changes had causal priority. The critical remarks that follow do not apply to Huey's kind of directional study, which should not be confused with the topic of our discussion. By a directional test, or hypothesis, we mean one that tests whether, or claims that, changes in one character (e.g. $a \rightarrow A$) precede and drive changes in another ($b \rightarrow B$).

There are two reasons we discern for preferring a directional test. The hypothesis may be explicitly historical and concerned with which character came first. We, however, are interested in functional hypotheses and not in history; indeed, it is doubtful whether the uncertainty in historical reconstructions, by their nature unique events, is really statistical in nature. The second reason is that a hypothesis may be inherently directional, claiming that one character causes a change in another, but not vice versa. Five points can be made.

1. A directional hypothesis predicts both an association and a direction of change; but a directional test is likely to be more powerful. It is therefore desirable, insofar as truly directional hypotheses exist, for people to work on developing directional comparative methods.

2. However, no formally justified directional test is available. No tests have, for example, reached the stage of justification that we set out above, and the existing tests trip up at earlier hurdles (see below, point 3). Until such a test is available, the fact that directional hypotheses also make associative predictions means that they can be tested with tests of association.

3. We expect that a valid and powerful directional test would be more difficult to construct than an associative test. One reason is the problem of phylogenetic overcounting, the difficulty of which is illustrated by the way Maddison's (1990) concentrated changes test — one of the most carefully constructed and thoroughly thought out of comparative methods — nevertheless suffers from the phylogenetic overcounting kind of nonindependence, as Maddison (1990) himself almost admitted (Grafen & Ridley 1995 show that his arguments on this point are in error) and Sanderson (1991) and Sillén-Tullberg (1993) have also discussed. Harvey and Pagel's (1991) method also phylogenetically overcounts, or seeks to compensate for overcounting in an arbitrary manner, and no one has yet devised a directional method that avoids the problem. That is not to say it is insurmountable.

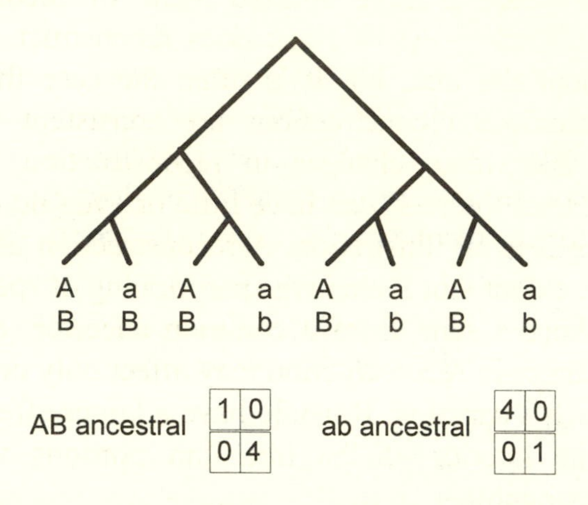

Figure 3. Influence of ancestral reconstructive error. Whether AB or ab is ancestral, the form and significance of the association is the same in Ridley's (1983) test. In this case there would be little or no effect on a directional test either, but in general directional tests are more sensitive to ancestral error than are associative tests.

The second reason is that directional methods will be more strongly affected by error in reconstructed ancestral states than are associative tests. This is essentially the criticism of Frumhoff and Reeve (1994), though they were not concerned to compare directional and associative tests. It also informs Höglund and Sillén-Tullberg's (in press) reply to Oakes (1992), in which they point out that Maddison's test, in Oakes' work, reconstructs many bird groups with lekking as the ancestral state even though it is a minority habit; such was Oakes' reason for concluding that sexual dimorphism often evolves after lekking. Höglund and Sillén-Tullberg suggest the reconstruction is erroneous and that sexual dimorphism is no more likely to evolve after lekking than after nonlekking.

Maddison's test reconstructs ancestral states in the phylogeny and counts the number of changes in the second character ($b \rightarrow B$ and $B \rightarrow b$) that occur in regions of the tree that are already a or A. Errors in the reconstruction of whether an ancestral state is A or a can have a large effect. The associative tests, however, take a more relaxed view of reconstructed ancestral error. Ridley's (1983) test does reconstruct ancestral states throughout the tree, but it is often the case that many equally parsimonious reconstructions are consistent with the observations; then large changes in reconstruction, such as altering the state at the root, can have little or even no effect on the test. Figure 3 is an illustration. A reconstruction affects the test only to the extent that it alters the partitioning of species into subsets that share a state from a common ancestor, and even quite large changes in reconstruction may affect only one or two elements of such a partition. It would have a larger effect on the division of the interior of the tree into portions with one character state or another.

The other associative tests, strictly speaking, do not rely at all on ancestral reconstruction. Grafen (1989, p. 148) argued that the phylogenetic regression, although superficially appearing to condition on ancestry, actually conditions on modern pattern alone; Pagel's (1994, p. 37) unargued assertion to the contrary is in error. The apparent ancestral reconstruction is best understood as simply being part of the working: the method compares weighted averages with other weighted averages, which are chosen by knowing the phylogeny. However, it is also worth noting that the sampling error in these weighted averages is taken into account in the test. Even, therefore, if the weighted averages were interpreted as ancestral reconstructions, the method would not be assuming their correctness. Likewise, the apparent ancestral reconstruction in the randomization test of Grafen and Ridley (1995) and in Burt's (1989) test is only needed to find partitions in the data for purposes of randomization; the methods do not assume anything in particular about the history of the character states in those regions.

We are thus distinguishing three degrees of ancestral reconstructive dependency. Directional tests are likely to have the strongest dependency, and be most vulnerable to error in ancestral reconstruction. They use inference as data, in Felsenstein's (1984) phrase, in a strong sense. Of the associative tests, Ridley's (1983) test has the stronger dependency, but it is weaker than the directional tests; and the other class of associative tests has only a weak and superficial dependency on ancestral reconstruction.

Simulations could reveal how much this argument matters. Maddison (1990, pp. 550–551) described some simulations to test the effect of parsimonious errors and found little evidence of damage, though there were imperfections. Maddison's simulations were a sensible method given the question he set out to answer; but two things are worth noting about them. One is that he held constant the areas of the phylogeny that had one or other state of the suppositious causal variable; he only randomized the suppositious caused variable. Reconstructive error is also possible for the causal variable, and it is error in that character that we are particularly concerned with in the argument given above. Secondly, Maddison's model of character change, like Harvey and Pagel's but unlike Grafen and Ridley's, had no "freezer" characters (like *C, D,* and *E* above). Every branch in Maddison's model had a separate independent chance of change, and therefore no taxa were frozen in a uniform state for an observed character. We should not expect problems with uniform taxa and reconstructive error in this model and doubt its usefulness for probing the problem. Thus Maddison's simulations are inadequate to contradict our logical point about the vulnerability of directional methods to reconstructive error. For the associative test of Ridley (1983), Grafen and Ridley (1995) obtained Type I error rates by simulation and, as we saw above, found that they were reasonably realistic in one phylogeny and conservative in another — which implies the method is robust to errors in ancestral inference. However, the relative vulnerability of directional and associative tests has not been measured in a direct comparison. Such an exercise would be premature at this stage because of the absence of a directional test that is free from non-independence due to phylogenetic overcounting.

4. Although many evolutionary hypotheses are expressed in a directional way and may seem at first sight to be directional, they may really be associative. That is, causation often works both ways and not just in one. Consider the link between gregariousness and aposematism among lepidopteran larvae,

analyzed by Sillén-Tullberg (1988) in a directional way, and discussed by Maddison (1990) and Harvey and Pagel (1991) as a directional hypothesis. The theory is that gregarious species may undergo selection for aposematism because predators that try one aposematic larva may then avoid its relatives.

However, there will be other selective forces at work. Aposematic species will sometimes undergo selection toward solitariness and gregariousness for other reasons. The aposematic species will have the balance tipped toward gregariousness because an aposematic species that is gregarious will experience less predation. The postulated selective force can therefore act in either direction and will be most appropriately investigated by testing for an association between gregariousness and aposematism. Many biological hypotheses are, we believe, of this form and — sometimes despite first appearances — make an associative and not a directional prediction.

This is a case in which the experimental analogy for comparative studies can mislead. In experiments there is usually unidirectional causality. If you add nitrogeneous fertilizer, the crops grow more — but that does not mean that if crops grow more for some other reason it will stimulate a nitrogeneous rain from the environment. Adaptive relations between two characters are rarely like that. If there is a selective equilibrium, in which two characters show an adaptive association, it will be quite peculiar for causality not to work in both directions. Sometimes a particular theory will suggest a hypothesis in a form that implies one causal direction rather than the other; but further thought (as in the gregariousness example) will often reveal that suggestion belongs to the context of discovery and not the logic of the hypothesis. Likewise, economy of exposition may constrain the proponents of a hypothesis to explain it as if it works in one causal direction; but an adaptive hypothesis in simplified expository form should not be confused with a truly directional hypothesis.

It may only be necessary to think about the two characters in the hypothesis by themselves in order to see that a truly directional hypotheses is implausible. That is, if *ab* and *AB* form adaptive associations, a little thought will usually show both that *a* tends to change to *A* in species that are *B* and that *b* tends to change to *B* in species that are *A*. But the multifactorial nature of selective influences on characters makes directional hypotheses even less likely. The characters studied in comparative research are typically influenced by a number of other factors. The gregariousness example drew on this point: here is a further example.

Harvey and Pagel (1988) discussed the relation between mating system and sexual dimorphism as a directional hypothesis. If a species is polygynous it is more likely to evolve sexual dimorphism. The same idea was at work in the exchange between Höglund (1989), Oakes (1992), and Höglund & Sillén-Tullberg (in press). The causal tendency for polygynous species to evolve sexual dimorphism is well known and follows simply from the theory of sexual selection. However, sexual dimorphism is influenced by other factors too, including species recognition and ecology. If sexual dimorphism evolves for some other reason, it would probably influence the future evolution of the mating system. If the sexual dimorphism were such that males had a higher mortality rate, that would automatically direct the species down the evolutionary road to polygyny, simply through its effect on sex ratio.

Maddison (1990) introduced directional tests from a different perspective, that of release from constraint. Imagine the *A/a* character exerts a constraint on the evolution of another character (*B/b*), such that the evolution of *B* is possible in an *A* species, but *a* constrains things and makes the evolution of *B* difficult. Then in *A* zones there will be more evolution from *b* to *B*. The constraint may operate directionally in this manner, but that does not show the evolutionary relation between the two characters is directional; it could again be that thought about change in the other direction too would reveal that the system can move in either direction. If *AB* is an adaptive association it is likely that *A/a* will change in *B* species.

A similar point can be made about "permissive" comparative hypotheses. Pagel and Johnstone (1992) tested the relation between the *C* value of the DNA of a species and the cell cycle time. The two variables are known to be correlated, but Pagel and Johnstone argued this is more likely to be a permissive than a causally adaptive relation: if the cell cycle is longer for other reasons, more junk DNA will accumulate (in contrast to the idea that DNA increases in order to slow the cell cycle). Again, it is also theoretically possible that the process can work either way. If some species have lower, and others higher, *C* values, and some are selected for longer and others shorter cell cycle times, then the species with higher *C* values will be more likely to be the ones that end up evolving longer cell cycles.

It is not our purpose to deny the logical possibility of directional adaptive hypotheses. In other areas of evolutionary biology, directional ideas do exist, such as in the literature about Dollo's Law (Bull & Charnov 1985), and Godfray's (1987) work on parasitoid clutch sizes. However, the hypotheses that have been discussed in the literature about directional comparative biology seem to us to be associative. There are good theoretical reasons to think that most hypotheses about the adaptive relations between characters will really be associative, and the theoretical discussions of directional comparative methods have underestimated the generality of bidirectional causal influences in adaptive associations.

5. Frumhoff and Reeve (1994), crediting Armbruster (1992), make a further point, that a directional causal process, even a one-way example, will often produce an association in the observed species with real data. If *B* is selected for after *A*, and the selection pressures are not very weak, then all observed species will be either *AB* or *ab*. There will be no telltale *Ab* group whose location could reveal in what direction evolution occurred.

We conclude that directional comparative methods have yet to be developed and justified, face formidable technical and inferential difficulties, and have a narrower range of application, relative to associative methods. Associative methods exist and should at present be the methods of choice in discrete comparative inquiry.

Other statistical approaches

The procedure we discuss above for investigating a test is to generate random hypothetical data sets using a "data-generation process," and to apply the proposed test to many such data sets to discover initially Type I and later Type II error rates. This is the standard statistical approach, though often analytical work can remove the need for actual simulation. In this section we discuss two other justificatory methods that might be thought to apply to comparative statistics (Harvey & Pagel 1991; Pagel 1994; Felsenstein 1988). We shall mainly be concerned with likelihood theory and briefly mention bootstrapping. We discuss their relationship with the fundamental criterion just outlined and offer reasons why we believe they can be of limited if any use in the present state of statistical theory.

The older of the two alternatives is the likelihood approach, essentially introduced by Fisher (1922) and later much developed and enhanced. [Cox & Hinkley 1974, Chap. 2, particularly section 4(vii) and the bibliographic notes, is a discussion intended for statisticians.] The likelihood approach begins by writing down a data-generation process or statistical model. Instead of actually generating data and applying a proposed test to it, it is possible by applying standard techniques to derive a test from the data generation process. If certain assumptions are upheld, this "likelihood-ratio test" enjoys many enviable properties. It might therefore be thought that a test could be written down directly, and simulations avoided, and this was the approach of Pagel (1994). However, this is possible only if those assumptions are indeed upheld nearly enough, and we shall come to the question of whether they are.

Most standard tests can be derived as likelihood tests. For example, multiple regression and analysis of variance, and indeed the "general linear model," can be derived as likelihood tests. It is natural, therefore, when devising new tests, to attempt the likelihood approach. Cox and Hinkley (1974, pp. 47–48) express a consensus that the ultimate criterion, to be applied to all tests including those suggested by likelihood theory, is the repeated trials criterion, which we have applied using our simulated data sets.

As new tests have been devised for more complex problems, statisticians have increasingly realized that the straightforward likelihood approach is no panacea. Special sub-branches of likelihood theory are devised to cope with specific problems: for example, partial likelihoods were invented by Cox (1972) for life table analysis (for a longer discussion, see Kalbfleisch & Prentice 1980). Let us turn to the kinds of difficulties that can lie in the way of standard likelihood theory. The following quotation from McCullagh (1991, p. 286) sets the scene:

> Classical likelihood-based analysis for complex problems may be intractable or impossible. In many instances this is due to the fact that a complete probabilistic specification of the model is not available or not suitable .... Related difficulties with standard likelihood-based analysis are the accommodation of possibly large numbers of nuisance parameters, and the choice of a reference set for computation of the relevant probabilities.

The first problem is the availability of a suitable complete specification of the model. In our case this means a data-generation process that we are prepared to accept as true. Thus acceptance of Pagel's (1994) likelihood method depends inter alia on acceptance of his model of character change. We gave reasons above for not agreeing with the model, so we would not accept his method. A broader point emerges from this need for an agreed model, and it is that we may never know enough to agree on a model. In simple statistical problems, the questions of interest can be answered by including only material of immediate interest. In more complex problems, answering a question (such as whether there is a functional relationship between two variables) may necessitate including in the model a lot of information of no direct interest (such as the phylogeny, its topology, and branch lengths). We may need to agree that the model for the whole lot is precisely right in order to agree that the test for the small part of interest is a good one. Any parameters of the model (unknown quantities that require estimation from the data, such as the intercept and slope in simple linear regression) that are not of direct interest are the "nuisance parameters" that McCullagh refers to in the quotation above.

In the case of discrete comparative methods that reconstruct character states at higher nodes, those reconstructions are nuisance parameters. Now the assumptions under which likelihood tests are justified do allow for nuisance parameters, but there is an important restriction. The general result about likelihood tests is asymptotic, and for our informal purposes we may think of it in the form "As the number of data points increases, the likelihood test becomes as close as you like to validity." (For a more technical treatment, see Stuart & Ord 1991, particularly pp. 658–661 and p. 870.) Now the restriction is that the number of nuisance parameters should be fixed as the number of data points increases. In a regression problem where an x-variable must be controlled for, this is easily accommodated. The one extra nuisance parameter is the coefficient for that extra variable, and the likelihood test for the coefficient of interest is thus justified. In phylogenetic problems, the number of ancestral states to be estimated increases as the tree increases. Indeed, the number of ancestral states to be estimated remains in all interesting cases a sizable fraction of the number of data points. This matters greatly. The essence of the justification of likelihood tests is that there is enough information to estimate the nuisance parameter as precisely as we like. We cannot assume in phylogenetic problems that the nuisance parameters of ancestral states can be estimated as precisely as we like. They will in general remain poorly estimated in data sets of any size.

How is Pagel's (1994) method affected by these considerations? His likelihood function does not have nuisance parameters, but a structural aspect of the likelihood function creates a parallel difficulty. A likelihood function in a typical simple case is a product of terms, with each term corresponding to one data point. Further, the terms are all defined in the same way, and differ from each other only because the values of the observations differ. Once logged, such a likelihood function is a sum of terms, where each term has the same relationship to its own data point. Readers may recall that the central limit theorem deals with sums of identically distributed random variables, and indeed the asymptotic behavior of likelihoods is derived from the central limit theorem by, for example, Stuart and Ord (1991, pp. 658–661).

Pagel's (1994) likelihood function is not a simple product. Instead it is recursive sum of products of sums of products and so on, in a structure conforming to the phylogeny. When we increase the size of a data set in an ordinary case, we simply add extra terms to the sum representing the log-likelihood. The asymptote, as the data set increases, then corresponds to the asymptote in the central limit theorem that guarantees normality of the sum. But when we increase the size of a phylogenetic data set, the structure of Pagel's likelihood changes and the analogy with the central limit theorem breaks down. Pagel envisages an asymptotic justification for the test, but it would need to be established what kind of asymptote was being considered. Specifically, what shape the phylogeny would have for each different-sized data set. Not only would standard results probably not apply, but the conclusion that the likelihood ratio has the required chi-squared distribution might well not be true. The recursive nature of Pagel's likelihood function therefore means that standard likelihood theory does not apply, and his method is unjustified.

Pagel suggests a simulation test whose validity he claims would not depend on asymptotic arguments. It involves estimating parameters for the characters on the basis of maximum likelihood and on the assumption of the null hypothesis of independent evolution. Then simulated data sets are created using those parameter values, and the distribution of the likelihood ratio test built up empirically by repetition of this process. We remark that this test, although it involves maximum likelihood and simulations, has no logical justification in terms of statistical theory, and Pagel offers no simulational evidence that it is a valid test. The reasonable-sounding procedure is dubious if the distribution of the likelihood ratio depends strongly on the parameter values. If it does not, it hardly matters whether they are fitted by maximum likelihood or not. If it does, it is not clear why the distribution generated by the maximum likelihood estimates is of sole concern. Nearby parameter values may well have given rise to the data, and may have importantly different distributions of the likelihood ratio. The question which needs to be answered to evaluate this simulation version of Pagel's test is, therefore, whether the distribution of the likelihood ratio under the null hypothesis depends much on the parameters.

The difficulties of handling likelihood methods in the presence of nuisance and incidental parameters are discussed by Wetherill (1986, sec. 13.4). On page 279, Wetherill singles out the related problem of reconstruction of phylogenies as being "even more slippery" than the already problematic example he is discussing. We did begin work on a likelihood based test. It included nuisance parameters representing freezing of character states in parts of the phylogeny. Two remarks are worth making about it. The nuisance parameters reintroduced into the problem all the complexities that have to be grappled with in the standard approach, so the choice of approach does not change the core of difficulties that have to be overcome. Indeed its behavior most closely resembled that of Ridley's (1983) method among those discussed in this paper. Second, the reason we abandoned this test, despite our attraction to its model of character change and freezing, was that there is no extant statistical justification for a likelihood test that necessarily involves so many nuisance parameters.

It follows that the likelihood approach to comparative statistics must await possible new theoretical developments before it can be relied upon. The possibility should not be discounted; as we mentioned above, the theory of partial likelihoods was developed by Cox (1972) to help construct methods for the now burgeoning field of tests for survival data, but at present it is only a possibility and we know of no relevant work on such an extension.

One possible attraction in using the likelihood approach is that the reconstruction of a phylogeny could be combined with testing a functional hypothesis in a grand likelihood scheme (e.g. Pagel 1994, pp. 38 and 42). We believe this would not be a helpful approach. Modern statistics does not tend to treat complicated problems with straightforward likelihood methods. Indeed an important strand in the recent history of statistics has been a sophistication of likelihood methods to handle special kinds of nuisance and incidental parameters that arise in particular kinds of applications. A grand likelihood test would be at least as technically dubious as its constituent parts are in the present state of likelihood theory.

Bootstrapping might be thought to be a second alternative approach. At present, the theory of bootstrapping is well worked out only for cases in which there is a sample whose members can be assumed to be independently drawn from some population (see for example Wu 1988). In inferring phylogenies, Felsenstein (1985) has taken the characters to be the random sample. The independence of molecular characters is certainly more plausible than the independence of ordinary phenotypic characters. In our case, however, the characters are fixed. The usual approach to bootstrapping is to resample the data points (i.e., species). However, it is not obvious what phylogeny to assume for a set of species, and furthermore the assumption of independence is not even close to being met. This probably explains why, to our knowledge, no one has yet suggested applying bootstrapping methods to the comparative method.

The approach we have taken to scrutinizing the behavior of a test is to form a data generation process or model, generate many data sets, and try the test out on those data sets. The alternative approaches of likelihood theory and bootstrapping lack statistical foundation in the complex setting of phylogenetic problems. The attraction of the methods would be superficial, for they gloss over rather than solve real statistical difficulties in the problem. At present, any test suggested by those two approaches needs to be scrutinized by the logically prior and statistically routine method which we adhere to in this chapter.

References

Armbruster, W. S. 1992. Phylogeny and the evolution of plant-animal interactions. *BioScience*, 42, 12–20.

Bull, J. J. & E. L. Charnov. 1985. On irreversible evolution. *Evolution*, 39, 1149–1155.

Burt, A. 1989. Comparative methods using phylogenetically independent contrasts. *Oxford Surv. Evol. Biol.*, 6, 33–53.

Clutton-Brock, T. H., & P. H. Harvey. 1977. Primate ecology and social organization. *J. Zool., Lond.* 183, 1–39.

Cox, D. R. 1972. Regression models and life-tables. (with discussion). *J. R. Stat. Soc. B*, 34, 187–220.

Cox, D. R., & Hinkley, D. V. 1974. *Theoretical Statistics*. London: Chapman and Hall.

Donoghue, M. J. 1989. Phylogenies and the analysis of evolutionary sequences, with examples from seed plants. *Evoution,* 43, 1137–1156.

Felsenstein, J. 1984. Review of "The explanation of organic diversity" by M. Ridley. *Nature*, 308, 565.

Felsenstein, J. 1985. Confidence limits on phylogenies: An approach using the bootstrap. *Evolution,* 39, 783–791.

Felsenstein, J. 1988. Phylogenies and quantitative characters. *Annu. Rev. Ecol. Syst.,* 19, 445–471.

Fisher, R. A. 1922 On the mathematical foundation of theoretical statistics. *Phil. Trans. R. Soc. Lond. A*, 222, 309–368.

Frumhoff, P. C., & Reeve, H. K. 1994. Using phylogenies to test hypotheses of adaptation: A critique of some current proposals. *Evolution.*, 48, 172–180.

Gittleman, J. 1981. The phylogeny of parental care in fishes. *Anim. Behav.*, 29, 936–941.

Godfray, H. C. J. 1987. The evolution of clutch size in parasitic wasps. *Am Nat..*, 129, 221–233.

Grafen, A. 1989. The phylogenetic regression. *Phil. Trans. R. Soc. Lond. B*, 326, 119–157.

Grafen, A., & M. Ridley. 1995. Statistical tests for discrete cross-species data. *J. Theor. Biol.* (submitted)

Harvey, P. H., & M. D. Pagel. 1991. *The Comparative Method in Evolutionary Biology*. Oxford, England: Oxford University Press.

Höglund, J. 1989. Size and plumage dimorphism in lek-breeding birds: a comparative analysis. *Am. Nat.*, 134, 72–87.

Höglund, J. & B. Sillén-Tullberg. (in press). Does lekking promote the evolution of male biased size dimorphism in birds? On the use of comparative approaches. *Am. Nat.*.

Huey, R. B. 1987. Phylogeny, history, and the comparative method. In: *New Directions in Ecological Physiology* (Ed. by M. E. Feder, A. F. Bennett, W. W. Burggren, & R. B. Huey), pp. 76–98. Cambridge, England: Cambridge University Press.

Kalbfleisch, J. D., & R. L. Prentice. 1980. *The Statistical Analysis of Failure Time Data*. New York: John Wiley & Sons.

Lande, R. 1979. Quantitative genetic analysis of multivariate evolution, applied to brain: body size allometry. *Evolution*, 33, 402–416.

Maddison, W. P. 1990. A method for testing the correlated evolution of two binary characters: Are gains or losses concentrated on certain branches of a phylogenetic tree? *Evolution*, 44, 539–557.

McCullagh, P. 1991. Quasi-likelihood and estimating functions. In: *Statistical Theory and Modeling* (Ed. by D. V. Hinkley, N. Reid, & E. J. Snell), pp. 265–286. London: Chapman and Hall.

Miles, D. B., & A. E. Dunham. 1993. Historical perspectives in ecology and evolutionary biology: the use of phylogenetic comparative analysis. *Annu. Rev. Ecol. Syst.*, 24, 587–619.

Møller, A. P., & T. R. Birkhead. 1992. A pairwise comparative method as illustrated by copulation frequency in birds. *Am. Nat.*, 139, 644–656.

Oakes, E. J. 1992. Lekking and the evolution of sexual dimorphism in birds: comparative approaches. *Am. Nat.*, 140, 665–694.

Pagel, M. 1994. Detecting correlated evolution on phylogenies: A general method for the comparative analysis of discrete characters. *Proc. R. Soc. Lond. B*, 255, 37–45.

Pagel, M. D., & R. A. Johnstone. 1992. Variation across species in the size of the nuclear genome supports the junk-DNA explanation for the *C*-value paradox. *Proc. R. Soc. Lond. B*, 249, 119–124.

Pagel, M. D., & P. H. Harvey. 1989. Comparative methods for examining adaptation depend on evolutionary models. *Folia Primatol.*, 53, 203–220.

Proctor, H. 1991. The evolution of copulation in water mites: a comparative test for nonreversing characters. *Evolution*, 45, 558–567.

Ridley, M. 1983. *The Explanation of Organic Diversity*. Oxford, England: Oxford University Press.

Sanderson, M. J. 1991. In search of homoplastic tendencies: statistical inference of topological patterns in homoplasy. *Evolution*, 45, 351–358.

Sillén-Tullberg, B. 1988. Evolution of gregariousness in aposematic butterfly larvae: a phylogenetic approach. *Evolution*, 42, 293–305.

Sillén-Tullberg, B. 1993. The effect of biased inclusion of taxa on the correlation between discrete characters in phylogenetic trees. *Evolution*, 47, 1182–1191.

Stuart, A., & J. K. Ord. 1991. *Kendall's Advanced Theory of Statistics*, 5th ed. London: Edward Arnold.

Wetherill, B. G. 1986. *Regression Analysis with Applications*. London: Chapman and Hall.

Wu, C. F. J. 1988. [Contributor to discussion of papers by Hinkley and Diciccio and Romano.] *J R. Stat. Soc. B*, 50, 338–354.

CHAPTER 4

The Mechanistic Bases of Behavioral Evolution: A Multivariate Analysis of Musculoskeletal Function

George V. Lauder and Stephen M. Reilly

Of the many approaches that one might take to studying animal behavior, the comparative and phylogenetic analysis of physiological traits has been one of the least utilized. On the one hand, the discipline of neuroethology has had spectacular success in investigating the neural basis of behavior in individual species, but only rarely have such studies ventured into phylogenetic territory (e.g., Arbas et al. 1991; Hoy et al. 1988; Katz 1991). For the most part, neuroethologists have focused on understanding the mechanistic basis of behavior in individual species and on clarifying the motor and sensory systems involved in generating behavior. However, some evolutionary biologists have used physiological and neural traits to understand evolutionary processes such as sexual selection and mate choice (Ryan & Keddyhector 1992; Ryan & Rand 1990).

On the other hand, comparative biologists have recently taken increased interest in studying behavioral evolution. Behavioral characters are again being used to generate phylogenies (De Queiroz & Wimberger 1993; Wenzel 1992), a return to earlier days in ethology when the phylogenetic study of behavior was common (Lauder 1986; McLennan et al. 1988). Behavioral traits have also been correlated with other characters such as body size, territory size, and brain size to clarify historical (phylogenetic) patterns of character coevolution (Brooks & McLennan 1991; Harvey & Keymer 1991; Harvey & Pagel 1991; Pagel & Harvey 1988), and many of these studies are using recently developed quantitative comparative methods (Felsenstein 1985; Garland et al. 1992, 1993; Grafen 1989; Harvey & Pagel 1991; Martins & Garland 1991). The central theme of Felsenstein's (1985) oft-cited paper, that species cannot be treated as statistically independent entities in comparative analysis due to their genealogical relationship to each other, has now permeated the comparative literature (e.g., Garland & Adolph 1994; Huey 1987; Miles & Dunham 1993).

However, only rarely have comparative analyses been extended to physiological traits that directly measure the function of phenotypic features. The relative lack of comparative analyses of physiological traits is understandable given that determining the value of physiological characters usually involves conducting laboratory experiments on multiple individuals in each of the taxa to be used in the phylogenetic analysis. Obtaining physiological or functional characters might involve using several different techniques to measure multiple traits, which greatly increases the effort needed to obtain comparative data. Furthermore, individuals in taxa critical for the phylogenetic analysis, such as outgroup clades, may not be readily available (or amenable) for laboratory study, making a complete phylogenetic analysis of physiological traits difficult.

Despite these difficulties, there is considerable value in analyzing physiological or functional traits that underlie behavior. First, there are few data on patterns of evolution in physiological or functional characters. Most comparative analyses are of structural, genetic (relying on DNA base or amino acid sequences), or behavioral data (Lauder 1990). Second, the analysis of physiological characters provides data on the proximate causes of behavior. For example, by studying the neuromuscular mechanisms that generate behavior in related species, we can understand the specific changes in neural activation and muscle physiology that cause novel behavior. Third, data on functional characteristics enable us to examine the relationship between form and function from a historical perspective. If we are to analyze the relationship between structure and function in a noncircular manner, then functional traits must not be inferred from structure but should be measured independently (Lauder 1990, 1995; Reilly & Wainwright 1994). Applying comparative phylogenetic methods to structural and functional traits will then allow us to examine coevolutionary patterns to structural and functional characters, much as host-parasite coevolutionary patterns may be studied phylogenetically (e.g., Brooks & McLennan 1993; Mitter et al. 1991).

The major aims of this chapter are, first, to explain in more detail the value of a comparative examination of the physiological bases of behavior as a means to understanding how behavior evolves; second, to illustrate one approach to examining the relationship among different classes of functional traits as a heuristic tool to aid in understanding the mechanistic bases of behavioral evolution; third, to provide a case study of aquatic feeding behavior in salamanders as an example of the interspecific analysis of behavioral differentiation.

Physiological bases of behavioral evolution

A. Levels of analysis

A key concept for the comparative analysis of mechanistic bases of behavioral differentiation among species is the heuristic separation of underlying physiological mechanisms into several different levels. As illustrated in Table 4.1, a behavioral difference between two species could be due to changes at one or more of a number of levels. Behavior itself (row 1 in Table 4.1) could be quantified by measuring the amplitude and direction of movement, and/or velocities or accelerations of bones using a high-speed film or video system. If two species differ in behavior, in the movements of the forelimb during a mating display for example, the differences we observe might be due to changes in one or more structural and physiological properties (rows 2 through 5 in Table 4.1). Species may differ in the topological arrangement of muscles and bones of the limb or in structural properties of the muscles causing the movement (such as muscle fiber type). In addition, observed interspecific differences in behavior might be due to alterations in one or more species of the physiological properties of peripheral musculature, such as the contraction time. Changes in the central nervous system, either in structure (such as the pattern of neuronal interconnection) or function (activation or modulation of neuronal circuits generating motor output) could also cause a novel behavior to be observed in one or more species. Table 4.1 certainly does not represent the only possible arrangement of causal levels; many other possible classes of traits could be chosen that might underlie behavioral variation. But whatever the specific levels or hierarchical organization chosen, some decomposition of causal mechanisms of behavior is likely to be a valuable heuristic for our attempts to understand behavioral evolution.

Table 4.1. Hierarchy of classes of data (characters or traits) that might be analyzed to understand the mechanistic bases of behavioral attributes.

Class of data	Example of an organismal trait that might be studied interspecifically
1. Behavioral	Pattern of forelimb movement during a mating display
2. Structural (at the level of peripheral tissues)	Topographic arrangement of muscles and bones; tissue histology; muscle fiber types
3. Functional/physiological (at the level of peripheral tissues)	Physiological properties of muscles; biomechanical tissue properties; pattern of muscle activation
4. Structural (at the level of the nervous system)	Neuronal morphology; topology of neuronal interconnection; wiring of sensory and motor pathways
5. Functional/physiological (at the level of the nervous system)	Neuronal spiking patterns; motor patterns; membrane properties; modulation by neurotransmitters

In this particular hierarchy, characters are grouped into either structural or functional/physiological classes to reflect potential proximate causes for variation at the behavioral level. Note that variation in a behavioral trait among species might be due to change at a number of possible levels. See also Fig. 1 for a schematic illustration of this idea.

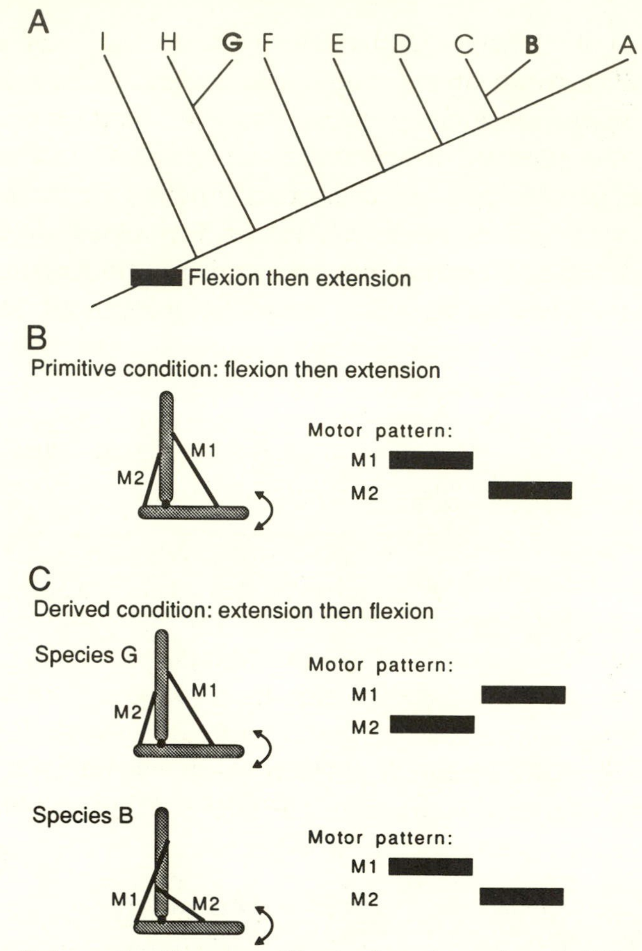

Figure 1. Schematic diagrams illustrating the role of different classes of characters in behavioral evolution. A: phylogenetic relationships of nine taxa. The ancestral condition for this clade is the behavior "flexion then extension" of the bony elements (shaded) shown in panel B. Taxa in bold (B and G) possess a derived behavior "extension then flexion" illustrated in panel C. B: morphological configuration of a limb in most taxa within the clade shown in panel A. Two muscles (M1 and M2) connect the bony elements that are attached at a joint (black dot). The pattern of muscle activation for these species is shown to the right. Black bars indicate the duration of muscle activity. Note that activation of M1 followed by M2 generates the behavior "flexion then extension." C: derived behavior, morphology, and motor pattern in species G and B. Both species convergently possess the derived behavior, but differ in the underlying mechanistic causes of the novel behavior. Species G retains the ancestral morphology for the clade but possesses a novel motor pattern. Species B exhibits a novel morphology (in which the muscles have changed their insertions) but retains the primitive motor pattern.

Figure 1 schematically illustrates how evolutionary changes at two of the levels discussed above might independently result in a novel behavior. A comparative behavioral investigation of the nine species in a clade (Fig. 1A) indicates that all taxa except B and G generate a threat behavior with the forelimb called "flexion followed by extension": the lower bone in Fig. 1B moves up in flexion, followed by a downward extension movement. Mapping this character onto the cladogram reveals that "flexion followed by extension" behavior is the ancestral condition for this clade. Further comparative investigation demonstrates that two species, B and G, possess a novel threat behavior that may be described as "extension followed by flexion." Within this clade these two species display convergently acquired behavioral novelties. How has this behavioral novelty been produced in these two species?

Investigation of the anatomy of species in this clade shows that all species possess a forelimb with a basic structure consisting of two bones connected at one joint which is spanned by two muscles (Fig. 1B), M1 and M2. However, in species B (and only in species B), M1 and M2 have changed position relative to the ancestral condition. An ontogenetic study shows that M1 grows posteriorly to attach behind the joint, while M2 grows anteriorly to insert anterior to the joint (Fig. 1C). This contrasts with the condition in all other species in which muscle M1 runs anteriorly to its attachment, while muscle M2 inserts posterior to the joint (Fig. 1B).

Measurement of the activation pattern of the two muscles in each species (by implanting small electrodes in the muscles and recording electromyograms during the threat behavior) reveals the motor pattern shown to the right of the morphology in panels B and C. In all species except species G, the motor activity pattern is that muscle M1 is activated first, followed by muscle M2. In species G, however, muscle M2 is activated first, followed by muscle M1. A simple biomechanical analysis confirms that given the morphology of the limb in species B, activating M1 causes extension of the horizontal bone, and that the subsequent activation of M2 causes flexion of that bone. In this species, then, the behavioral novelty ("extension followed by flexion") results from a topological rearrangement at the level of musculoskeletal structure and not from any change in neural output to the musculature. In fact, species B shares the ancestral motor pattern for the clade as a whole. Recordings of the motor pattern in species G show a different result. Unique to all species in the clade, species G possesses a novel motor pattern in which muscle M2 is activated prior to muscle M1. Given the topology of the musculoskeletal system, such a motor pattern will produce a movement of "extension followed by flexion".

Even though the behavior might appear to be similar in species B and G, the underlying mechanistic bases of the behavior are different. The novel behavior in B resulted from changes at the level of peripheral gross morphology, while the novel behavior in species G resulted from evolution in the motor pattern. If we do not investigate the mechanistic basis of behavioral evolution, then we will be unable to focus on the appropriate hierarchical level at which behavioral novelty is generated (Lauder 1994).

One benefit to such a schema for considering behavioral evolution is the Newtonian mechanical basis for musculoskeletal and nervous function that permits increased confidence in attributions of cause and effect. If two taxa differ in behavior and motor pattern and yet are similar in morphology, a biomechanical analysis of the effects of those differences in motor pattern enables functional morphologists to predict differences in behavior. Behavioral differences are caused (in the mechanical sense) by novelties at some level (Table 4.1), and evolutionary patterns at each level may not be congruent. In our view, a key question in the analysis of behavioral evolution is the identification of the appropriate hierarchical level or levels that account causally for behavioral differences among species.

B. Visualizing multivariate levels

One limitation to the approach discussed above, of dissecting the causal basis of behavioral differences among species, is that too often such analyses are univariate in nature and focus on only a few variables. Yet all levels presented in Table 4.1, from behavior to neuronal physiology, are intrinsically multivariate; the analysis of many attributes of each level is necessary to adequately capture interspecific variation. Univariate approaches have been of considerable value in the past, used extensively in biomechanical analyses of musculoskeletal function where the influence of specific features of muscle structure and function on movement are of interest. In an example taken from our own work, we analyzed the mechanism by which salamanders feeding on land project their tongues toward the prey during feeding (Reilly & Lauder 1991). We studied the effect of one muscle (the subarcualis rectus one, SAR1) on feeding performance and behavior. As this muscle was believed to provide the main motive force projecting the tongue from the mouth, our focus was on the effect of structure and function of the SAR1 on behavior.

However, in order to adequately describe interspecific variation in any class of traits, a multivariate approach allows many different measured attributes at each level to contribute to decisions as to which species differ from others. While use of multivariate analyses is hardly new in most areas of biology (Bookstein et al. 1985; Rohlf & Bookstein 1990), in the interspecific study of organismal function and physiology multivariate studies of variation are relatively rare.

One way in which we might extend the depiction of different mechanistic levels underlying behavior (Table 4.1) to a multivariate framework is to use a basic multivariate technique such as principal components analysis to describe variation at each level. For example, measurement of a variety of different aspects of movement of the hindlimb during locomotion (such as the amplitude of excursion of each of the bones in the limb and their peak velocities and accelerations) for several individuals in each species of a clade could be used to measure variation at the behavioral level. This would generate a data set of several kinematic variables for each species, and such a data set could be subjected to principal components analysis to summarize variation among the species.

A principal components analysis (PCA) generates a set of new variables that are linear combinations of the original variables in the data set. Coefficients of variables contributing to the first PC are chosen to maximize the variance for PC1, and PC1 thus represents the vector of greatest variation in the data set. PC2 is calculated similarly, but the PC scores calculated from the linear combination of variables and coefficients must be uncorrelated with the PC1 scores. The variation in the data set explained by PC2 is thus uncorrelated with that explained by PC1. The number of principal components calculated is equal to the number of original variables, with each successive component accounting for successively less variation; the first four PCs commonly account for greater than 75% of variation in the data set. Useful descriptions of PCA are given in Dunteman (1989) and Harris (1975). An important use of PCA is to reduce a data set with a large number of intercorrelated variables to a smaller set of uncorrelated variables. Analyzing these uncorrelated PCs avoids the problem of interpreting separate univariate analyses that are correlated with each other in a complex fashion (Bray & Maxwell 1985; Willig et al. 1986). Multivariate variation in a large number of variables can be summarized succinctly. Each variable can also be examined for its contribution to the overall PC score on each component, so that the specific pattern of variation in measured variables can be determined.

Figure 2 illustrates schematically one possible result from such a PCA. Given five taxa (A to E), we could analyze behavioral variation by conducting a PCA on the set of kinematic variables and then plot the mean position of each species in the resulting multivariate space (Fig. 2B). Only principal components 1 and 2 are shown, as these represent the greatest percentage of variation within the data set. While the technique of principal components analysis itself does not account for phylogenetic relationships among species in calculating the component scores, these scores can be calculated using phylogenetically standardized contrasts of the original data matrix (if a corroborated phylogeny of the species is available, Felsenstein 1985). Martins (1993) provides an example of this procedure in an analysis of behavioral characters. In order to define differences among species, we could also analyze principal component scores to test for significant groupings among the taxa. Figure 2B also illustrates an example of a multivariate analysis of variance on the principal component scores (accounting for individuals within each taxon). Taxa A and B are not significantly different in behavior from each other (and are thus grouped by a circle), while these two taxa together do occupy a significantly different portion of the behavioral space from taxa C, D, and E, all of which are behaviorally distinct from each other as well as from taxa A and B together.

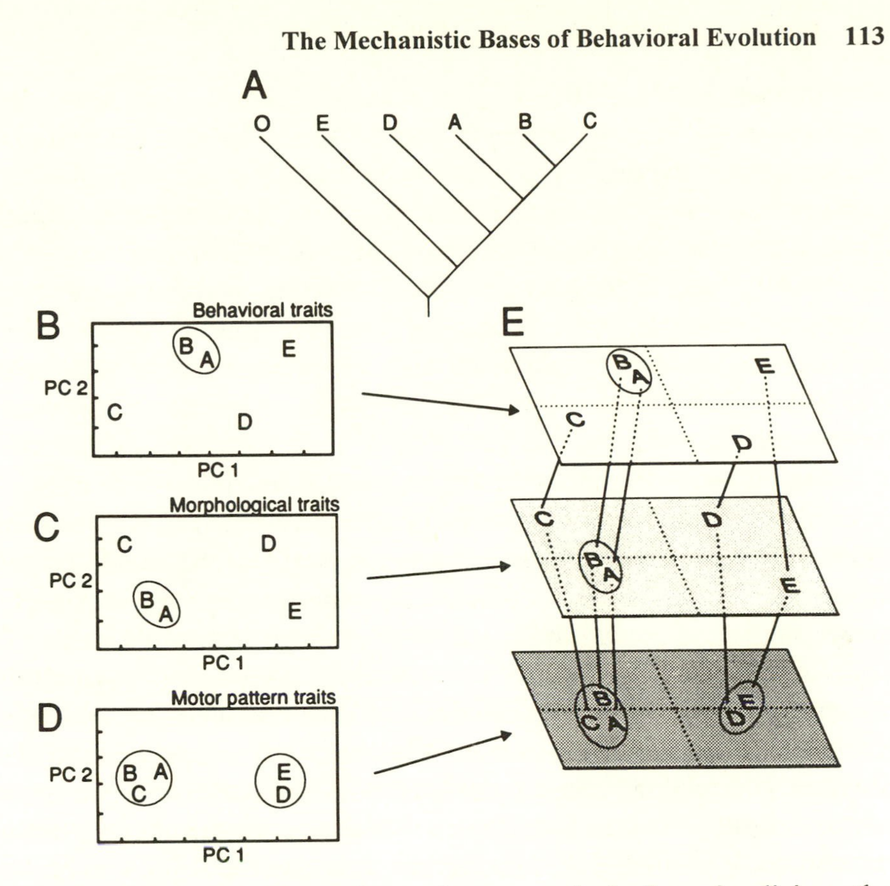

Figure 2. Schematic illustration of one method for visualizing the relationships among classes of characters for an interspecific analysis of the mechanistic bases of behavioral evolution. A: Phylogenetic relationships of five taxa (A to E) with an outgroup taxon (nearest phylogenetic relative) O. B: Principal components analysis of behavioral traits measured, for example, from high-speed video records of movement. Letters indicate the mean for each taxon of values for all the individuals studied in that taxon. Taxa enclosed by a circle are not significantly different from each other. Taxa not enclosed by a circle are significantly different in principal component 1 and 2 scores from all other taxa. C: Principal components analysis of morphological traits measured for the same five taxa. D: Principal components analysis of muscle activity traits. Note the changing pattern of differences among taxa in the different classes of characters. E: Schematic overview of the relationships among the three classes of characters. Lines between adjacent levels connect taxa and show the mapping of character variation; see text for further discussion.

For the same species one might also measure a variety of structural attributes of the musculoskeletal system that are mechanically involved in generating the observed locomotor behavior: e.g., mass of leg muscles and bones, lever arms, or muscle fiber lengths and angles. These data could also be subjected to a PCA (Fig. 2C), and the resulting plot might reveal a different array of taxa in multivariate space. In this schematic example, taxa A and B occupy similar regions of the multivariate morphological space and are similar structurally. These taxa together differ from taxa C, D, and E, all of which occupy distinct morphological positions.

Finally, one might conduct a similar analysis for a set of physiological/functional characters of the limb muscles. For example, study of the pattern of activation (motor pattern) of the limb muscles is of interest in understanding how the nervous system drives limb morphology to generate the locomotor behavior quantified in Fig. 2B. For each species we record the electrical patterns in several limb muscles involved in generating locomotor behavior and quantify such patterns by measuring the duration of electromyographic bursts, peak amplitudes, and relative onset times of activity. These data can then be subjected to a PCA and the position of taxa in the muscle activity pattern space plotted (Fig. 2D). In this schematic example, taxa A, B, and C are similar in motor pattern, while taxa D and E share similar motor patterns to each other and differ from the A, B, and C group.

The most heuristic feature of an analysis of interspecific variation among character classes is evident when we consider the relationships among levels. What is the mapping among taxa from one level to another? When a taxon possesses a novel morphology, does it tend to also possess a novel motor pattern? To what extent are behavioral novelties in a taxon generated by both morphological and neuromuscular changes? Of particular interest is the changing pattern of differences among taxa as we move among levels. This mapping may be visualized simply by orienting the results of each PCA as a plane and connecting taxa with lines (Fig. 2E). The significant feature of this visualization is not changes in absolute locations of each taxon at each level: the positions of each taxon are calculated from a different data matrix at each level, and shifts in absolute position reflect changes in variable correlations within each data set. Rather, it is the changes that occur in groupings of significant differences among taxa that provide the information of interest. Thus, taxon C possesses a behavior distinct from other members of this clade (Fig. 2E). In addition, this taxon possesses novel features of morphology as compared to outgroup taxa (close relatives) A and B (the phylogeny of this clade is shown in panel A). Taxon C shares a similar motor pattern with taxa A and B. Given a primitively similar motor pattern, a reasonable mechanical/evolutionary hypothesis is that taxon C possesses a derived behavior because of morphological novelties. Taxa A and B share similar motor patterns and morphologies and thus must possess a similar kinematic (behavioral) pattern. Taxa D and E share similar patterns of muscle activation but differ in morphology and behavior. A similar causal evolutionary hypothesis would indicate that the behavioral differences observed between these two taxa result from changes in morphology (such as bone lengths, muscle lever arms and masses) which produce a different behavior given a similar motor pattern.

A variety of theoretically possible patterns of interrelationships among taxa and levels can be envisioned. Figure 3 shows three such patterns in association with the phylogenetic relationships of the five taxa (A to E) and the closest relative of this clade, outgroup taxon O. Figure 3A shows a situation in which there is no general pattern to behavioral, morphological, and motor pattern evolution: each taxon is distinct from the others and possesses derived traits at each level. This pattern is not expected, given the mechanically causal relationships that must exist among levels and our knowledge that many traits are relatively conservative phylogenetically, but it is possible that each taxon could possess autapomorphies (uniquely derived features) for each class of traits.

Figure 3B shows a clade in which all ingroup taxa share a common motor pattern but show interspecific differentiation at the morphological and behavioral levels. In this case, taxa D and E share novelties in morphology that map onto shared behavioral novelties. Variation in morphology is a likely causal explanation of behavioral differentiation among these taxa, since similarities in motor pattern are a plesiomorphic (ancestral) characteristic of the clade and cannot explain behavioral differentiation among taxa. Thus, taxa A, B, and C differ morphologically, and these differences are mirrored by differentiation at the behavioral level.

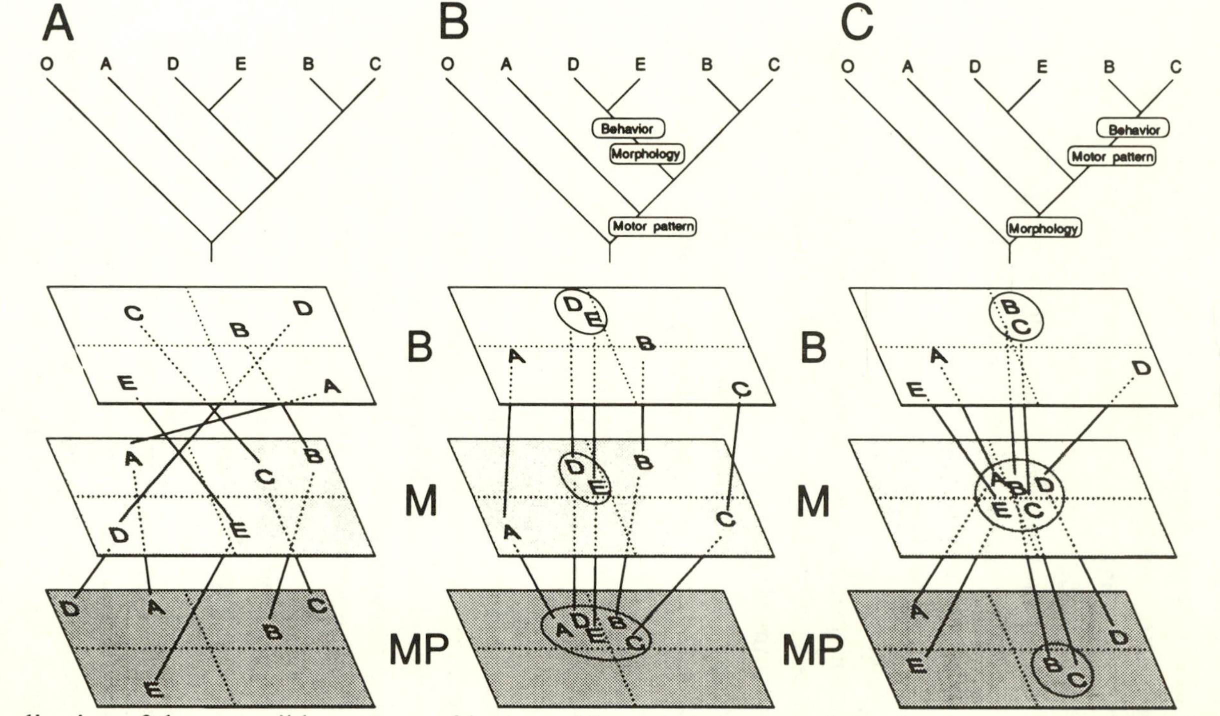

Figure 3. Visualization of three possible patterns of interspecific variation among behavioral, morphological, and motor pattern characters. Each plane represents a plot of principal component 2 (y axis) versus principal component 1 (x axis). Conventions for taxa and groupings follow Fig. 2. Phylogenetic interpretations of each pattern are given at the top of each panel. Although other possibilities exist, these three panels show a range of possible results. A: Each taxon possesses uniquely derived (autapomorphic) traits at each level, and there is no clear mapping pattern among levels. B: All five taxa share a common motor pattern. Differentiation at the morphological and behavioral levels is similar. Two taxa (D and E) share common derived morphology and behavior. C: All five taxa share similar morphology, but show divergent motor patterns and behavior. Taxa B and C share novel muscle activity and behavior. Abbreviations: B, Behavioral level; M, morphological level; MP, motor pattern level.

The phylogenetic pattern depicted in Fig. 3B, with taxa showing relatively little differentiation in motor pattern and behavioral differentiation resulting from differences among taxa in morphology, has been the focus of considerable research in the field of functional and evolutionary morphology. A number of authors have suggested that homologous muscles may retain primitive activity patterns, and that behavioral evolution may be a consequence of changes at the morphological level alone Dial et al. 1991; Gatesy 1994; Goslow et al. 1989; Jenkins & Goslow 1983; Lauder 1990; Lauder & Shaffer 1988). Evidence both for and against ontogenetic and phylogenetic conservation of motor patterns has been presented, and as yet no general conclusions are possible (Smith 1994).

Figure 3C illustrates taxa in a clade sharing a similar plesiomorphic (ancestral) morphological configuration, while exhibiting diversity at the motor pattern and kinematic levels. Taxa B and C share derived conditions of both motor output and kinematics (behavior), while taxa A, D, and E possess novel motor patterns and kinematics that are not shared with other taxa.

There are two useful features of such an analysis for understanding causal relationships among the levels deriving from the Newtonian mechanical relationships that connect variation among classes of characters. First, a kinematic differentiation between two taxa that share a common structural plan (e.g., taxa A and E in Fig. 3C) allows the general prediction that A and E must differ in motor pattern. Only by differentiation in motor output could two taxa with similar morphology display different behavior patterns. Second, and more specifically, the individual traits that make up the components can be recovered from the analysis and evaluated against a biomechanical model. Thus, taxon A differs from taxon C in morphology and kinematics (Fig. 3B). An examination of the variables loading highly on morphological PC1 might show that taxon C possesses a longer lower jaw and hyoid linkage in the skull as compared to taxon A. Given the shared ancestral activation pattern of muscles controlling these elements (Fig. 3B), the type of behavioral difference between these taxa can be predicted: a longer time course for mouth movements, for example. This prediction can be checked against the variable loadings for those movements on principal components one and two at the kinematic level.

A case study: aquatic feeding behavior in salamanders

A. Background

The feeding mechanism of salamanders has been the subject of numerous studies over the last 20 years (Cundall 1983; Erdman & Cundall 1984; Larsen et al. 1989; Lauder & Reilly 1990; Lauder & Shaffer 1985, 1988; Lombard & Wake 1976, 1977; Miller & Larsen 1990; Reilly & Lauder 1989a, 1990; Roth 1976; Schwenk & Wake 1993; Thexton et al. 1977; recent reviews in Lauder & Reilly 1994; Lauder & Shaffer 1993; Reilly 1994). These analyses have served to characterize basic biomechanical features of the musculoskeletal system and many aspects of the diversity in salamander skull structure and function. Thus, there is moderately extensive knowledge of behavioral variation, morphology, and muscle function during a behavior that is important to fitness: food acquisition. This background on skull biomechanics facilitates a comparative analysis of feeding behavior (kinematics), morphology of the skull, and motor patterns of cranial musculature that can be used to help analyze patterns of differentiation among taxa in these classes of characters.

Here we present a case study of aquatic feeding behavior in six taxa of salamanders that represent most of the salamander clades exhibiting aquatic prey capture. In this analysis we used our previous functional research on both aquatic and terrestrial prey capture in ambystomatid salamanders (Lauder & Reilly 1988, 1990, 1994; Lauder & Shaffer 1985, 1988; Reilly & Lauder 1988, 1989a, 1989b, 1990; Shaffer & Lauder 1985a, 1985b) to define relevant biomechanical variables at the morphological and motor pattern levels (Table 4.1). The analysis of behavior and morphology has been published elsewhere (Reilly & Lauder 1992), but the electromyographic data on motor patterns used to generate data for the third class of characters (Table 4.1) has not been previously presented. We will demonstrate that interesting historical questions about the evolution of behavior and its underlying physiological bases may emerge from a comparative analysis of different classes of characters that are causally related to the generation of behavior.

B. Methods and data

The six taxa of salamanders used in this analysis are shown in Figure 4 along with their phylogenetic relationships as most recently analyzed by Larson and Dimmick (1993). One difficulty inherent in a comparative analysis involving different salamander families at this time is that current views of the phylogenetic relationships of these families are still the subject of some controversy and ambiguity. We utilized the phylogenetic relationships resulting from a combined analysis of molecular and morphological data sets as summarized by Larson and Dimmick (1993), but other morphological data (which include fossil taxa) support a different topology (Cloutier in press; Trueb & Cloutier

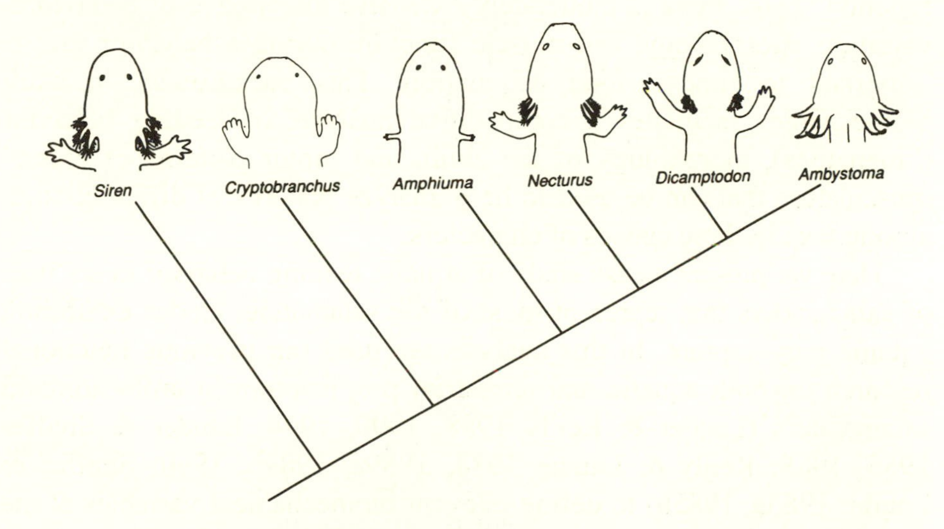

Figure 4. Phylogenetic relationships of the six taxa of salamanders used in this study. Behavioral and morphological data sets were generated for all six taxa, but only four taxa were studied in the analysis of motor patterns (*Siren*, *Cryptobranchus*, *Necturus*, and *Ambystoma*). Phylogenetic relationships of the taxa are from Larson and Dimmick (1993), and the branching pattern reflects only relative time of divergence, not amount of phenotypic change. The external morphology of the head (in dorsal view) is shown for each taxon to illustrate head shape and the size and location of external gill filaments; figures are not drawn to the same scale.

1991). The morphological characters of the feeding mechanism analyzed here are not part of the data used by Larson and Dimmick (1993) and Cloutier (in press) to construct their phylogeny. Prey capture behavior in a total of six families was analyzed: *Siren intermedia* (Sirenidae), *Cryptobranchus alleghenensis* (Cryptobranchidae), *Amphiuma means* (Amphiumidae), *Necturus maculosus* (Proteidae), *Dicamptodon tenebrosus* (Dicamptodontidae), and *Ambystoma mexicanum* (Ambystomatidae). The number of individuals studied varied between 2 and 10 in each species.

Prey capture in salamanders is rapid (taking between 30 and 120 ms), requiring high-speed films or video recordings (at 200 frames per second) for accurate measurement of cranial bone movements. Individual salamanders were trained to feed in an aquarium under filming lights, and 1- to 2-cm-long pieces of earthworms (*Lumbricus*) were used as prey. Video or film recordings of prey capture were analyzed frame by frame to quantify head movements. From each frame we digitized movements of the head, hyoid region, and jaws. Profiles of the movements of each set of structures were plotted against time (a sample kinematic profile for *Necturus* is shown in Figure 5) and from these profiles seven kinematic variables were derived that describe the behavior of the head during feeding. Behavioral variables measured both the timing of movement (such as the time to maximal head angle and the duration of the gape cycle from the onset of mouth opening to mouth closing) and the amplitude of bone movement. At least five feedings were quantified from each individual.

Cranial morphology of the six taxa was characterized by measuring five structural variables that described features of head shape relevant to prey capture behavior. Morphological variables included the number of open gill slits at the back of the buccal cavity (a hydrodynamically important factor in the feeding mechanism, Lauder & Shaffer 1986; Reilly & Lauder 1988), and measurements of lower jaw length and head size.

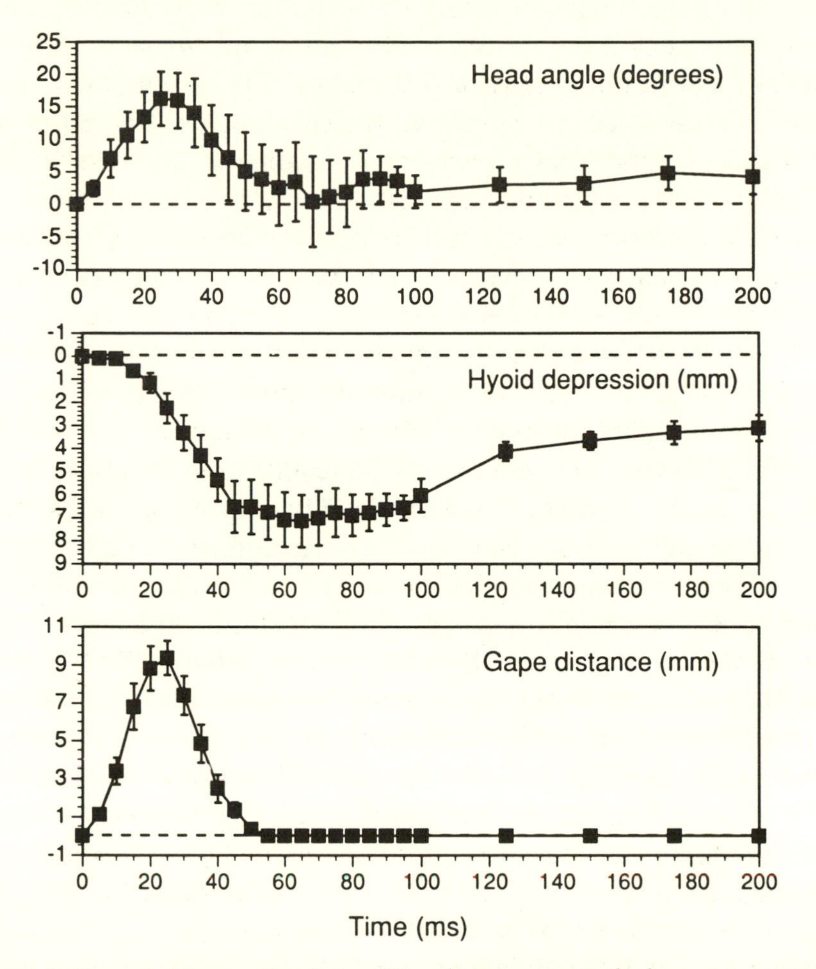

Figure 5. Sample kinematic plots of head movements during prey capture by *Necturus*. Points shown are means (*N*=5) from one individual with the standard error about the mean. Head angle increases from 0 to 25 ms indicating elevation of the skull on the vertebral column. Peak gape is also reached at 25 ms, while hyoid depression (reflecting expansion of the throat) begins a plateau at 45 ms. Prey capture occurs within 50 ms. Plots from individual feedings such as this were used to generate the behavioral variables for the principal components analysis.

Motor output to the cranial musculature was measured by implanting fine-wire electrodes into five muscles involved in generating movements of the head, jaws, and hyoid during prey capture. These muscles were chosen to reflect likely causal relationships between morphology and feeding behavior. For example, head angle was one of the kinematic variables measured, and the epaxial muscles are the only set of muscles in the head capable of generating an increase in head angle; thus, activity in the epaxial musculature was quantified. Individual salamanders were anaesthetized and electrodes implanted percutaneously into cranial muscles as in previous research (Lauder & Shaffer 1985; Reilly & Lauder 1991). The five homologous muscles studied in each taxon were (a) the epaxial muscles, which act to elevate the head and thus contribute to opening the mouth during feeding; (b) the depressor mandibulae muscle, which acts to move the lower jaw ventrally and thus open the mouth; (c) the rectus cervicis muscle, which moves the hyoid apparatus posteroventrally expanding the volume inside the mouth cavity and creating suction; (d) the adductor mandibulae externus muscle, which is one of two major mouth closing muscles; (e) the branchiohyoideus muscle, which acts to abduct the gill arches during prey capture [Lauder and Shaffer (1985) provide morphological descriptions of these muscles in *Ambystoma*]. *Cryptobranchus* possesses the capability of generating asymmetrical movements of the right and left sides of the head during feeding (Cundall et al. 1987), and we implanted bilateral electrodes (for a total of 10 channels) to assess asymmetry in one individual. Electrodes from each muscle were bundled together into a common cable that led from an attachment on the animal's back to AC preamplifiers (amplification was 5000× to 10,000×) and then into an FM tape recorder. All channels were recorded simultaneously. The analog tape recordings of electromyographic data were digitized at 12-bit resolution (8000-Hz sample rate) into a binary data file from which individual motor pattern variables were measured (to the nearest 0.1 ms resolution). Due to limited animal availability and performance with implanted electrodes, electromyographic data could be obtained from only four taxa: *Siren*, *Cryptobranchus*, *Necturus*, and *Ambystoma*.

Electromyographic data from a single prey capture event by *Cryptobranchus* are shown in Fig. 6. Most muscles show a rapid onset of activity (within 10 ms of each other) except for the adductor mandibulae, which occasionally showed a considerable delay in onset. From these data, 14 motor pattern variables were measured. The onset

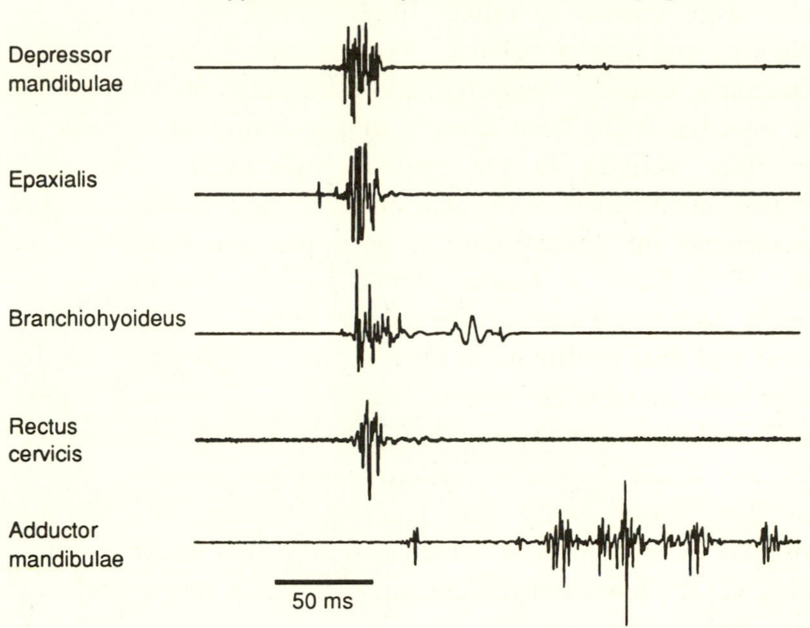

Figure 6. Representative pattern of muscle activity for five cranial muscles during prey capture by *Cryptobranchus*. Note the nearly simultaneous onset time in all muscles except for the adductor mandibulae. Recordings such as this were used to generate the motor pattern variables for the principal components analysis.

of activity in four muscles was measured relative to the onset time of the rectus cervicis muscle. The rectus cervicis is a major mouth-opening muscle, often used as a marker of the onset of the feeding motor pattern. We measured the duration of each muscle burst and the total rectified integrated electrical activity within each burst.

Separate statistical analyses on the three data sets (behavioral, morphological, and motor pattern) included standard two-level nested analysis of variance (with individuals nested within taxon), PCA on the correlation matrix of each data set (since variables differed in measurement scale; Bookstein et al. 1985), and multivariate analysis of variance on the principal component scores to define groups of taxa within each class of data.

C. Results and interpretation

Results from the analysis of three classes of data for the four taxa common to all three levels are presented in Fig. 7. This figure schematically represents three plots of principal components one (horizontal or *x* axis) and two (*y* axis) resulting from analyses of the three separate data sets. The first two levels represent a subset of the behavioral and morphological results presented by Reilly and Lauder (1992) for all six taxa. Each taxon is represented by a letter reflecting the mean position of the values for all individuals measured in that .taxon, and taxa that are not significantly different from each other are grouped by circles. Taxa that are not located within a circle are significantly different from other groupings within each level when all characters are considered together.

Behaviorally, *Ambystoma* and *Necturus* share common features of prey capture movements that result in both taxa grouping together in the principal components analysis of feeding behavior (Fig. 7). *Siren* and *Cryptobranchus* are located in different regions of the multivariate space that describes prey capture and are distinct both from each other and from the *Ambystoma* + *Necturus* group. High values of principal component 1 reflect larger values for maximal mouth opening and hyoid expansion, while larger values of principal component 2 indicate more rapid feedings (shorter gape cycle times and time to maximal gape). Thus, *Cryptobranchus* possesses slower prey capture and larger mouth and hyoid excursions than the other taxa. Morphologically, *Siren* and *Ambystoma* possess similar cranial structures (Fig. 7), while *Necturus* and *Cryptobranchus* occupy divergent areas of morphological space.

Mappings between the behavioral and morphological levels reveal several interesting patterns, and this visualization facilitates consideration of the extent to which behavioral variation among taxa may be due to changes in morphology. First, taxa such as *Ambystoma* and *Siren*, which share similar morphology, show divergent behavior. This result must be due to differences between these two taxa at the motor pattern level. The mapping between the behavioral and morphological levels can thus be used to predict patterns of variation in motor pattern. Second, since *Ambystoma* and *Necturus* are divergent in morphology but not in behavior, one cannot predict the pattern of interspecific variation in motor pattern. Morphological variation alone could cause behavioral differences, as could differences in both morphology and motor pattern. A similar result obtains for the mappings between levels for *Cryptobranchus*. Third, taxa may show similar behavior and divergent morphologies, or the converse. No overall pattern corresponding to the examples in Fig. 3 was found.

Examination of the motor pattern data for the four taxa (Fig. 7) shows that *Necturus* and *Cryptobranchus* share similar regions of motor pattern space, while *Ambystoma* and *Siren* are divergent. High values on principal component 1 primarily reflect greater activity (longer duration and higher rectified integrated area) for the rectus cervicis muscle, longer duration activity in the branchiohyoideus and epaxial muscles, and shorter relative onset times of the depressor mandibulae and epaxial muscles. High values of principal component 2 indicate greater activity in the depressor mandibulae and adductor mandibulae muscles and longer relative onset times of the depressor mandibulae and epaxial muscles. Thus, compared to *Siren*, the motor pattern used by *Ambystoma* during prey capture involves relatively less activity in the rectus cervicis and longer times between rectus cervicis onset and the start of activity in the depressor and adductor mandibulae muscles.

Overall, the mappings among the three levels demonstrate three significant patterns. First, two of the taxa, *Ambystoma* and *Necturus*, possess divergent motor patterns and morphology but share similar prey capture behavior patterns. These two taxa have taken different mechanistic "paths" to arrive at a similar behavioral phenotype; differences in motor pattern and morphology are interacting to generate similar bone movements. Second, the prediction discussed above (based on the analysis of data from the behavioral and morphological levels), that *Ambystoma* and *Siren* must show divergent motor patterns is correct. These two taxa show significantly different patterns of muscle activity. Causally, differences at the motor pattern level must account for differences at the behavioral level. Third, *Cryptobranchus*, is distinct in morphology and behavior but possesses a similar motor pattern to *Necturus*. This result indicates that morphological and behavioral divergence in *Cryptobranchus* is not mirrored at the motor pattern level. Overall, the relationships observed among taxa and levels in this study are complex, and no general patterns are apparent.

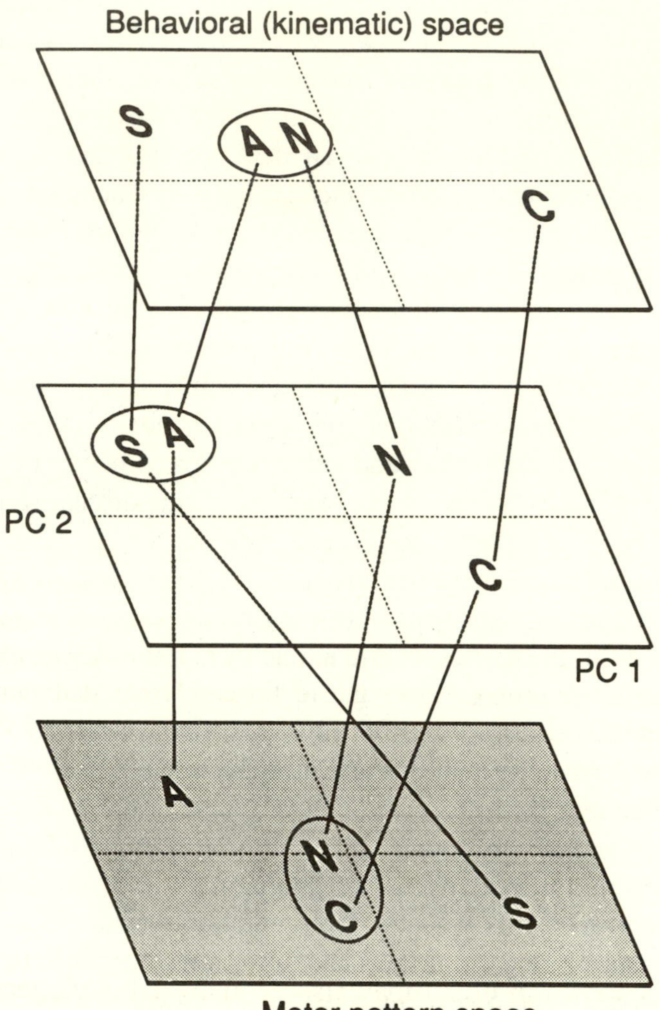

Figure 7. Results from the principal components analysis of three data sets: behavioral, at the top; morphological, in the middle; motor pattern, at bottom. Each plane represents principal component 2 (PC2) plotted against principal component 1 (PC1). Following the conventions established in Figure 2, lines connect taxa on adjacent planes, while taxa that are not significantly different from each other are circled. Abbreviations: A, *Ambystoma*; C, *Cryptobranchus*; N, *Necturus*; S, *Siren*.

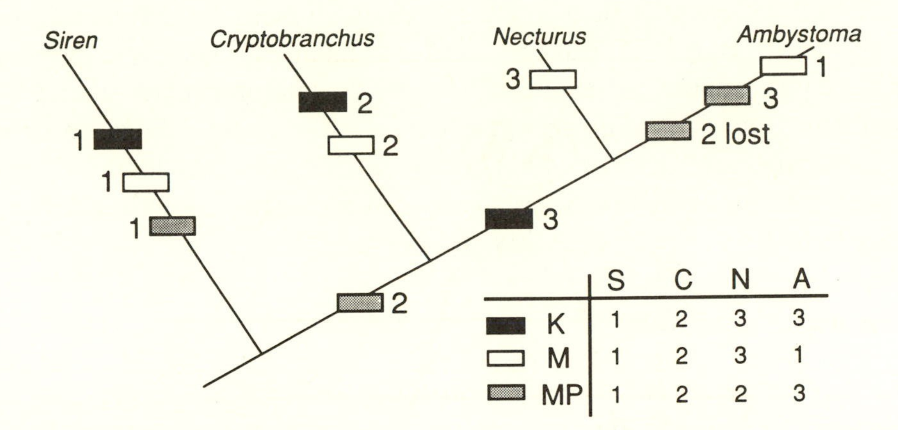

Figure 8. Phylogenetic interpretation of the changes in traits at each of three levels: K, kinematic (behavioral), M, morphological, and MP, motor pattern. The data matrix in the lower right codes the differences among taxa for each class of characters based on the results of statistical tests on principal component 1 and 2 scores. Taxa with the same numbered state are not significantly different from each other in this trait. The resulting characters in the data matrix are mapped onto the phylogeny. Note that motor pattern character state 2 is illustrated as having been gained once and then lost in *Ambystoma*. An equally parsimonious explanation of this character is that it was gained independently in *Cryptobranchus* and *Necturus*.

The data shown in Fig. 7 may also be analyzed phylogenetically as discrete character states. In Fig. 8 the variation among the four taxa for each of the three classes of characters is coded into discrete states (1, 2, or 3) based on statistical differences among taxa at each level. For example, at the kinematic level (K), *Necturus* and *Ambystoma* are coded as state 3 based on similar patterns of bone movement, while the other two taxa receive character state codes of 1 and 2. Perhaps the most striking feature of the phylogenetic pattern depicted in Fig. 8 is the extent to which the four taxa possess uniquely derived (autapomorphic) features of behavior, morphology, and motor pattern: there has been considerable phylogenetic divergence in all three classes of characters. Only one character (the shared kinematic pattern of *Necturus* and *Ambystoma*, coded 3) can be used to unambiguously diagnose a monophyletic clade. A second character, the motor pattern shared by *Necturus* and *Cryptobranchus* (coded 2) could diagnose the *Cryptobranchus+Necturus+Ambystoma* clade (with a loss of this character in *Ambystoma*). Alternatively, this character could have arisen independently in *Cryptobranchus* and *Necturus*.

General discussion

The method presented here for visualizing differences among taxa in behavior focuses on the mechanical interrelationships among bone movement (behavior), morphology, and output from the central nervous system. But not all interspecific differences in behavior are the result of changes at these levels. Learned behavior patterns, for example, do not result from changes in morphology (although learning may change motor output; Wainwright 1986). Differences among taxa in biochemistry or in organismal properties not addressed by the levels in Table 4.1 may also contribute to interspecific behavioral differentiation. Furthermore, changes in behavior may have arisen phylogenetically prior to changes in morphology. Indeed, there has been considerable discussion in past ethological literature as to the primacy of morphological or behavioral change (Lauder 1986). The visualization presented in Fig. 2 suggests that changes in behavior are due (and caused by) changes in morphology or motor pattern, and thus that behavioral changes occur either along the same branch or at a later time phylogenetically than changes in morphology. A phylogenetic analysis of behavioral and morphological characters might show that a change in behavior preceded the occurrence of novel morphology, thus changing the direction of causality implied by Fig. 2. In such cases, the analysis of characters at the motor pattern level is critical, as alterations in motor pattern may well have caused behavioral change prior to alterations in musculoskeletal structure. For example, neuromodulators within the nervous system may change and produce novel behavior in a species without any change in peripheral morphology. Neuromodulators can alter motor output directly by their effect on central neurons and thus generate a novel behavior (Katz 1991 discusses examples). Change in neuromodulators and the consequent change in behavior may then characterize a clade which is similar in morphology. This scenario would produce a pattern similar to that shown in Fig. 3C.

Answering the question "Which changes first, morphology or behavior?" depends on (1) a phylogenetic analysis of morphological and behavioral characters and (2) an assessment of motor patterns produced by the nervous system. We predict that when phylogenetic analysis demonstrates novelties in behavior preceding morphology, motor output from the nervous system will be found to have changed concordantly with behavior, and novelties in both these classes of characters will precede morphology.

Although we believe that there is much to be gained by the comparative mechanistic analysis of interspecific differences in behavior, this is one aspect of ethology that has received relatively little attention. Compared to progress that has been recently made in the comparative (phylogenetic) study of behavioral ecology (e.g., Brooks & McLennan 1991; Harvey & Pagel 1991; Harvey et al. 1990), our knowledge of behavioral, structural, and functional variation among taxa is minimal. In discussing the method of analyzing interspecific behavioral differentiation presented here, we will first consider practical issues that bear on the development of case studies, then discuss analytical considerations, and finally address strengths of this approach. There are a number of reasons why comparative functional data are relatively scarce. Obtaining both physiological and morphological data on a variety of taxa is extremely time consuming, particularly if complex physiological procedures must be performed on multiple individuals within each taxon. Measuring physiological traits in living animals often requires a number of invasive experimental techniques. This may limit the size range of animals that can be studied, restricting comparative functional data to those taxa in which adults are a suitable size. In addition, species vary in their ability to tolerate experimental procedures while still exhibiting normal behavioral responses. Some species simply do not respond well to laboratory conditions, and yet such species may be critical for testing a given comparative hypothesis. Unless physiological data can be obtained along with behavioral characters, we will not be able to make the link between functional and behavioral variation. Of course, the ability to obtain functional data depends critically on the availability of species in the chosen clade in the first place. While structural data may be obtained from museum specimens, physiological and behavioral data clearly cannot. In this study we began with a behavioral and morphological analysis of six taxa (Fig. 4; Reilly & Lauder 1992), but data on muscle activity patterns could be obtained for just four taxa.

A significant problem may arise when taxa studied are reduced to a small number. With such a small number of taxa, the opportunity for finding cases of convergent character acquisition (which may themselves provide tests of causal hypotheses) is small, and the possibility of conducting statistical tests of character association greatly reduced. The availability of a small number of taxa within a larger clade also increases the possibility of identifying some traits as uniquely derived characters of a single taxon (autapomorphies) when in reality these traits might characterize a larger of grouping of taxa.

The conventional approach to analyzing traits of this kind has been to map characters onto a phylogeny of the studied taxa (e.g., Lauder 1990; Brooks & McLennan 1991; Maddison & Maddison 1992). Such mappings allow the sequential acquisition of characters in a clade to be determined. Given a sufficiently large number of taxa and a well-corroborated phylogeny, relationships among characters can be assessed by using quantitative phylogenetic techniques for the analysis of correlated characters (e.g., Pagel 1994). But if the number of taxa is small, then, as is evident from the character mappings presented in Fig. 8, relatively few novel insights are gained into the data using a mapping type of analysis. On the other hand, when the number of character classes equals or exceeds three and the number of taxa is also large, simply gaining a basic understanding of the pattern of variation among characters can be difficult.

One of the reasons for presenting the multivariate visualization illustrated in this paper is that it facilitates analysis of variation among several taxa as well as among multiple classes of characters simultaneously. But as yet there are no analytical techniques that have been applied to data such as those presented here which would allow a quantitative assessment of mapping among levels. A significant advance would be the direct incorporation of information on the phylogenetic history of the taxa into the analysis of the mappings themselves. Use of phylogenetically standardized contrasts (based on a well-corroborated phylogeny) at each level as data for the principal components analysis (Felsenstein 1985; Martins 1993) would incorporate information on phylogeny into each plane of Fig. 2, but would not resolve the problem of analyzing mappings among levels. Path analytic techniques might provide a partial solution to this problem, as might analyses of data matrices for each class of characters as suggested by Douglas and Endler (1982). Alternatively, the statistical relationships among taxa at each level could be transformed into alternative hypotheses for phylogenetic relationships among taxa. Each class of characters provides evidence of genealogical relationships among taxa, and these cladograms (one for each class of characters) could be examined for regions of congruence and incongruence.

There are considerable benefits to be gained from a comparative analysis of structural and functional traits that underlie behavioral variation among taxa. Comparative analyses are needed to provide proximate explanations of behavioral differentiation, and such explanations are critical for a comprehensive understanding of behavioral evolution. We present this study as an example of one simple approach to visualizing the relationships among both taxa and classes of characters that underlie behavioral diversification. By considering the causal bases of behavioral differentiation among taxa as consisting of several different classes of characters (Table 4.1), we are able to hypothesize how behavior and its underlying mechanistic bases have evolved. Is there congruence in patterns of variation among types of characters? Does the existence of behavioral novelties in a clade necessarily imply novelties at both the morphological and motor pattern levels? Do motor pattern and behavioral characters tend to coevolve prior to morphological differentiation? Such questions may be approached, at least to a first approximation, by describing the variation within each class of characters using principal components analysis, and then considering the taxonomic mappings among levels.

One advantage of this approach is the generation of an understandable scheme for visualizing both the changing relationships among characters as well as among classes of traits. Figure 7, for example, reduces a considerable amount of information into a manageable format which highlights key changes across levels. As a visualization and description of the data, such figures point to areas where further investigation would be profitable (such as a detailed study of the comparative feeding biomechanics of *Ambystoma* and *Necturus*). In addition, there is a predictive aspect to such visualizations that extends the approach beyond the merely descriptive. For example, analysis of behavior and morphology in salamander feeding mechanisms led Reilly and Lauder (1992) to make several predictions about results at the motor pattern level prior to collecting the muscle activity data for this chapter. A prediction was made that *Cryptobranchus* and *Siren* would likely possess significantly different motor patterns, as should *Siren* and *Ambystoma*. Because of the mechanically causal relationships among levels, differences in behavior possessed by taxa that are similar in morphology must be generated by differences in muscle activity. The motor pattern data gathered for this study confirm both predictions.

Although the approach described here is but one avenue by which the analysis of behavioral evolution might proceed, in conjunction with other research directions and an increased focus on the evolution of physiological and functional traits in general (Garland & Carter 1994; Huey 1987; Huey & Bennett 1987; Lauder 1991a, b, 1994; Wainwright & Reilly 1994), we are likely to see much new information over the next few years on the evolution of the mechanistic bases of animal behavior.

Acknowledgments

We thank Emília Martins, Al Bennett, Ted Garland, Don Miles, Julian Humphries, John Endler, Barry Chernoff, Dick Hudson, and Steve Frank for helpful discussions concerning the data and concepts in this paper. Two anonymous reviewers provided many helpful comments. This research was supported by NSF IBN 91-19502 to GVL, and Ohio University Research Challenge Grant RC94-028 to SMR.

References

Arbas, E. A., I. A. Meinertzhagen, & S. R. Shaw. 1991. Evolution in nervous systems. *Annu. Rev. Neurosci.*, 14, 9–38.

Bookstein, F., B. Chernoff, R. Elder, J. Humphries, G. Smith, & R. Strauss. 1985. *Morphometrics in Evolutionary Biology*. Philadelphia: Academy of Natural Sciences.

Bray, J. H., & S. E. Maxwell. 1985. *Multivariate Analysis of Variance*. Beverly Hills, California: Sage Publications.

Brooks, D. R., & D. A. McLennan,. 1991. *Phylogeny, Ecology and Behavior*. Chicago: University of Chicago Press.

Brooks, D. R., & D. A. McLennan. 1993. *Parascript. Parasites and the Language of Evolution*. Washington, D. C.: Smithsonian Institution Press.

Cloutier, R. (in press). Phylogenetic status, basal taxa, and interrelationships of lower sarcopterygian groups. *Zool. J. Linn. Soc. Lond.*

Cundall, D. 1983. Activity of head muscles during feeding by snakes: A comparative study. *Am. Zool.*, 23, 383–396.

Cundall, D., J. Lorenz-Elwood, & J. D. Groves. 1987. Asymmetric suction feeding in primitive salamanders. *Experientia*, 43, 1229–1231.

Dial, K. P., G. E. Goslow, & F. A. Jenkins. 1991. The functional anatomy of the shoulder in the European starling (*Sturnus vulgaris*). *J. Morphol.*, 207, 327–344.

Douglas, M. E., & J. A. Endler. 1982. Quantitative matrix comparisons in ecological and evolutionary investigations. *J. Theor. Biol.*, 99, 777–795.

Dunteman, G. H. 1989. *Principal Components Analysis*. Newbury Park, California: Sage Publications.

Erdman, S., & D. Cundall. 1984. The feeding apparatus of the salamander *Amphiuma tridactylum*: morphology and behavior. *J. Morphol.*, 181, 175–204.

Felsenstein, J. 1985. Phylogenies and the comparative method. *Am. Nat.*, 125, 1–15.

Garland, T., & S. C. Adolph. 1994. Why not to do 2-species comparisons: Limitations on inferring adaptation. *Physiol. Zool.*, 67, 797–828.

Garland, T., & P. A. Carter. 1994. Evolutionary physiology. *Annu. Rev. Physiol.*, 56, 579–621.

Garland, T., A. W. Dickerman, C. M. Janis, & J. A. Jones. 1993. Phylogenetic analysis of covariance by computer simulation. *Syst. Biol.*, 42, 265–292.

Garland, T., P. H. Harvey, & A. R. Ives. 1992. Procedures for the analysis of comparative data using phylogenetically independent contrasts. *Syst. Biol.*, 41, 18–32.

Gatesy, S. M. 1994. Neuromuscular diversity in archosaur deep dorsal thigh muscles. *Brain Behav. Evol.*, 43, 1–14.

Goslow, G. E., K. P. Dial, & F. A. Jenkins. 1989. The avian shoulder: an experimental approach. *Am. Zool.*, 29, 287–301.

Grafen, A. 1989. Phylogenetic regression. *Phil. Trans. R. Soc. Lond., Series B.*, 326, 119–157.

Harris, R. J. 1975. *A Primer of Multivariate Statistics*. New York: Academic Press.

Harvey, P. H., & A. E. Keymer. 1991. Comparing life histories using phylogenies. *Phil. Trans. R. Soc. Lond., Series B.*, 332, 31–39.

Harvey, P. H., & M. D. Pagel. 1991. *The Comparative Method in Evolutionary Biology*. Oxford, England: Oxford University Press.

Harvey, P. H., A. F. Read, & D. E. Promislow. 1990. Life history variation in placental mammals: unifying the data with theory. *Oxford Surv. Evol. Biol.*, 6, 13–31.

Hoy, R. R., A. Hoikkala, & K. Kaneshiro. 1988. Hawaiian courtship songs: Evolutionary innovation in communication signals of *Drosophila*. *Science*, 240, 217–219.

Huey, R. 1987. Phylogeny, history, and the comparative method. In: *New Directions in Ecological Physiology* (Ed. by M. Feder, A. F. Bennett, W. W. Burggren, and R. B. Huey), pp. 76–101. Cambridge, England: Cambridge University Press.

Huey, R. B., & A. F. Bennett 1987. Phylogenetic studies of coadaptation: preferred temperature versus optimal performance temperature of lizards. *Evolution*, 41, 1098–1115.

Jenkins, F. A., & G. E. Goslow. 1983. The functional anatomy of the shoulder of the Savannah Monitor Lizard (*Varanus exanthematicus*). *J. Morphol.*, 175, 195–216.

Katz, P. S. 1991. Neuromodulation and the evolution of a simple motor system. *Semin. Neurosci.*, 3, 379–389.

Katz, P. S., & K. Tazaki. 1992. Comparative and evolutionary aspects of the crustacean stomatogastric system. In: *Dynamic Biological Networks: The Stomatogastric Nervous System* (Ed. by R. M. Harris-Warrick, E. Marder, A. I. Selverston, & M. Moulins), pp. 221–261. Cambridge, Massachusetts: MIT Press.

Larsen, J. H., J. T. Beneski, & D. B. Wake. 1989. Hyolingual feeding systems of the Plethodontidae: Comparative prey capture kinematics of free projectile, attached projectile and protrusible tongued salamanders. *J. Exp. Zool.*, 252, 25–33.

Larson, A., & W. W. Dimmick. 1993. Phylogenetic relationships of the salamander families: An analysis of congruence among morphological and molecular characters. *Herpetol. Monogr.*, 7, 77–93.

Lauder, G. V. 1986. Homology, analogy, and the evolution of behavior. In: *The Evolution of Behavior* (Ed. by M. Nitecki, & J. Kitchell), pp. 9–40. Oxford, England: Oxford University Press.

Lauder, G. V. 1990. Functional morphology and systematics: studying functional patterns in an historical context. *Annu. Rev. Ecol. Syst.*, 21, 317–340.

Lauder, G. V. 1991a. Biomechanics and evolution: integrating physical and historical biology in the study of complex systems. In: *Biomechanics in Evolution* (Ed. by J. M. V. Rayner, & R. J. Wootton), pp. 1–19. Cambridge, England: Cambridge University Press.

Lauder, G. V. 1991b. An evolutionary perspective on the concept of efficiency: How does function evolve? In: *Efficiency and Economy in Animal Physiology* (Ed. by R. W. Blake), pp. 169–184. Cambridge, England: Cambridge University Press.

Lauder, G. V. 1994. Homology, form, and function. In: *Homology: The Hierarchical Basis of Comparative Biology* (Ed. by B. K. Hall), pp. 151–196. San Diego, California: Academic Press.

Lauder, G. V. 1995. On the inference of function from structure. In: *Functional Morphology in Vertebrate Paleontology* (Ed. by J. Thomason), pp. 1–18. Cambridge, England: Cambridge University Press.

Lauder, G. V., & S. M. Reilly. 1988. Functional design of the feeding mechanism in salamanders: causal bases of ontogenetic changes in function. *J. Exp. Biol.*, 134, 219–233.

Lauder, G. V., & S. M. Reilly. 1990. Metamorphosis of the feeding mechanism in tiger salamanders (*Ambystoma tigrinum*): the ontogeny of cranial muscle mass. *J. Zool.*, 222, 59–74.

Lauder, G. V., & S. M. Reilly. 1994. Amphibian feeding behavior: comparative biomechanics and evolution. In: *Advances in Comparative and Environmental Physiology: Biomechanics of Feeding in Vertebrates* (Ed. by V. L. Bels, M. Chardon, & P. Vandewalle), pp. 163–195. Berlin: Springer-Verlag.

Lauder, G. V., & H. B. Shaffer. 1985. Functional morphology of the feeding mechanism in aquatic ambystomatid salamanders. *J. Morphol.*, 185, 297–326.

Lauder, G. V., & H. B. Shaffer. 1986. Functional design of the feeding mechanism in lower vertebrates: Unidirectional and bidirectional flow systems in the tiger salamander. *Zool. J. Linn. Soc. Lond.*, 88, 277–290.

Lauder, G. V., & H. B. Shaffer. 1988. The ontogeny of functional design in tiger salamanders (*Ambystoma tigrinum*): Are motor patterns conserved during major morphological transformations? *J. Morphol.*, 197, 249–268.

Lauder, G. V., & H. B. Shaffer. 1993. Design of feeding systems in aquatic vertebrates: Major patterns and their evolutionary interpretations. In: *The Skull*, Vol. 3*: Functional and Evolutionary Mechanisms* (Ed. by J. Hanken and B. K. Hall), pp. 113–149. Chicago: University of Chicago Press.

Lombard, R. E., & D. B. Wake. 1976. Tongue evolution in the lungless salamanders, family Plethodontidae: I. Introduction, theory, and a general model of dynamics. *J. Morphol.*, 148, 265–286.

Lombard, R. E., & D. B. Wake. 1977. Tongue evolution in the lungless salamanders, family Plethodontidae: II. Function and evolutionary diversity. *J. Morphol.*, 153, 39–80.

Maddison, W. P., & D. R. Maddison. 1992. *MacClade Version 3: Analysis of Phylogeny and Character Evolution.* Sunderland, Massachusetts: Sinauer and Associates.

Martins, E. P. 1993. A comparative study of the evolution of *Sceloporus* push-up displays. *Am. Nat.*, 142, 994–1018.

Martins, E. P., & T. Garland, Jr. 1991. Phylogenetic analysis of the correlated evolution of continuous characters: A simulation study. *Evolution*, 45, 534–557.

McLennan, D. A., D. R. Brooks, & J. D. McPhail. 1988. The benefit of communication between comparative ethology and phylogenetic systematics: A case study using gasterosteid fishes. *Can. J. Zool.*, 66, 2177–2190.

Miles, D. B., & A. E. Dunham. 1993. Historical perspectives in ecology and evolutionary biology: The use of phylogenetic comparative analyses. *Ann. Rev. Ecol. Syst.*, 24, 587–619.

Miller, B. T., & J. H. Larsen. 1990. Comparative kinematics of terrestrial prey capture in salamanders and newts (Amphibia: Urodela: Salamandridae). *J. Exp. Zool.*, 256, 135–153.

Mitter, C., B. Farrell, & D. J. Futuyma. 1991. Phylogenetic studies of insect-plant interactions: insights into the genesis of diversity. *TREE*, 6, 290–293.

Pagel, M. 1994. Detecting correlated evolution on phylogenies: a general method for the comparative analysis of discrete characters. *Proc. R. Soc. Lond., Series B*, 255, 37–45.

Pagel, M. D., & P. H. Harvey. 1988. How mammals produce large-brained offspring. *Evolution*, 42, 948–957.

Reilly, S. M. 1994. The ecological morphology of metamorphosis: heterochrony and the evolution of feeding mechanisms in salamanders. In: *Ecological Morphology: Integrative Organismal Biology* (Ed. by P. C. Wainwright, & S. M. Reilly), pp. 319-338. Chicago: University of Chicago Press.

Reilly, S. M., & G. V. Lauder. 1988. Ontogeny of aquatic feeding performance in the eastern newt *Notophthalmus viridescens* (Salamandridae). *Copeia*, 1988, 87–91.

Reilly, S. M., & G. V. Lauder. 1989a. Kinetics of tongue projection in *Ambystoma tigrinum*: quantitative kinematics, muscle function, and evolutionary hypotheses. *J. Morphol.*, 199, 223–243.

Reilly, S. M., & G. V. Lauder. 1989b. Physiological bases of feeding behavior in salamanders: do motor patterns vary with prey type? *J. Exp. Biol.*, 141, 343–358.

Reilly, S. M., & G. V. Lauder. 1990. The strike of the tiger salamander: quantitative electromyography and muscle function during prey capture. *J. Comp. Physiol.*, 167, 827–839.

Reilly, S. M., & G. V. Lauder. 1991. Experimental morphology of the feeding mechanism in salamanders. *J. Morphol.*, 210, 33–44.

Reilly, S. M., & G. V. Lauder. 1992. Morphology, behavior, and evolution: comparative kinematics of aquatic feeding in salamanders. *Brain Behav. Evol.*, 40, 182–196.

Reilly, S. M., & P. C. Wainwright. 1994. Ecological morphology and the power of integration. In: *Ecological Morphology:Iintegrative Organismal Biology* (Ed. by P. C. Wainwright, & S. M. Reilly), pp. 339–354. Chicago: University of Chicago Press.

Rohlf, F. J., & F. L. Bookstein, editors. 1990. *Proceedings of the Michigan Morphometrics Workshop*. Ann Arbor: University of Michigan Museum of Zoology.

Roth, G. 1976. Experimental analysis of the prey catching behavior of *Hydromantes italicus* Dunn (Amphibia, Plethodontidae). *J. Comp. Physiol.*, 109, 47–58.

Ryan, M. J., & A. Keddyhector. 1992. Directional patterns of female mate choice and the role of sensory biases. *Am. Nat.*, 139(S), S4–S35.

Ryan, M. J., & A. S. Rand. 1990. The sensory basis of sexual selection for complex calls in the Túngara frog, *Physalaemus pustulosus* (sexual selection for sensory exploitation). *Evol.*, 44(2), 305–314.

Schwenk, K., & D. B. Wake. 1993. Prey processing in *Leurognathus marmoratus* and the evolution of form and function in desmognathine salamanders (Plethodontidae). *Biol. J. Linn. Soc.*, 49, 141–162.

Shaffer, H. B., & G. V. Lauder. 1985a. Aquatic prey capture in ambystomatid salamanders: patterns of variation in muscle activity. *J. Morphol.*, 183, 273–326.

Shaffer, H. B., & G. V. Lauder. 1985b. Patterns of variation in aquatic ambystomatid salamanders: kinematics of the feeding mechanism. *Evol.*, 39, 83–92.

Smith, K. K. 1994. Are neuromotor systems conserved in evolution? *Brain Behav. Evol.*, 43, 293–305.

Thexton, A. J., D. B. Wake, & M. H. Wake. 1977. Tongue function in the salamander *Bolitoglossa occidentalis*. *Arch. Oral Biol.*, 22, 361–366.

Trueb, L., & R. Cloutier. 1991. A phylogenetic investigation of the inter- and intrarelationships of the Lissamphibia (Amphibia: Temnospondyli). In: *Origins of the Higher Groups of Tetrapods: Controversy and Consensus* (Ed. by H.-P. Schultze, & L. Trueb), pp. 223–313. Ithaca, New York: Comstock Publishing Associates.

Wainwright, P. C. 1986. Motor correlates of learning behavior: feeding on novel prey by the pumpkinseed sunfish (*Lepomis gibbosus*). *J. Exp. Biol.*, 126, 237–247.

Wainwright, P. C., & S. M. Reilly, editors. 1994. *Ecological Morphology: Integrative Organismal Biology*. Chicago: University of Chicago Press.

Willig, M. R., R. D. Owen, & R. L. Colbert. 1986. Assessment of morphometric variation in natural populations: the inadequacy of the univariate approach. *Syst. Zool.*, 35, 195–203.

CHAPTER 5

Geographic Variation in Behavior: A Phylogenetic Framework for Comparative Studies

Susan A. Foster and Sydney A. Cameron

Comparative studies have been used for more than a century to explore the origins and evolutionary histories of behavior patterns. Darwin employed comparison extensively in his discussion of instincts in *The Origin of Species* (1859) and clearly recognized that behavior patterns were the products of both natural selection and phylogeny. This insight formed the foundation upon which Whitman (1899), Heinroth (1911, 1930), and Lorenz (1935, 1941) expanded in their pioneering writings on the application of the comparative method to the study of behavioral evolution. During the first half of this century, these and other ethologists developed a remarkable diversity of hypotheses concerning the evolution of behavioral patterns, particularly display behavior, using the comparative method (reviewed in Hinde & Tinbergen 1958; Blest 1961; Foster 1995). Although this approach to the study of behavioral evolution fell into disfavor during the third quarter of this century, the development of more rigorous analytical methods and faster algorithms for inferring phylogenies (Farris 1983, 1988; Felsenstein 1993; Hendy & Penny 1982; Swofford 1993) and for estimating their reliability (Templeton 1983; Felsenstein 1985a,b; Faith 1991; Donoghue et al. 1992; Hillis & Huelsenbeck 1992) have promoted the production and application of phylogenetic hypotheses. In the last several years, rigorous comparative studies of behavior evolution using phylogenies have appeared increasingly often (e.g., Greene 1986; Coddington 1988; Sillén-Tullberg 1988; Carpenter 1989; McLennan et al. 1988; Cameron 1993).

When the comparative method has been used to infer the evolutionary histories of behavior patterns, comparisons have usually been made among species or higher-order taxa (Burghardt & Gittleman 1990; Brooks & McLennan 1991; Harvey & Pagel 1991; Foster 1995). The method has required that individual trait values be assigned to each species (or higher taxonomic grouping) for each behavior of interest (Hinde & Tinbergen 1958). Typically, trait values have been based on the behavior of individuals from single populations. Implicit in this procedure is the assumption that behavior patterns may vary insignificantly, if at all, among populations within a species, an assumption that is proving suspect at best.

At the time ethologists were developing the comparative method to explore the evolution of behavior patterns which they deemed "species-typical," or characteristic of all members of a species, population and quantitative geneticists were focusing on variation within species. This approach again could be traced to Darwin's (1859) writings, in this instance to his recognition of the importance of variation in the evolutionary process. Although behavior was rarely the focus of early research on geographic variation (Endler 1977, 1986; Zink & Remsen 1986), this has changed in the last two decades. Recent studies have demonstrated unequivocally that geographic variation in behavior is common, that geographic differences often are heritable, and that such variation is not limited taxonomically (reviewed in Boake 1994; Pomiankowski & Sheridan 1994; Foster & Endler 1995).

A fusion of these two areas of endeavor is clearly necessary. In order to understand the evolution of behavioral phenotypes, we must examine variation in behavior at the appropriate hierarchical (taxonomic) level. Ignoring intraspecific variation can obscure evolutionary insights or, worse, lead to errors in historical inference. For all comparative studies of behavior, the availability of a well-supported phylogeny is crucial because populations are connected to one another via ancestor-descendant relationships, as are species (Avise 1989).

Below the species level, however, the patterns of relationship can be difficult to interpret due to the complication of gene flow, which can lead to reticulate evolution (considered below). Until recently, phylogeographic (*sensu* Avise et al. 1987; Avise 1994) reconstruction was hindered by the paucity of phylogenetically informative morphological characters at the population level. The application of molecular characters, especially those of mitochondrial DNA (mtDNA), has virtually revolutionized our ability to detect patterns of phylogeographic structure (reviewed in Avise 1994). However, few of these phylogeographic studies have considered patterns of behavioral evolution.

A consequence of the resurgence of interest in a phylogenetic perspective is a new focus on the use of phylogenies to test hypotheses of behavioral evolution (e.g., Carpenter 1989; Donoghue 1989) and to gain new insights into the relative roles of historical and environmental (local adaptive) factors in the process of behavioral evolution (Brooks & McLennan 1991). Population comparisons can provide particularly powerful means of evaluating adaptive hypotheses for two reasons. The first is that there tend to be fewer differences between populations than between species. Consequently, there are fewer covarying traits to confound analyses (Lott 1991; Arnold 1992). Second, divergent populations are often relatively young and may be more likely to reside in the habitats in which their derived character states evolved than is the case for divergent species with potentially longer intervening histories (Foster et al. 1992).

Insights from population comparisons are not limited to adaptive inference. When population phylogenies are available, and appropriate outgroups can be identified, derived states and the evolutionary transformation of characters can be assessed via outgroup analysis (Watrous & Wheeler 1981; Maddison et al. 1984). Thus, the number of independent evolutionary events involved in the transformation of a behavior can be inferred and acquisition of novel behavior patterns can be distinguished from loss of ancestral patterns (e.g., Lauder 1986; McLennan et al 1988; Brooks & McLennan 1991).

In this chapter we first provide a brief summary of the evidence indicating that geographic variation in behavior is common. Our purpose is to demonstrate not only that one must account for possible intraspecific geographic variation in interspecific analyses but also that the existing geographic variation in behavior is likely to be of interest to those wishing to understand behavioral evolution. We then present a hierarchical framework within which to examine variation in behavioral traits at any taxonomic level, emphasizing the importance of incorporating knowledge of both within- and between-population variation in the analyses. This is followed by a summary of the phylogenetic methodology that can be used to examine the causes and patterns of behavioral evolution, along with some illustrative examples.

Geographic variation in behavior

As field and laboratory research on behavior has become increasingly common in recent years, multiple populations of a number of species have been studied. Comparison of the results of these studies has revealed that geographic variation is not only common but is perhaps to be expected in most taxa (papers in Foster & Endler 1995). Moreover, the kinds of behavior patterns that have been seen to vary within taxa include those often thought to be "species-typical" or invariant within a species. Courtship and other aspects of reproductive behavior exemplify this class of behavior patterns (Hinde & Tinbergen 1958; Lorenz 1970; King & West 1990). Hence, we will begin our discussion with reproductive behavior.

The earliest comprehensive efforts to evaluate the extent of geographic variation in behavior within species had the goal of identifying and studying the earliest stages in speciation. This approach was pioneered by Dobzhansky, but the research program was carried out by many (reviewed in Chatterjee & Singh 1989). Flies of the genus *Drosophila* were the typical subjects of choice (but see Bush 1994). Flies were collected from geographically disparate populations and homotypic and heterotypic matings were attempted. In this way, sexually incompatible populations could be identified. Little attention was paid to possible underlying differences in behavior, although it was perhaps reasonable to assume that they existed in many cases. Similar studies of geographic differences in compatibility, combined with ethological research on associated behavioral differences, have revealed geographic variation in the sexual behavior of, for example, insects (Duijm 1990; Ross & Shoemaker 1993; Wilkinson & Reillo 1994;

Carroll & Corneli 1995), guppies (Luyten & Liley 1985, 1991; Houde & Endler 1990), frogs (e.g. Gerhardt 1974; Ryan & Wilczynski 1991; Littlejohn 1993), salamanders (Verrell & Arnold 1989; Tilley et al. 1990), and birds (Baker & Cunningham 1985; King & West 1990).

Other elements of social behavior also vary geographically within species, although they have been less well studied in this context. Examples include variation in territorial and aggressive behavior in spiders (Riechert 1987, 1993a,b) and fish (Huntingford 1976; Bakker 1986). A more complex aspect of social behavior, colony social structure, also varies across the range of the fire ant, *Solenopsis invicta* (Ross & Fletcher 1985; Ross & Trager 1990; Ross & Shoemaker 1993) and the sweat bee, *Dialictus lineatulus* (Eickwort 1986). Finally, the migratory behavior of insects (Dingle 1988), fishes (Snyder & Dingle 1989) and birds (Berthold et al. 1992) also varies geographically, and has been shown to evolve remarkably rapidly (Berthold et al. 1992).

Geographic variation is by no means limited to social behavior, a finding that is not surprising in that food resources and predators often vary in type and abundance across the ranges of widespread species. Among the best-studied examples of ecotypic differences in foraging behavior are those in the desert spider, *Agelenopsis aperta* (Hedrick & Riechert 1989), the garter snakes, *Thamnophis elegans* (Arnold 1977) and *T. ordinoides* (Brodie 1989), and the nymphalid butterflies, *Euphydryas editha* (Ehrlich 1965; Ehrlich et al. 1975) and *E. chalcedona* (Bowers 1986). Differences in predation regime are also reflected in ecotypic differences in antipredator behavior in a number of species of fish (Seghers 1974; Magurran 1986; Foster & Ploch 1990; Magurran & Seghers 1990a,b; Huntingford et al. 1994), in ground squirrels (Goldthwaite et al. 1990; Towers & Coss 1990) and in the spider, *A. aperta* (Riechert & Hedrick 1990).

A final, elegant set of studies can serve to remind students of animal behavior of the importance of studying behavior at the appropriate hierarchical level. Recent research by Thompson & Pellmyr (1991) and others (Singer et al. 1992) have revealed that a variety of phytophagous butterfly species originally characterized as polyphagous, or generalist herbivores, are actually collections of populations, each population specializing on a different plant species found within its home range. Thompson (1988a) initially documented distinct differences in host plant preference among closely related species of *Papilio*. He showed that a difference in oviposition preference was the result of reorganization in the genes associated with preference ranking (Thompson 1988b). However, these initial studies were based on analyses of single populations representing each species. A subsequent study of several populations of *Papilio zelicaon* revealed high levels of geographic variation in oviposition preference in different parts of the species' range. In this case, however, the preference for different hosts was not the result of genetic changes in preference, indicating an evolutionary conservatism in the preference hierarchy among the populations, even after hundreds of generations (Thompson 1993). Thus, at higher levels, oviposition preference appears to be influenced by genetic reorganization but at lower levels it is not.

This example provides compelling evidence that behavior must be examined at various hierarchical levels if the quality of evolutionary inference is to be maximized. In the next sections we summarize methods for achieving this goal.

A hierarchical approach to the study of behavioral variation

The first step in a comparative study of different taxa (i.e., within species, genera, families, etc.) is to assess whether the trait of interest is invariant within each unit of comparison or whether it is variable, precluding the assignment of a single character state to a taxon. For instance, if the comparative analysis is a between-species comparison but intraspecific variation is found in the character, the question of whether the species have been recognized correctly must be addressed. Answering this question requires concurrent analyses of behavior and phylogeny. Behavioral variation must be analyzed within each species (i.e., among different populations and within populations) to identify the nature of the intraspecific variation.

In conjunction with the behavioral assessment, an independent phylogenetic hypothesis of relationships among the taxa is required to test the associations between the phylogeny and the behavioral traits. If the variation in behavior occurs primarily among species with little or no variation within each species, the appropriate level of analysis is a species comparison using statistical methods designed to incorporate phylogenetic information (see in this volume, Martins & Hansen, Chap. 2 for review). The same methods can be applied when there is substantial variation among populations but little within and when barriers to dispersal are thought to have severely curtailed gene flow across populations. In this instance, population level analysis using standard phylogenetic methods would be appropriate. If, on the other hand, the behavior exhibits substantial but continuous variation within the species with little or no structured population or geographic variation, the appropriate level of analysis is a species comparison using statistical methods designed to accommodate unstructured within-species variation (e.g., Lynch 1991; Martins 1994).

A more complex problem arises when, as is probably typical, substantial gene flow has occurred among populations. Gene exchange leads to reticulate relationships among taxa rather than to simple branching trees. Recent developments in coalescent theory have provided the basis for reconciling gene trees with population histories, taking into account the problems of ancestral polymorphism (e.g. Ewens 1990; Hudson 1990; Kingman 1982) and gene flow (Slatkin & Maddison 1989, 1990). As with species trees, ancestral behavioral traits can be reconstructed using a reconciled population tree and interpreted in a phylogenetic context. A second set of methods for handling reticulations involves removing "phylogenetic" effects (correlations of trait values or genealogical correlations among individuals) using an autocorrelational approach (Cheverud et al. 1985; Gittleman & Kot 1990; Edwards & Kot 1995).

In the ensuing sections, we will provide an analysis of behavioral evolution in which gene flow is sufficiently curtailed among populations that a branching tree is an appropriate phylogenetic model for relationships among the populations of interest. We then discuss the problem of reticulate evolution, providing a more detailed explanation of the methods used to address phylogenetic issues when gene flow has blurred population boundaries. In all of these discussions we focus on cases in which there exists geographic or population-level differentiation.

To reiterate in closing this section, identification of the proper hierarchical level in the phylogeny is, we feel, a crucial procedure in the comparative analysis of behavioral traits. This procedure may require reciprocal illumination between the development of phylogenies and surveys of behavioral variation. Once the proper hierarchical level for analysis has been identified, and a phylogeny has been developed, a number of evolutionary questions may be addressed (e.g., Ridley 1983; Lauder 1986, 1990; McLennan et al. 1988; Carpenter 1989; Brooks & McLennan 1991; Harvey & Pagel 1991).

When a branching tree is an appropriate model: low gene flow

When populations are separated by discontinuities that restrict gene flow or the taxa of interest are poor dispersers, cladistic methods can sometimes be used to reconstruct phylogenetic relationships among populations (see Avise 1994 for review). If a strongly supported tree topology can be obtained using an independent data set (usually morphological and/or DNA data), it can serve as a hypothesis of evolutionary relationships among the populations that can be used in examining evolutionary transformations in behavioral traits. Ancestral behavioral phenotypes can then be reconstructed to examine patterns of evolutionary change in the traits, and if other characters are also reconstructed, associations among behavioral and other variables can be explored (Maddison 1990).

The most commonly used method of reconstructing evolutionary changes in characters is character optimization, a method that relies on parsimony (Swofford & Maddison 1987; 1992; Maddison 1990). Although the method is readily accessible (Maddison & Maddison 1992; Swofford 1993), it should be employed with caution in analyses of population data. Parsimony is a method of inferring phylogenies, or overlaying character evolution on a tree, that assumes evolutionary change in the character is relatively rare. A character is "optimized" onto a tree by finding the transformational sequence that involves the smallest number of evolutionary changes capable of explaining the observed pattern (e.g., Felsenstein 1983; Swofford & Maddison 1987, 1992).

The problem with this method in comparative studies at the population level is that the characters we seek to examine often differ among populations that have become separated from one another relatively recently (e.g., Cruz & Wiley 1989; Goldthwaite et al. 1990;

Foster 1994a,b). By definition then, they are not evolving slowly (i.e., changes are not rare) and might not fit the assumptions of parsimony analysis. The problem is further aggravated if the characters are chosen because they are thought to be targets of natural selection, because such traits are certainly likely to evolve rapidly and perhaps nonparsimoniously. Recently, alternative methods have been developed to account for attributes of intraspecific data ignored by traditional parsimony methods (e.g., Templeton 1992; Crandall 1994). These methods are still being evaluated. Other methods of analysis do not assume low rates of evolutionary change, although they assume similar rates of evolution throughout the tree (for review see, in this volume, Martins & Hansen, Chap. 2) and may be more appropriate for population-level comparative studies. In the example we discuss here, one of the few phylogenetic studies of behavioral evolution conducted at the population level, parsimony analysis provided an appropriate estimation of the tree topology.

The example is derived from a phylogenetic analysis of relationships among populations in a diploid-tetraploid cryptic species pair of gray tree frogs *Hyla chrysoscelis* and *H. versicolor*. Wasserman (1970) first suggested that the tetraploid species *H. versicolor* ($2n = 48$) had arisen from *H. chrysoscelis* ($2n = 24$) by autopolyploidy. Since that time, both protein electrophoretic data (Ralin et al. 1983) and allele frequency data (Romano et al. 1987) have been argued to best support the hypothesis of a single origin of the tetraploid frogs from the diploid. Two lineages of the diploid *H. chrysoscelis* have been recognized based on allozyme divergence and chromosome polymorphisms (Gerhardt 1974; Ralin 1977; Wiley 1983; Wiley et al. 1989). Both species share the same breeding sites in many parts of the central and mid-Atlantic regions of the United States (Bogart 1980; Fig. 1). Only the diploid *H. chrysoscelis* is found in most of the southeastern United States (Fig. 1).

Inferences based on molecular data appeared to be well supported by advertisement call divergence among the groups. Call differences among diploid populations (fast pulse rate in the west and slow pulse rate along the east coast) provided the first evidence that there were at least two lineages of *H. chrysoscelis* (Gerhardt 1974). The tetraploid, *H. versicolor*, was also first distinguished from its diploid progenitor on the basis of advertisement call pulse rate (Johnson 1966, Wasserman 1970). All tetraploid pulse rates are 50 to 60 percent lower than those of the

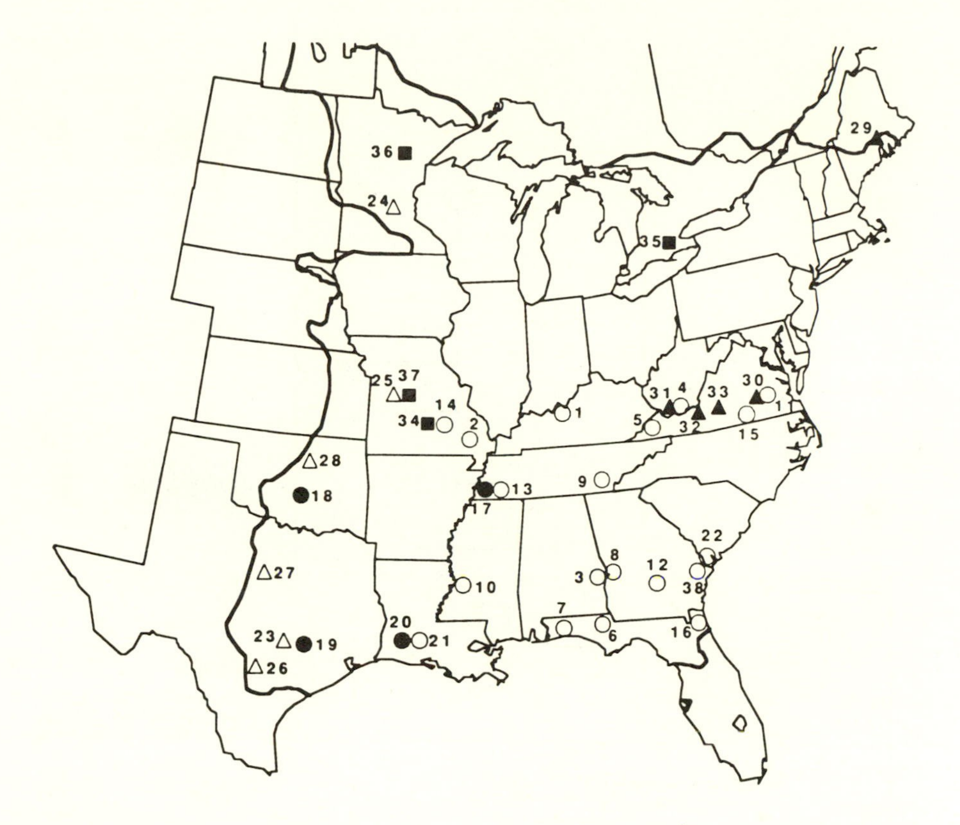

Figure 1. Collecting sites for *Hyla chrysoscelis* (open symbols) and *H. versicolor* (solid symbols). The solid lines indicate the approximate boundaries of the distribution of the gray tree frog complex. Individuals in the same clade in Fig. 2 share symbol types. Collecting site information is in Ptacek et al. (1994). Figure from Ptacek et al. (1994).

diploids that occur in sympatry or parapatry and the temperature-corrected pulse rates of tetraploid populations vary by no more than 10 percent (Gerhardt 1994; personal communication).

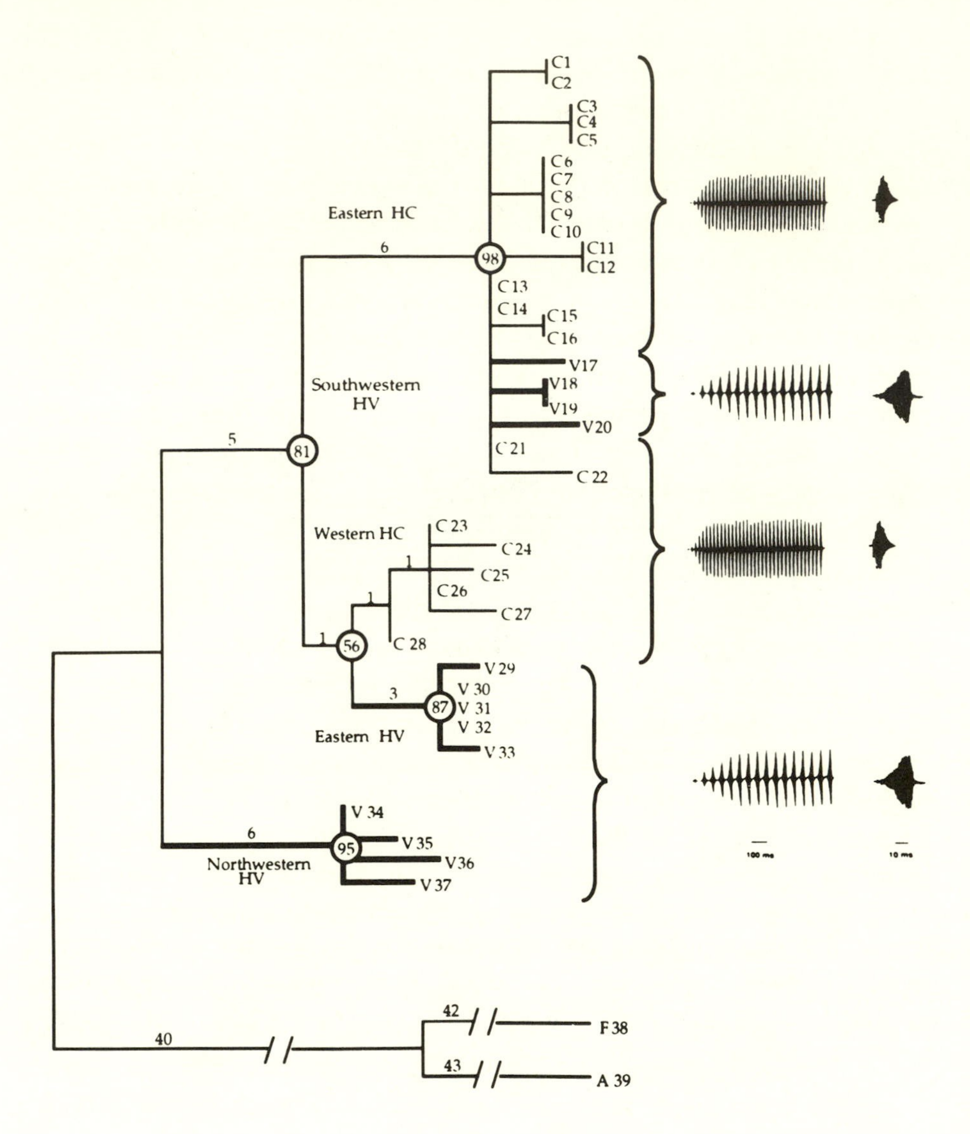

Figure 2. The strict consensus tree for the cytochrome *b* gene for populations of *Hyla chrysoscelis* (labeled C) and *H. versicolor* (V). Branch lengths are indicated above the branches and circled numbers are bootstrap values from 1000 bootstrap replications. See Ptacek et al. (1994) for collecting locations and sample sizes. Redrawn from Ptacek et al. (1994).

A recent phylogenetic analysis of the cytochrome *b* gene of the mitochondrial DNA of individuals from 24 populations of *H. chrysoscelis* and 13 populations of *H. versicolor* has challenged the hypothesis of a single origin for the tetraploid gray tree frog (Ptacek et al 1994). Instead, this analysis provides evidence of at least three separate, independent origins for the tetraploid (Fig. 2). More than one independently derived tetraploid lineage may exist in the eastern clade (upper clade, Fig. 1) which is an unresolved polytomy that almost certainly contains at least two diploid lineages as well. Thus, the tetraploid could have arisen more than three times (Ptacek et al. 1994). From a behavioral perspective, the result is intriguing because it suggests remarkable similarity in the advertisement calls of the independently derived tetraploid lineages and raises the question of how the calls came to be so similar in those lineages.

Because the two species are syntopic through much of their ranges, several explanations for the consistent differences in call between the two species and for the comparative similarity of calls among lineages within species are possible. These include initial shifts to lower pulse rates (Ueda 1993) and shifts in female preference (e.g., Bogart & Wasserman 1972) associated with polyploidy, and possibly, character displacement in zones of sympatry (Gerhardt 1994). What is known at present is that more of the geographical variation in pulse rate in the diploid is explained by these ill defined lineages (e.g., the chromosome morphotype/mtDNA tree) than by whether or not the diploid population is sympatric with tetraploids. Nevertheless, there are within-lineage differences in female selectivity in the diploid that are consistent with the pattern expected of character displacement (Gerhardt, pers. comm.). The value of the phylogenetic analysis is that it will permit researchers to identify appropriate sympatric and allopatric populations within a lineage to test for character displacement in calls and differences in female selectivity. Without this information differences due to character displacement could readily be confounded by those due to historical differentiation (pleiotropy, etc.) caused by unknown factors (Gerhardt 1994).

When a reticulate tree is an appropriate model: high gene flow

Standard cladistic methods are unlikely to provide an appropriate evaluation of relationships among populations with no clear geographic subdivisions or between which there may be substantial gene flow. The resolution of this problem may lie in the recently developed coalescent approach (Hudson 1990) to inferring population structure in such populations. These methods not only reveal the pattern of gene flow within geographically structured populations (Slatkin & Maddison 1989) but can potentially be used to distinguish current gene flow from genetic similarity due to common ancestry, and hence to infer population phylogenies (Templeton et al. 1995).

Estimating within-species phylogenetic relationships is particularly difficult due to the paucity of characters associated with the taxonomic units (OTUs). Traditional methods of phylogeny reconstruction (e.g., maximum parsimony) have their statistical power when large numbers of characters differentiate OTUs (Huelsenbeck and Hillis 1993; Hillis et al. 1994). The cladogram estimation procedure of Templeton et al. (1992), however, has its statistical power when few characters differentiate OTUs (Crandall et al. 1994). Additionally, a comparative method has been developed based on the intraspecific cladogram to statistically test for associations between phenotype and genotype (Templeton et al. 1987). The phenotype in question can be either continuous (Templeton et al. 1987) or categorical (Templeton & Sing 1993). One great advantage of this comparative method is that it allows for uncertainty in the cladogram estimate (Templeton & Sing 1993).

This approach is essentially cladistic in nature, in that it uses a character-state based haplotype evolutionary tree, or essentially, a gene genealogy. Once a tree is produced, analyses of the way that characters overlay on the tree can yield insights into natural selection (e.g., Golding 1987; Golding & Felsenstein 1990; Hartl & Sawyer 1991) and evolutionary transformations in character states (Donoghue 1989; Maddison 1990). Analyses that involve overlay of geography upon the gene tree have been termed "intraspecific phylogeography" (Avise 1989). This method often reveals strong associations between the position of haplotypes on a tree and their geographical location, allowing inference of gene flow in population level analyses.

Although applications of this approach to the study of geographically variable traits hold great promise, to our knowledge there are no behavioral examples to date. The methodology is in its developmental stages, and questions still need to be answered about the accuracy, robustness and applicability of the methods to different kinds of data and questions. Without question, when geographical issues are to be addressed, as will always be the case in population comparisons, appropriate population sampling designs are essential. For discussions of sampling design requirements and appropriate algorithms for analyses, see Templeton et al. (1992; 1995)

The effects of population history and structure on geographical distributions of alleles have also been explored using spatial autocorrelation methods (e.g., Sokal et al. 1989; Slatkin & Arter 1991; Epperson 1993), principal component analyses (e.g., Ammerman & Cavalli-Sforza 1984), and multidimensional scaling (Lessa 1990). These methods are based on an identity measure or a measure of genetic distance rather than on cladistic analyses of genetic data and thus differ substantially from other methods of population level analyses described in this paper. Under the assumption that such genealogical similarities reflect common ancestry, methods have been developed to statistically remove the portions of intraspecific variation in geographically variable traits that can be explained by genetic correlations ("phylogenetic component"), leaving only the "specific component" (Cheverud et al. 1985; Gittleman & Kot 1990; Gittleman & Luh 1992). Because this method is not based on cladistic (i.e., hierarchical) analyses of haplotypes, nonhierarchical relationships among conspecific individuals may be better taken into account by this method.

Autocorrelation statistics have been used to estimate, or "remove," the phylogenetic component of variation in several morphological features and in group size in 10 populations of the gray-crowned babbler (*Pomatostomus temporalis*), a cooperatively breeding songbird (Edwards & Kot 1995). The autoregressive model described in this work effectively removed intraspecific correlations for the morphological variables but did not do so for group size. This result was inferred to reflect the greater lability of group size than of morphology (Edwards & Kot 1995). Other conclusions were more speculative, in part because, as is the case with coalescent methodology, the limitations and advantages of the technique have not been fully explored (but see Gittleman & Kot 1990; Edwards & Kot 1995).

Evolutionary inference without a complete phylogeny

Comparative approaches to the study of behavioral evolution require some knowledge of the phylogenetic relationships among the taxa of interest. Often, complete, well-supported phylogenies do not exist for the taxa of interest, in many cases because too few morphological differences exist among them for phylogenetic reconstruction. This problem is likely to be alleviated at all hierarchical levels as regions of the DNA are discovered that evolve at an appropriate rate, providing molecular characters for use in phylogenetic analysis (e.g. Avise 1989; Cameron et al. 1992; Simon et al. 1994; Edwards & Kot 1995). Nevertheless, many researchers do not yet possess the requisite molecular data for phylogenetic analyses of population relationships, and other means of assessing these relationships must be employed where possible.

One method involves combining geological, distributional and biological data on the species of interest to infer historical relationships among populations. This approach is likely to be especially rewarding in regions where Pleistocene and post-Pleistocene changes in the environment are well enough understood that historical patterns of migration and population subdivision and distribution can be ascertained with a reasonable degree of certainty. In some instances the resolution may be good enough that the polarity of character evolution can be inferred and adaptive processes can be revealed through population comparison (e.g. Endler 1986; Goldthwaite et al. 1990; Bell & Foster 1994; Coss in press).

A particularly good example of this approach is found in research on the evolutionary diversification of the threespine stickleback fish *Gasterosteus aculeatus*, in northwestern North America. This small fish must have been absent from much of the northern part of its modern freshwater range at the time of the last glacial maximum (for review, Bell & Foster 1994). Marine and anadromous forms which are ancient (Orti et al. 1994), and have changed little morphologically during the last 10 million years (Bell 1977, 1994), are thought to have become landlocked in freshwater habitats repeatedly and independently in recently glaciated coastal regions following the last glacial maximum. Character states common to marine populations throughout their range can be considered ancestral to divergent states that have arisen in the freshwater populations, making inference of character polarity possible. Additionally, similar, derived character states found in populations in separate drainages and on islands can be assumed to have had independent origins derived from the marine form as dispersal of freshwater fish between such sites is unlikely (Bell & Foster 1994).

In lieu of the necessary genetic data, this biogeographical information can be used to infer the evolutionary histories of behavioral phenotypes (Foster 1994a,b). For example, female stickleback in some freshwater populations form large groups and attack and consume eggs guarded by paternal conspecifics. Males are unable to defend their nests against the groups, but can use a complex diversionary display to divert the groups from approaching the nest. In several noncannibalistic populations this display is missing. Both cannibalism and the diversionary display are common to all marine populations studied to date (Foster 1994a,b, in press) indicating that they are ancestral character states. As both phenotypes have been lost in at least four lakes in different drainages, and ecological conditions appear to favor the loss in these lakes but not in others that have been studied, adaptive differentiation is inferred (Foster 1994a,b, in press).

A similar approach, using knowledge of Pleistocene and post-Pleistocene changes in habitat distribution, has been used to infer ancestral and derived behavioral character states in ground squirrels (Goldthwaite et al. 1990; Coss in press). Although this kind of information can permit inference of character polarity and in some cases permits adaptive inference (Bell & Foster 1994; Foster 1994a,b), complex relationships among multiple populations will rarely be revealed. The same can be said of evolutionary inference based on introductions in recent history (Cruz & Wiley 1989; Davies & Brooke 1989). Nonetheless, though of limited usefulness compared to rigorous phylogenetic reconstruction, these approaches can provide working hypotheses, ultimately testable with phylogenetic information.

Riechert (1993a,b) has taken an unusual approach to interpreting the cause of apparently maladaptive territorial, aggressive, and antipredator behavior in one riparian woodland population of the spider *Agelenopsis aperta*. The members of this population possess behavior patterns similar to those in a surrounding "aridlands" population rather than those predicted by theory. To determine whether the spiders were simply unable to reach the predicted optimum in this population because of a lack of appropriate genetic variation in the species or whether gene flow prevented local adaptation, Riechert used a combination of experimental approaches to measure gene flow and the response to selection. She uncovered strong evidence of gene flow, and found evidence in an experimental setting, of a rapid response to selection in the absence of gene flow. Finally, she examined an isolated riparian population near which there was no source of aridland adapted spiders and found that this population possessed behavior patterns appropriate to the habitat in which it was found. She therefore concluded that gene flow and not phylogenetic constraint prevented the deviant riparian population from evolving adaptive behavior patterns.

These additional approaches have provided creative methods of circumventing the absence of complete population phylogenies, and both have provided unique evolutionary insights. Neither is likely to be as widely applicable as will be phylogenetic analyses once the molecular and statistical limitations can be eliminated. Nevertheless, they permit evaluation of evolutionary hypotheses in appropriate ways. In the latter case, specific hypotheses generated from theory and population comparison were explicitly tested.

Population comparison and adaptive inference

Ideally, experimentation should follow the generation of evolutionary hypotheses using the comparative method regardless of approach (Endler 1986). This is a point that has often been overlooked in the recent rush to use phylogenetic information to evaluate evolutionary scenarios (e.g., Brooks & McLennan 1991; Harvey & Pagel 1991). It is a problem that is particularly serious when the comparative method is used to test adaptive hypotheses. As pointed out by Endler (1986), population comparisons should be used only to generate adaptive hypotheses, and then only when the populations have been carefully selected to avoid confounding factors.

Lauder et al. (1993) and Leroi et al. (1994) have recently discussed the limitations of adaptive inference when an explicitly phylogenetic approach is taken. Phylogenetic associations may be due to selection on correlated characters rather than on the characters of interest and correlated environmental variables may actually be responsible for differences in selective regimes. Also, the environmental variables that can be compared may be too general to provide adequate resolution of the responsible selective agent (Endler 1986; Lauder et al. 1993; Leroi et al. 1994). In summary, population comparisons alone provide poor resolution of evolutionary mechanism and must be complemented with experimental tests of adaptive hypotheses, or with carefully designed fitness measurements across the populations being compared (Endler 1986; Lauder et al. 1993; Leroi et al. 1994).

Conclusion

Comparative studies of behavior, or of any phenotype, are only as accurate as the data upon which they are based. If character states are assigned at an inappropriate hierarchical level, the analyses will be flawed, however complex the statistical procedures. This kind of error seems to be most common when comparisons are made among species. This is probably because historically, the view has been taken that many, if not most, behavior patterns are invariant within species, or are "species-typical." However, there now exists ample evidence of behavioral variation within and among populations of many kinds of animals (Lott 1991, Foster & Endler in press). Such variation must be taken into account in future comparative studies. We have described methods for doing so that should prove to be of general applicability once population phylogenies are obtained, and have described other methods that are likely to be appropriate for fewer taxa but that can provide evolutionary insights when a well-supported population phylogeny is not available.

The reasons for examining intraspecific variation are not limited to enhancement of the quality of higher-order comparisons. Comparative analysis of such variation can provide unique insights into microevolutionary processes. Not only are comparative studies of intraspecific variation uniquely suited to the study of microevolutionary processes, they can often provide better insight into the causes of adaptive diversification than can higher order comparisons (Arnold

1992; Foster et al. 1992). Through such research we can hope to discern, for example, whether micro- and macroevolutionary processes are a consequence of similar or different evolutionary mechanisms. Equally, it is only through the comparative study of behavioral evolution at multiple hierarchical levels, in combination with experimental tests of hypotheses generated through such comparative studies, that we will come to understand the mechanisms by which behavioral phenotypes evolve.

Acknowledgments

We thank Emília Martins for organizing the symposium that stimulated the writing of this paper and K. A. Crandall, W. A. Cresko, J. Losos, J. B. Whitfield and two anonymous reviewers for comments that substantially improved the manuscript. H. C. Gerhardt and M. B. Ptacek provided invaluable help with the example from hylid tree frogs. We thank R. W. King for figure preparation. Writing of the manuscript was supported by a National Science Foundation Presidential Faculty Fellowship (DEB-9253718) to S. A. Foster and National Science Foundation grants (DEB-9396006 and GER-9450117) to S. A. Cameron.

References

Ammerman, J., & E. Cavalli-Sforza. 1984. *The Neolithic Transition and the Genetics of Populations in Europe.* Princeton New Jersey: Princeton University Press.

Arnold, S. J. 1977. Polymorphism and geographic variation in the feeding behavior of the garter snake, *Thamnophis elegans. Science*, 197, 676–678.

Arnold, S. J. 1992. Behavioural variation in natural populations. VI. Prey responses by two species of garter snakes in three regions of sympatry. *Anim. Behav.*, 44, 705–719.

Avise, J. C. 1989. Gene trees and organismal histories: a phylogenetic approach to population biology. *Evolution*, 43, 1192–1208.

Avise, J. C. 1994. *Molecular Markers, Natural History, and Evolution.* New York: Chapman and Hall.

Avise, J. C., J. Arnold, R. M. Ball, E. Bermingham, T. Lamb, J. E. Neigel, C. A. Reeb, & N. C. Saunders. 1987. Intraspecific phylogeography: The mitochondrial DNA bridge between population genetics and systematics. *Annu. Rev. Ecol. Syst.*, 18, 489–522.

Baker, M. C., & M. A. Cunningham. 1985. The biology of bird song dialects. *Behav. Brain. Sci.*, 8, 85–133.

Bakker, T. C. M. 1986. Aggressiveness in sticklebacks (*Gasterosteus aculeatus* L.): A behaviour-genetic study. *Behaviour*, 98, 1–144.

Bell, M. A. 1977. A late Miocene marine threespine stickleback, *Gasterosteus aculeatus aculeatus*, and its zoogeographic and evolutionary significance. *Copeia*, 1977, 277–282.

Bell, M. A. 1994. Paleobiology and the evolution of threespine stickleback. In: *The Evolutionary Biology of the Threespine Stickleback* (Ed. by M. A. Bell, & S. A. Foster) pp. 438–471. Oxford, England: Oxford University Press.

Bell, M. A., & S. A. Foster. 1994. Introduction to the evolutionary biology of the threespine stickleback. In: *The Evolutionary Biology of the Threespine Stickleback* (Ed. by M. A. Bell & S. A. Foster) pp. 1–27. Oxford, England: Oxford University Press.

Berthold, P., A. J. Helbig, G. Mohr, & U. Querner. 1992. Rapid microevolution of migratory behaviour in a wild bird species. *Nature*, 360, 668–670.

Blest, A. D. 1961. The concept of ritualization. In: *Current Problems in Animal Behaviour*. (Ed. by W. H. Thorpe, & O. L. Zangwill) pp. 102–124. Cambridge, England: Cambridge University Press.

Boake, C. R. B. 1994. *Quantitative Genetic Studies of Behavioral Evolution*. Chicago: University of Chicago Press.

Bogart, J. P. 1980. Evolutionary implications of polyploidy in amphibians and reptiles. In: *Polyploidy: Biological Relevance* (Ed. by W. H. Lewis), pp. 341–376. New York: Plenum Press.

Bogart, J. P., & A. O. Wasserman. 1972. Diploid-tetraploid species pairs: A possible clue to evolution by polyploidization in anuran amphibians. *Cytogenetics*, 11, 7–24.

Bowers, M. D. 1986. Population differences in larval host plant use in the checkerspot butterfly, *Euphydryas chalcedona*. *Entomol. Exp. Appl.*, 40, 61–69.

Brodie, E. D. III. 1989. Genetic correlations between morphology and antipredator behaviour in natural populations of the garter snake *Thamnophis ordinoides*. *Nature*, 342, 542–543.

Brooks, D. R. & D. A. McLennan. 1991. *Phylogeny, Ecology, and Behavior: A Research Program in Comparative Biology*. Chicago: University of Chicago Press.

Burghardt, G. M. & J. L. Gittleman. 1990. Comparative behavior and phylogenetic analyses: new wine, old bottles. In: *Interpretation and Explanation in the Study of Animal Behavior*. Vol. 2: *Explanation, Evolution, and Adaptation*. (Ed. by M. Bekoff & D. Jamieson) pp. 192–225. San Francisco: Westview Press.

Bush, G. L. 1994. Sympatric speciation in animals: new wine in old bottles. *Trends Ecol. Evol.*, 9, 285–288.

Cameron, S. A. 1993. Multiple origins of advanced eusociality in bees inferred from mitochondrial DNA sequences. *Proc. Natl. Acad. Sci. USA* , 90, 8687–8691.

Cameron, S. A., J. N. Derr, A. D. Austin, J. B. Woolley & R. A. Wharton. 1992. The application of nucleotide sequence data to phylogeny of the Hymenoptera: A review. *J. Hym. Res.*, 1, 63–79.

Carpenter, J. M. 1989. Testing scenarios: Wasp social behavior. *Cladistics*, 5, 131–144.

Carroll, S. P., & P. S. Corneli. 1995. The evolution of behavioral norms of reaction as a problem in ecological genetics: Theory, methods and data. In: *Geographic Variation in Behavior: An Evolutionary Perspective* (Ed. by S. A. Foster, & J. A. Endler). Oxford, England: Oxford University Press.

Chatterjee, S., & B. N. Singh. 1989. Sexual isolation in Drosophila. *Indian Rev. Life Sci.*, 9, 101–135.

Cheverud, J. M., M. M. Dow, & W. Leutenegger. 1985. The quantitative assessment of phylogenetic constraints in comparative analyses: sexual dimorphism in body weights among primates. *Evolution,* 39, 1335–1351.

Coddington, J. A. 1988. Cladistic tests of adaptational hypotheses. *Cladistics*, 4, 3–22.

Coss, R. G., (in press). Evolutionary persistence of behavior: restraints on geographic variation and phenotypic plasticity. In: *Geographic Variation in Behavior, an Evolutionary Perspective.* (Ed. by S. A. Foster, & J. A. Endler). Oxford, England: Oxford University Press.

Crandall, K. A. 1994. Intraspecific cladogram estimation: Accuracy at higher levels of divergence. *Syst. Biol.*, 43, 222–235.

Crandall, K. A., A. R. Templeton, & C. F. Sing. 1994. Intraspecific phylogenetics: Problems and solutions. In: *Models in Phylogeny Reconstruction.* (Ed. by R. W. Scotland, D. J. Siebert, & D. M. Williams), pp. in press. Oxford, England: Clarendon Press.

Cruz, A., & J. W. Wiley. 1989. The decline of an adaptation in the absence of a presumed selection pressure. *Evolution*, 43, 55–62.

Darwin, C. 1859. *The Origin of Species by Means of Natural Selection.* London: Murray.

Davies, N. B., & M. De L. Brooke. 1989. An experimental study of co-evolution between the cuckoo, *Cuculus canorus*, and its hosts: II. Host egg markings, chick discrimination and general discussion. *J. Anim. Ecol.*, 58, 225–236.

Dingle, H. 1988. Quantitative genetics of life history evolution in a migrant insect. In: *Population Genetics and Evolution* (Ed. by G. de Jong), pp. 83–93. Berlin: Springer-Verlag.

Donoghue, M. J. 1989. Phylogenies and the analysis of evolutionary sequences, with examples from seed plants. *Evolution*, 43, 1137–1156.

Donoghue, M. J., R. G. Olmstead, J. F. Smith, & J. D Palmer. 1992. Phylogenetic relationships of Dipsacales based on rbcL sequences. *Ann. Missouri Bot. Gardens*, 79, 333–345.

Duijm, M. 1990. On some song characteristics in *Ephippiger* (Orthoptera: Tettigoniidae) and their geographic variation. *Neth. J. Zool.*, 40, 428–453.

Edwards, S. V., & M. Kot. 1995. Comparative methods at the species level: Geographic variation in morphology and group size in Gray-crowned Babblers (*Pomatostomus temporalis*). *Evolution*, in press.

Ehrlich, P. R. 1965. The population biology of the butterfly, *Euphydryas editha*: II. The structure of the Jasper Ridge colony. *Evolution*, 19, 327–336.

Ehrlich, P. R., R. R. White, M. C. Singer, S. W. McKechnie, & L. E. Gilbert. 1975. Checkerspot butterflies: A historical perspective. *Science*, 188, 221–228.

Eickwort, G. C. 1986. First steps into eusociality: The sweat bee *Dialictus lineatulus*. *Fla. Ent.*, 69, 742–754.

Endler, J. A. 1977. *Geographic Variation, Speciation, and Clines.* Princeton, New Jersey: Princeton University Press.

Endler, J. A. 1986. *Natural Selection in the Wild.* Princeton, New Jersey: Princeton University Press.

Epperson, B. K. 1993. Recent advances in correlation studies of spatial patterns of genetic variation. *Evol. Biol.*, 27, 95–155.

Ewens, W. J. 1990. Population genetics theory — the past and the future. In: *Mathematical and Statistical Developments of Evolutionary Theory* (Ed. by S. Lessard), pp. 177–227. Dordecht, Netherlands: Kluwer Academic Publishers.

Faith, D. P. 1991. Cladistic permutation tests for monophyly and nonmonophyly. *Syst. Zool.*, 40, 366–375.

Farris, J. S. 1983. The logical basis of phylogenetic analysis. In: *Advances in Cladistics* Vol. 2: *Proc. 2nd Meeting Willi Hennig Society* (Ed. by N. I. Platnick, & V. A. Funk), New York: Columbia University Press.

Farris, J. S. 1988. *Hennig86: Version 1.5.* Stony Brook, New York: State University of New York.

Felsenstein, J. 1983. Parsimony in systematics: Biological and statistical issues. *Annu. Rev. Ecol. Syst.*, 14, 313–333.

Felsenstein, J. 1985a. Confidence limits on phylogenies with a molecular clock. *Syst. Zool.*, 34, 152–161.

Felsenstein, J. 1985b. Confidence limits on phylogenies: An approach using the bootstrap. *Evolution*, 39, 783–791.

Felsenstein, J. 1993. *PHYLIP (Phylogeny Inference Package) version 3.5.* Distributed by the author. Department of Genetics, University of Washington, Seattle.

Foster, S. A. 1994a. Evolution of the reproductive behavior of threespine stickleback. In: *The Evolutionary Biology of the Threespine Stickleback.* (Ed. by M. A. Bell, & S. A. Foster), pp. 381–398. Oxford, England: Oxford University Press.

Foster, S. A. 1994b. Inference of evolutionary pattern: diversionary displays of three-spined sticklebacks. *Behav. Ecol.*, 5, 114–121.

Foster, S. A. (in press). Constraint, adaptation, and opportunism in the design of behavioral phenotypes. In: *Perspectives in Ethology*, Vol. 11 (Ed. by N. Thompson), pp. 61–81. New York: Plenum Press.

Foster, S. A. (in press). Understanding the evolution of behaviour in threespine stickleback: The value of geographic variation.. *Behaviour.*

Foster, S. A., & Endler, J. A. (Eds.). (in press). *Geographic Variation in Behavior: An Evolutionary Perspective.* Oxford, England: Oxford University Press.

Foster, S. A., & S. A. Ploch. 1990. Determinants of variation in antipredator behavior of territorial male threespine stickleback in the wild. *Ethology*, 84, 292–313.

Foster, S. A., J. A. Baker, & M. A. Bell. 1992. Phenotypic integration of life history and morphology: An example from the three-spined stickleback, *Gasterosteus aculeatus* L. *J. Fish Biol.*, 41 (suppl.), 21–35.

Gerhardt, H. C. 1974. Mating call differences between eastern and western populations of the gray tree frog *Hyla chrysoscelis. Copeia*, 1974, 534–536.

Gerhardt, H. C. 1994. Reproductive character displacement of female mate choice in the gray tree frog, *Hyla chrysoscelis. Anim. Behav.*, 47, 959–969.

Gerhardt, H. C., M. B. Ptacek, L. Barnett, & K. G. Torke. 1994. Hybridization between the diploid-tetraploid tree frogs *Hyla chrysoscelis* and *Hyla versicolor*. *Copeia*, 1994, 51–59.

Gittleman, J. L., & M. Kot. 1990. Adaptation statistics and a null model for estimating phylogenetic effects. *Syst. Zool*, 39, 227–241.

Gittleman, J. L., & H-K. Luh. 1992. On comparing comparative methods. *Annu. Rev. Ecol. Syst.*, 23, 383–404.

Golding, G. B. 1987. The detection of deleterious selection using ancestors inferred from a phylogenetic history. *Genet. Res.*, 49, 71–82.

Golding, G. B., & J. Felsenstein. 1990. A maximum likelihood approach to the detection of selection from a phylogeny. *J. Mol. Evol.*, 31, 511–523.

Goldthwaite, R. O., R. G. Coss, & D. H. Owings. 1990. Evolutionary dissipation of an antisnake system: differential behavior by California and arctic ground squirrels in above- and below-ground contexts. *Behaviour*, 112, 246–269.

Greene, H. W. 1986. Diet and arboreality in the emerald monitor, *Varanus prasinus*, with comments on the study of adaptation. *Fieldiana, Zool.* (NS), 31, 1–12.

Greene, H. W., & G. M. Burghardt. 1978. Behavior and phylogeny: Constriction in ancient and modern snakes. *Science*, 200, 74–77.

Hartl, D. L., & S. A. Sawyer. 1991. Inference of selection and recombination from nucleotide sequence data. *J. Evol. Biol.*, 4, 519–532.

Harvey, P. H., & M. D. Pagel. 1991. *The Comparative Method in Evolutionary Biology*. Oxford, England: Oxford University Press.

Hedrick, A. V., & S. E. Riechert. 1989. Population variation in the foraging behavior of a spider: The role of genetics. *Oecologia*, 80, 533–539.

Heinroth, O. 1911. Beitrage zur biologie, insbesondere psychologie und ethologie der Anatiden. *Verh. 5 Int. Ornith. Kongr.*, pp. 589–702, Berlin.

Heinroth, O. 1930. Über bestimmte Bewegungsweisen der Wirbeltiere. *Sitzber. Ges. Naturforsch.* Freunde, Berlin, pp. 333–342.

Hendy, M. D., & D. Penny. 1982. Branch and bound algorithms to determine minimal evolutionary trees. *Math. Biosci.*, 59, 277–290.

Hillis, D. M., & J. P. Huelsenbeck. 1992. Signal, noise and reliability in molecular phylogenetic analyses. *J. Hered.*, 83, 189–195.

Hillis, D. M., J. P. Huelsenbeck, & C. W. Cunningham. 1994. Application and accuracy of molecular phylogenies. *Science*, 264, 671–677.

Hinde, R. A., & N. Tinbergen. 1958. The comparative study of species-specific behavior. In: *Behavior and Evolution* (Ed. by A. Roe & G. G. Simpson), pp. 251–268. New Haven, Connecticut: Yale University Press.

Houde, A. E., & J. A. Endler. 1990. Correlated evolution of female mating preferences and male color patterns in the guppy *Poecilia reticulata*. *Science*, 248, 1405–1408.

Hudson, R. R. 1990. Gene genealogies and the coalescent process. *Oxf. Surv. Evol. Biol.*, 7, 1–44.

Huelsenbeck, J. P., & D. M. Hillis. 1993. Success of phylogenetic methods in the four-taxon case. *Syst. Biol.*, 42, 247–264.

Huntingford, F. A. 1976. The relationship between anti-predator behaviour and aggression among conspecifics in the three-spined stickleback, *Gasterosteus aculeatus*. *Anim. Behav.*, 24, 245–260.

Huntingford, F. A., P. J. Wright, & J. F. Tierney. 1994. Adaptive variation in antipredator behaviour in threespine stickleback. In: *The Evolutionary Biology of the Threespine Stickleback*. (Ed. by M. A. Bell, & S. A. Foster), pp. 277–296. Oxford, England: Oxford University Press.

Johnson, C. F. 1966. Species recognition in the *Hyla versicolor* complex. *Tex. J. Sci.*, 18, 361–364.

King, A. P., & M. J. West. 1990. Variation in species-typical behavior: A contemporary issue for comparative psychology. In: *Contemporary Issues in Comparative Psychology*. (Ed. by D. A. Dewsbury), pp. 321–339. Sunderland, Massachusetts: Sinauer and Associates.

Kingman, J. F. C. 1982. On the genealogy of large populations. *J. Appl. Prob.,* 19A, 27–

Lauder, G. V. 1986. Homology, analogy and the evolution of behavior. In: *The Evolution of Behavior* (Ed. by M. Nitecki & J. Kitchell) pp. 9–40. Oxford, England: Oxford University Press.

Lauder, G. V. 1990. Functional morphology and systematics: studying functional patterns in an historical context. *Annu. Rev. Ecol. Syst.*, 21, 317–340.

Lauder, G. V., A. M. Leroi, & M. R. Rose. 1993. Adaptations and History. *Trends Ecol. Evol.*, 8, 294–297.

Leroi, A. M., M. R. Rose, & G. V. Lauder. 1994. What does the comparative method reveal about adaptation? *Am. Nat.,* 143, 381–402.

Lessa, E. P., 1990. Multidimensional analysis of geographic genetic structure. *Syst. Zool.*, 39, 242–252.

Littlejohn, M. J. 1993. Homogamy and speciation: a reappraisal. In *Oxford Surveys in Evolutionary Biology*, Vol. 9 (Eds. D. J. Futuyma, & J. Antonovics), pp. 135–165. Oxford, England: Oxford University Press.

Lorenz, K. 1935. Der Kumpan in der Umwelt des Vogels. *J. Ornithol.*, 83, 137–213, 289–413.

Lorenz, K. 1941. Vergleichende Bewegungsstudien an Anatinen. *J. Ornithol.*, 89, Sonderheft, 19–29.

Lorenz, K. 1970. *Studies in Animal and Human Behaviour*. London: Methuen Press.

Lott, D. F. 1991. *Intraspecific Variation in the Social Systems of Wild Vertebrates.* Cambridge, England: Cambridge University Press.

Luyten, P. H., & N. R. Liley. 1985. Geographic variation in the sexual behaviour of the guppy, *Poecilia reticulata* (Peters) *Behaviour*, 95, 164–179.

Luyten, P. H., & N. R. Liley. 1991. Sexual selection and competitive mating success of male guppies (*Poecilia reticulata*) from four Trinidad populations. *Behav. Ecol. Sociobiol.*, 28, 329–336.

Lynch, M. 1991. Methods of the analysis of comparative data in evolutionary biology. *Evol.*, 45, 1065–1080.

Maddison, W. P. 1989. Reconstructing character evolution on polytomous cladograms. *Cladistics*, 5, 365–377.

Maddison, W. P. 1990. A method for testing the correlated evolution of two binary characters: Are gains or losses concentrated on certain branches of a phylogenetic tree? *Evolution*, 44, 539–557.

Maddison, W. P., & D. R. Maddison. 1992. *MacClade: Analysis of Phylogeny and Character Evolution, Version 3*. Sunderland, Massachusetts: Sinauer and Associates.

Maddison, W. P., M. J. Donoghue, & D. R. Maddison. 1984. Outgroup analysis and parsimony. *Syst. Zool.*, 33, 83–103.

Magurran, A. E. 1986. Predator inspection behaviour in minnow shoals: Differences between populations and individuals. *Behav. Ecol. Sociobiol.*, 19, 267–273.

Magurran, A. E., & B. H. Seghers. 1990a. Population differences in predator recognition and attack cone avoidance in the guppy, *Poecilia reticulata*. *Anim. Behav.*, 40, 443–452.

Magurran, A. E., & B. H. Seghers. 1990b. Population differences in the schooling behaviour of newborn guppies, *Poecilia reticulata*. *Ethology*, 84, 334–342.

Martins, E. P. 1994. Estimating the rate of phenotypic evolution from comparative data. *Am. Nat.*, 144, 193–209.

McLennan, D. A., D. R. Brooks, & J. D. McPhail. 1988. The benefits of communication between comparative ethology and phylogenetic systematics: A case study using gasterosteid fishes. *Can. J. Zool.*, 66, 2177–2190.

McPhail, J. D. 1984. Ecology and evolution of sympatric sticklebacks (Gasterosteus): Morphological and genetic evidence for a species pair in Enos Lake, British Columbia. *Can. J. Zool.*, 62, 1402–1408.

Orti, G., M. A. Bell, T. E. Reimchen, & A. Meyer. 1994. Global survey of mitochondrial DNA sequences in the threespine stickleback: Evidence for recent migrations. *Evolution*, 48, 608–622.

Pomiankowski, A., & L. Sheridan. 1994. Linked sexiness and choosiness. *Trends Ecol. Evol.*, 9, 242–244.

Ptacek, M. B., H. C. Gerhardt, & R. D. Sage. 1994. Speciation by polyploidy in tree frogs: Multiple origins of the tetraploid, *Hyla versicolor*. *Evolution*, 48, 898–908.

Ralin, D. B. 1977. Evolutionary aspects of mating call variation in a diploid-tetraploid species complex of tree frogs (Anura). *Evolution,* 31, 721–736.

Ralin, D. B., M. A. Romano, & C. W. Kilpatrick. 1983. The tetraploid tree frog *Hyla versicolor*: Evidence for a single origin from the diploid *H. chrysoscelis*. *Herpetologica*, 39, 212–225.

Ridley, M. 1983. *The Explanation of Organic Diversity*. Oxford, England: Clarendon Press.

Riechert, S. E. 1987. Between-population variation in spider territorial behavior: Hybrid-pure population line comparisons. In: *Evolutionary Genetics of Invertebrate Behavior*. (Ed. by M. Huettel), pp. 33–42. New York: Plenum Press.

Riechert, S. E. 1993a. A test for phylogenetic constraints on behavioral adaptation in a spider system. *Behav. Ecol. Sociobiol.*, 32, 343–348.

Riechert, S. E. 1993b. Investigation of potential gene flow limitation of behavioral adaptation in an aridlands spider. *Behav. Ecol. Sociobiol.*, 32, 355–363.

Riechert, S. E., & A. V. Hedrick. 1990. Levels of predation and genetically based anti-predatory behavior in the spider, *Agelenopsis aperta*. *Anim. Behav.*, 40, 679–687.

Romano, M. A., D. B. Ralin, S. I. Guttman, & J. H. Skillings. 1987. Parallel electromorph variation in the diploid-tetraploid gray tree frog complex, *Hyla chrysoscelis* and *Hyla versicolor*. *Am. Nat,* 130, 864–878.

Ross, K. G., & D. J. C. Fletcher. 1985. Comparative study of genetic and social structure in two forms of the fire ant, *Solenopsis invicta* (Hymenoptera: Formicidae). *Behav. Ecol. Sociobiol.*, 17, 349–356.

Ross, K. G., & D. D. Shoemaker. 1993. An unusual pattern of gene flow between the two social forms of the fire ant *Solenopsis invicta*. *Evolution*, 47, 1595–1605.

Ross, K. G., & J. C. Trager. 1990. Systematics and population genetics of fire ants (*Solenopsis saevissima* complex) from Argentina. *Evolution*, 44, 2113–2134.

Ryan, M. J., & W. Wilczynski. 1991. Evolution of intraspecific variation in the advertisement call of a cricket frog (*Acris crepitans*, Hylidae). *Biol. J. Linn. Soc.*,

Schluter, D., & J. D. McPhail. 1992. Ecological character displacement and speciation in sticklebacks. *Am. Nat.*, 140, 85–108.

Seghers, B. H. 1974. Schooling behavior in the guppy (*Poecilia reticulata*): An evolutionary response to predation. *Evolution*, 28, 486–489.

Sillén-Tullberg, B. 1988. Evolution of gregariousness in aposematic butterfly larvae: A phylogenetic analysis. *Evolution*, 42, 293–305.

Simon, C., F. Frati, P. Flook, A. Beckenbach, B. Crespi, & H. Liu. 1995. Assessing the phylogenetic usefulness of specific mitochondrial genes and a compilation of conserved PCR primers. *Ann. Entomol. Soc. Am.*, in press.

Singer, M. C., D. Ng, D. Vasco, & C. D. Thomas. 1992. Rapidly evolving associations among oviposition preferences fail to constrain evolution of insect diet. *Am. Nat.*, 139, 9–20.

Slatkin, M., & H. E. Arter. 1991. Spatial autocorrelation methods in population genetics. *Am. Nat.*, 138, 499–517.

Slatkin, M., & W. P. Maddison. 1989. A cladistic measure of gene flow from the phylogenies of alleles. *Genetics*, 123, 603–613.

Slatkin, M., & W. P. Maddison. 1990. Detecting isolation by distance using phylogenies of genes. *Genetics*, 126, 249–260.

Snyder, R. J., & H. Dingle. 1989. Adaptive, genetically based differences in life history between estuary and freshwater threespine sticklebacks (*Gasterosteus aculeatus* L.). *Can J. Zool.*, 67, 2448–2454.

Sokal, R. R., G. M. Jacquez, & M. C. Wooten. 1989. Spatial autocorrelation analysis of migration and selection. *Genetics,* 121, 845–855.

Swofford, D. L. 1993. PAUP: *Phylogenetic Analysis Using Parsimony, Version 3.1.1.* Champaign, Illinois: Illinois Natural History Survey.

Swofford, D. L., & W. P. Maddison. 1987. Reconstructing ancestral character states under Wagner parsimony. *Math. Biosci.*, 87, 199–229.

Swofford, D. L., & W. P. Maddison. 1992. Parsimony, character-state reconstructions, and evolutionary inferences. In: *Systematics, Historical Ecology, and North American Freshwater Fishes*. (Ed. by R. L. Mayden), pp. 186–223. Stanford, California: Stanford University Press.

Templeton, A. R. 1983. Convergent evolution and non-parametric inferences from restriction fragment and DNA sequence data. In: *Statistical Analysis of DNA Sequence Data* (Ed. by B. Weir.), pp. 151–179. New York: Marcel Dekker.

Templeton, A. R. 1992. Human origins and analysis of mitochondrial DNA sequences. *Science*, 255, 737.

Templeton, A. R., E. Boerwinkle, & C. F. Sing. 1987. A cladistic analysis of phenotypic associations with haplotypes inferred from restriction endonuclease mapping: I. Basic theory and an analysis of alcohol dehydrogenase activity in *Drosophila*. *Genetics*, 117, 343–351.

Templeton, A. R., K. A. Crandall, & C. F. Sing. 1992. A cladistic analysis of phenotypic association with haplotypes inferred from restriction endonuclease mapping and DNA sequence data: III. Cladogram estimation. *Genetics*, 132, 619–633.

Templeton, A. R., E. Routman, & C. A. Phillips. 1995. Separating population structure from population history: A cladistic analysis of the geographical distribution of mitochondrial DNA haplotypes in the tiger salamander, *Ambystoma tigrinum*. *Genetics*, in press.

Templeton, A. R., & C. F. Sing. 1993. A cladistic analysis of phenotypic associations with haplotypes inferred from restriction endonuclease mapping. IV. Nested analyses with cladogram uncertainty and recombination. *Genetics*, 134, 659–669.

Thompson, J. N. 1988a. Variation in preference and specificity in monophagous and oligophagous swallowtail butterflies. *Evolution*, 42, 118–128.

Thompson, J. N. 1988b. Evolutionary genetics of oviposition preference in swallowtail butterflies. *Evolution*, 42, 1223–1234.

Thompson, J. N. 1993. Preference hierarchies and the origin of geographic specialization in host use in swallowtail butterflies. *Evolution*, 47, 1585–1594.

Thompson, J. N., & O. Pellmyr. 1991. Evolution of oviposition behavior and host preference in Lepidoptera. *Annu. Rev. Entomol.*, 36, 65–89.

Tilley, S. G., P. A. Verrell, & S. J. Arnold. 1990. Correspondence between sexual isolation and allozyme differentiation: A test in the salamander *Desmognathus ochrophaeus*. *Proc. Natl. Acad. Sci. USA*, 87, 2715–2719.

Towers, S. R., & R. G. Coss. 1990. Confronting snakes in the burrow: Snake-species discrimination and antisnake tactics of two California ground squirrel populations. *Ethology*, 84, 177–192.

Ueda, H. 1993. Mating calls of auto-triploid and autotetraploid males in *Hyla japonica, Sci. Rep. Lab. Amphibian Biol., Hiroshima Univ.*, 12, 177–189.

Verrell, P. A., & S. J. Arnold. 1989. Behavioral observations of sexual isolation among allopatric populations of the mountain dusky salamander, *Desmognathus ochrophaeus*. *Evol.*, 43, 745–755.

Wasserman, A. O. 1970. Polyploidy in the common tree toad, *Hyla versicolor* Le Conte. *Science,* 167, 385–386.

Watrous, L. E., & Q. D. Wheeler. 1981. The out-group comparison method of character analysis. *Syst. Zool.*, 30, 1–11.

Whitman, C. O. 1899. Animal Behavior. *Woods Hole Mar. Biol. Lect.*, 6, 285–338.

Wilkinson, G. S., & P. R. Reillo. 1994. Female choice response to artificial selection on an exaggerated male trait in a stalk-eyed fly. *Proc. R. Soc. Lond., Series B*, 255, 1–6.

Wiley, J. E. 1983. Chromosome polymorphism in *Hyla chrysoscelis. Copeia*, 1983, 273–275.

Wiley, J. E., M. A. Little, M. A. Romano, D. A. Blount, & G. R. Cline. 1989. Polymorphism in the location of the 18s and 28s rRNA genes on the chromosomes of the diploid tetraploid tree frogs *Hyla chrysoscelis* and *H. versicolor. Chromosoma*, 97, 481–487.

Zink, R. M., & J. V. Remsen, Jr. 1986. Evolutionary processes and patterns of geographic variation in birds. *Curr. Ornithol.*, 4, 1–69.

CHAPTER 6

Phylogenetic Lability and Rates of Evolution: A Comparison of Behavioral, Morphological and Life History Traits

John L. Gittleman, C. Gregory Anderson, Mark Kot, and Hang-Kwang Luh

"How fast, as a matter of fact, do animals evolve in nature?" Fifty years ago Simpson (1944) posed this deceptively simple question. Even though it lies at the heart of evolutionary biology, many conceptual, methodological, and theoretical problems remain unstudied in our understanding of the lability and rate of trait evolution. Many advances have arisen as the result of direct estimates of evolutionary rates from the fossil record. However, most such studies focus on morphological characters (Fenster & Sorhannus 1991) that are easily preserved in the fossil record. For many behavioral, ecological, and other poorly preserved traits we must instead depend on indirect measurements and on inferences from cross-taxonomic (comparative) study (Gittleman & Decker 1994). Even though numerous theoretical statements argue that behavioral and ecological characters evolve at different and faster rates, this is a relatively untouched line of inquiry. Now, as this volume reflects, we have better tools to get at such problems: an explosion of phylogenetic information (both molecular and morphological) coupled with new statistical methodologies and large comparative data bases are all readily available.

In this paper, we address three interrelated questions concerning species differences in the phylogeny, lability, and rate of evolution:

1. Are quantitative traits (behavioral and morphological) strongly correlated with phylogeny?
2. Are behavioral traits more plastic or labile through evolutionary time than morphological and life history traits?
3. Do behavioral traits evolve at different, presumably faster, rates than morphological and life history traits?

To answer these questions, we present various statistical methods and apply them to molecular phylogenies and comparative data on behavior, life histories, and morphology in mammals. Our study parallels and is intended to complement cladistic-type analyses of the origin, transitions, congruence, and historical constraints of behavior and ecology (e.g., Coddington 1988; de Queiroz & Wimberger 1993; Greene 1994; Lauder 1986, 1990, 1994; McLennan 1991; for review, see Brooks & McLennan 1991). However, in contrast to this body of work, we focus exclusively on continuous traits, statistical comparative methods, and molecular phylogenies for measuring taxonomic divergence and rate. That is, we co-opt some of the now massive literature on comparative methods for analyzing quantitative traits (Harvey & Pagel 1991) to pursue a new line of inquiry. Ideally, using quantitative techniques may permit a broader comparative sweep (both in traits and taxa) for analyzing evolutionary trends, and more importantly, fully use the perhaps more precise nature (branch lengths and divergence times) of molecular phylogenies; after all, molecular systematics has given the promise of increasingly accurate phylogenies and it is now time to use them. We emphasize, however, that the present work is inherently descriptive, not of a causal form. We are primarily concerned with showing, via modern comparative methods, how quantitative traits evolve through phylogenetic time; although some speculation is given for why traits evolve in a specific manner, this is not the focus of the present work. Prior to the empirical analysis, we first outline some relevant conceptual issues.

Phylogenetic correlation: a null hypothesis of trait evolution

In order to understand why the issue of "phylogenetic correlation" of traits is important, it is necessary to trace some history. Early statements in many areas of organismal biology firmly indicate that phylogeny was viewed as important for explaining trait variation. In animal behavior, both Lorenz (1965) and Tinbergen (1951) asserted that comparative studies would show evolutionary transitions in behavior just as classical work had shown in comparative morphology (Burghardt & Gittleman 1990; Gittleman & Decker 1994; Greene 1994). In ecology, Elton (1927) discussed how ecological differences among species reflect historical patterns rather than environmental selection. And, in evolutionary biology, particularly during formative stages of the modern synthesis, adaptive trait evolution was considered to be constrained by historical antecedents (Darwin 1859; Bateson 1894; Morgan 1903; Robson & Richards 1936; Huxley 1942; Simpson 1944; for review see, Provine 1985). Such emphasis on phylogenetic history was followed by a heyday of adaptation: traits evolve exclusively in response to environmental problems. Undoubtedly the watershed of this development was publication of Wilson's (1975) *Sociobiology: A Modern Synthesis*. Although this volume ushered in a generation of students interested in the ecology and evolution of behavior, emphasis was clearly placed on explaining trait variability within an adaptive framework. The phrase, "phylogenetic inertia," though introduced in Wilson's book, is only mentioned one other time in 575 pages.

Seemingly, the pendulum has swung back in the other direction, perhaps even a bit too far to the extent that phylogeny is now assumed to relate to trait variation irrespective of any empirical support. At least four general advances, both conceptual and methodological, have resulted from resurrecting phylogeny: (a) Greater establishment of homologous relations (e.g., Roth 1988; Hall 1994) and inclusion of a variety of traits in erecting phylogenies (e.g., Sanderson & Donoghue 1989); (b) Following Gould and Lewontin's (1979) influential paper, which was equally extreme in arguing for constraints as Wilson was in arguing for adaptation, phylogeny is now considered an important alternative explanation to strict adaptation (McKitrick 1993; Gould 1993); (c) The acceptance of the idea that when shared evolutionary history of different taxa results in points that are not statistically independent, hypothesis testing may miscalculate degrees of freedom and reach incorrect results unless phylogeny is incorporated into comparative tests (Clutton-Brock & Harvey 1977; Felsenstein 1985; Huey 1987; Martins & Garland 1991 for reviews, see Harvey & Pagel

1991; Gittleman & Luh 1992; Miles & Dunham 1993; and, in this volume, Martins & Hansen, Chap. 2); (d) Accumulation of morphological/molecular phylogenies which provide a rich source of phylogenetic hypotheses to explain species differences in traits.

All of these advances rest on an important assumption: trait variation indeed is "phylogenetically correlated," defined as a statistically significant relationship between trait variation and a given taxonomy or phylogeny (Gittleman & Luh 1992; Martins & Hansen, Chap. 2 in this volume). Many comparative studies show such phylogenetic correlation; these include a wide variety of traits across both broad taxonomic groups as well as relatively small taxa or guilds. Examples include social organization in mammals (Berger 1988); morphology, behavior, ecology, life histories and social organization in primates (Clutton-Brock & Harvey 1977; Harvey & Clutton-Brock 1985), carnivores (Gittleman 1991, 1993, 1994), small mammals (Mace et al. 1981) and birds (Bennett & Harvey 1985a, b); cooperative breeding in birds (Edwards & Naeem 1993); life history traits in reptiles (Miles & Dunham 1992). Importantly, each of these studies is distinguished from other comparative work by diagnosing phylogenetic correlation *prior to* hypothesis testing (Miles & Dunham 1993). As a result of these studies and the proliferation of comparative tools to incorporate and/or remove such phylogenetic effects, many comparative researchers are now approaching comparative problems with the assumption that traits are always phylogenetically correlated, without any statistical diagnosis. An array of comparative studies now reflect this presumptive approach, analyzing such diverse problems as mammalian physiology (Promislow 1991), sexual dimorphism in birds (Promislow et al. 1992), bird song (Read & Weary 1992), life history patterns (Berrigan et al. 1993), maternal investment (Pontier et al. 1993), and sperm competition (Gomendio & Roldan 1993), to cite only a few recent examples. To quote from one, "It is well known that species cannot be considered as independent points in comparative analyses" (Read & Weary, p. 169). Further, in a review on comparative methods, Harvey & Purvis (1991) begin by describing why phylogeny is (theoretically) critical to hypothesis testing: "Species values do not provide independent data points for comparative analyses because species share characteristics through descent from common ancestors. And the more related that species are, the more similar they tend to be"

(p. 619). But, no mention is made of whether trait variation indeed is correlated with, or in any way related to, phylogeny. That is, no basis is given for affirming this (empirical) assumption.

This problem is not trivial. In recent simulation studies assessing the error rates of different comparative methods for incorporating phylogeny in comparative tests, it was found that statistical error may in fact be created by applying these methods when no phylogenetic patterns are in the data. Gittleman & Luh (1992, 1994) simulated trait evolution along various phylogenies and found that in the case where there is no relationship between the traits and the phylogeny, different comparative methods may generate unacceptable error rates. Further, error rates may actually be increased by imposing a comparative method under a condition of no phylogenetic correlation, especially with small sample sizes (e.g., less than 10 species) and inaccurate phylogenetic information. Indeed, from an empirical perspective, traits may not relate to a phylogeny for a variety of reasons including strong stabilizing selection, evolutionary lability, or even inaccurate phylogenetic information.

Because of these issues, we strongly encourage diagnosing the presence and distribution of phylogenetic correlation prior to any comparative test (Gittleman & Kot 1990; Gittleman 1991, 1993; Gittleman & Luh 1992, 1994; Miles & Dunham 1993). One should consider the possibility that there is no phylogenetic correlation as well as the possibility that all traits exhibit significant phylogenetic correlation. More realistically, one should determine whether some traits are more significantly correlated with phylogeny than others and assume phylogenetic correlation only if there is a strong *a priori* reason (Gittleman & Decker 1994; Felsenstein 1985, p. 6). Thus, it is important to test the hypothesis, "Most comparative data are correlated with phylogeny, as currently assumed in the literature, and some traits are more phylogenetically correlated than others?" In the following, we present a statistical method and sample data sets to examine this hypothesis. Assuming different traits (behavioral, morphological, life history) correlate with phylogeny differently, as certainly anticipated, this sets forth interesting questions about relative lability and evolutionary rate through phylogenetic time.

Evolutionary lability

Much speculation surrounds the idea that different kinds of traits are relatively more labile than others through evolutionary time (Gause 1947; Schmalhausen 1949; Bradshaw 1965; Mayr 1974). Generally, it is thought that behavioral traits are more labile than morphological or physiological traits. Indeed, West-Eberhard (1987, 1989) suggests that four mechanisms, genetic (allelic-switch) polymorphism, disruptive selection, pleiotropic effects, or origin via contextual shift (preadaptation), may causally allow for greater lability in a trait.

Early statements indicate that different kinds of traits are equally conservative or labile among taxa. Hinde and Tinbergen (1958) argued that behavioral differences among species follow phyletic lines just as morphological differences do and therefore "...the behavior of closely related species is more similar than that of distantly related ones" (p. 261); Colbert (1958) and Simpson (1958) expressed similar views about equal evolutionary divergence between behavioral and morphological traits. Lorenz (1965) emphasized that some behavior patterns would be more plastic than others, but behavior would be more influential in bringing about evolutionary change; likewise, Mayr (1963) stated, "A shift into a new niche or adaptive zone is, almost without exception, initiated by a change in behavior. The other adaptations to the new niche, particularly the structural ones, are acquired secondarily." (1963, p. 604). Thus, a fundamental assumption in the behavioral and evolutionary literature, either explicit or implicit, is that behavior is more evolutionarily labile than other traits: behavior is malleable to respond to immediate ecological, environmental, and/or social problems.

Not only is it asserted that behavioral traits are more plastic than others, but it is explicitly stated that behavioral change usually precedes morphological evolution; behavior is frequently considered to be an inducer of evolutionary change (Wcislo 1989), speciation events (Hinde 1959), or "breaker" of phenotypic constraint (Arnold 1992). Some phylogenetic comparative studies have indeed shown that behavior can effect morphology (e.g., Basolo 1990). At least five causal factors can bring about change via behavioral plasticity (Bateson 1988; Wcislo 1989): (a) inherited capacities allow animals to learn to exploit new environments; (b) parental behavior elicit new behavior(s) in young; (c) behavioral invasion to a new environment expresses greater genetic variability via progeny; (d) behavioral plasticity compensates for structural or physiological constraints; (e) behavioral traits develop later in ontogeny than structural traits.

Two observations further emphasize the need to analyze relative lability between behavior and other types of traits. First, some morphological traits change significantly and rapidly with environmental change similar to behavioral change (James 1983). For example, size, anatomical structure, and chemistry of the cerebral cortex are rapidly altered with environmental enrichment (Rosenzweig 1971; Rosenzweig & Bennett 1977; Greene 1994). Second, in some organisms (e.g., ectotherms) physiological capacities can fluctuate more than behavior and, in fact, reveal that behavior actually constrains adjustment to new habitats (Huey & Bennett 1987; Huey 1991). Thus, from a comparative perspective, it is unclear whether behavioral and morphological traits show phylogenetic divergence in lability. As West-Eberhard (1987) asserts, "The comparative study of flexible strategies in related species is crucial to understanding the evolution of social phenotypes, which combine ancient and widespread flexible characters to give new results in new circumstances" (p. 48).

To address the problem of relative lability in phenotypic traits, we describe a methodology for examining the degree to which different kinds of traits change (observed lability) through phylogenetic time. In contrast to many analogous questions pertaining to phenotypic plasticity (Scheiner 1993), we are concerned with phylogenetic history and its relationship with trait variation. It is also important to note that our interest is not in behavioral vs. morphological variability. Others (e.g., Slater 1986) have shown that behavior is generally more variable than morphology, as measured by coefficients of variability (problems of this measure notwithstanding). We are explicitly assessing trait change in relation to phylogenetic pattern.

Rates of evolution

Problems related to evolutionary rate have always formed a centerpiece in evolutionary biology (Simpson 1944, 1953; Haldane 1949). Rates of evolution are acknowledged to influence clade diversity, levels of speciation, and evolutionary origin (Simpson 1944; Eldredge & Gould

1972). Studies of evolutionary rate often involve comparative methods: the measurement of rates (univariate or multivariate) within a single phylogenetic lineage or across diverging lineages during speciation. Thus, in proposing to study phylogeny and lability of trait evolution, problems of evolutionary rate are involved. Because the issue of evolutionary rate is so large (Hecht 1965; Simpson 1953; Fenster & Sorhannus 1991; Gingerich 1983, 1993), we mainly focus our attention on: (a) differential rate change among traits; (b) methods for estimating evolutionary rate; and (c) reconstructing phylogenies to measure rate change.

A. Differential rate change among traits

Many authors have suggested that rates of evolution are at least partly determined by overall behavioral complexity or that behavior acts as a "pacemaker" of evolution (Schmalhauson 1949; Mayr 1963). Typically, it is argued, behavioral plasticity allows for individuals or populations to enter new niches. This invasion then increases variability of other traits for selection to act upon and, in so doing, reveals that behavior must evolve at a more rapid pace. Although well-studied examples of behavioral evolution are available at the population level (Endler 1986; Endler & McLellan 1988; Scheiner 1993), cross-species studies are restricted to cladistic type analyses which show transformational change among discrete traits (e.g., Greene & Burghardt 1978; Lauder 1986; McLennan 1991; for review, see Brooks & McLennan 1991). These studies do not directly assess rate in a quantitative sense, similar to examples (see below) in macroevolution and paleontology.

Moreover, in a recent review on rates of evolution (Fenster & Sorhannus 1991) four broad (trait) categories were listed as sources of data: (a) nucleotide, protein, or biochemical; (b) morphological; (c) taxonomic; and (d) behavioral. Of these, behavior received no discussion. Indeed, despite repeated references to behavior evolving at different and/or faster rates than other phenotypic traits, we know of no comparative (cross-taxonomic) analysis that analytically evaluates this view. Using the methods described below we will assess relative evolutionary rates among behavioral, life history and morphological traits.

B. Measuring evolutionary rate

Gingerich (1993) recently reviewed and critiqued three methods for quantitatively measuring evolutionary rates. (a) Haldane (1949) proposed a unit of proportional change, the *darwin*, which he defined as evolutionary change by a factor of *e* (the base of natural logarithms) per million years. On a semilogarithmic scale (time versus *log* length), a rate of one *darwin* would correspond to a difference of one logarithmic unit per million years. (b) Haldane further suggested that the phenotypic standard deviation is an alternative to measuring *e*, pointing out that population variation is the target of selection and thus a more appropriate measure; Gingerich notes that rates of evolution calculated in standard deviations per million years may be referred to as the *simpson*. (c) Haldane also suggested that evolutionary rate be measured in generations rather than years because this would incorporate the reproductive cycling underlying evolutionary change; Gingerich deems this unit of measure the *haldane*. There are various strengths and weaknesses of each measure which primarily involve the traits in question, quality of data, and what inferences are being made. The present study relies on extant taxa which do not contain standard measures of time scale or generation time; variances, and hence standard deviations, of morphological and behavioral traits are not always known, thus precluding the use of simpsons or haldanes. Therefore, given constraints imposed by the comparative data, *darwins* will be used in the present work. Measurement of evolutionary rate in darwins provides a useful and analytically appropriate estimate for descriptive comparisons (Gingerich 1993, p. 458).

C. Phylogenetic information

Essential to measuring evolutionary rates is that time in terms of ancestral-descendant (i.e., phylogenetic) relationships either be known or inferred. Without reliable phylogenies, rate of evolution cannot be determined for any taxa or traits (Cracraft 1984). As with most problems of evolutionary history, issues of rate change are hampered by the fact that we rarely can observe evolution at the species level. Except for unusually complete fossil records, our primary evidence for evolutionary change is via inferences about sequences of taxonomic change based on end-point analysis. To study times of change and consequential rate patterns it is necessary to reconstruct past events, as in standard phylogenetic reconstruction. The present work accepts that we are at a stage in which working phylogenies are realistic estimates of real phylogenies (Grafen 1989; Purvis et al. 1994). That is, phylogenies should provide information about topological changes among taxa: points of divergence (nodes) and branch lengths (time) in phylogenies can be used to estimate rates of trait evolution (Maddison & Maddison 1992). For this information, we do not support the view that molecular information is superior to morphological. Rather, phylogenetic information must be selected for the specific questions at hand (Hillis 1987; Donoghue & Sanderson 1992). Because we are explicitly interested in comparing phylogenetic differences among behavioral and other types of traits (morphological; life history), using phylogenies that are independent of the traits being studied reduces risk of circularity. We therefore exclusively use molecular phylogenies. Here, we restrict our analysis to the Mammalia because we are familiar with this group.

Methods

To examine the three questions posed at the outset, we quantitatively use phylogenies to examine phylogenetic correlation, lability, and rate of evolution in different kinds of traits. The study involves three elements: (a) molecular phylogenies; (b) quantitative species values of morphological, life history (ecological), and behavioral traits; (c) comparative statistical methods. All of this information is drawn from the literature and is available from the authors. Admittedly, there are weaknesses and restrictive assumptions in each element of the study. We present these as they become apparent and potentially influence comparative results; future work will involve computer simulation to place statistical boundaries on our methods and results.

In the following, we describe methodological aspects pertaining to (a) establishing a terminology for comparing, and characterizing the evolution of, different kinds of traits; (b) each hypothesis and the statistical procedures used to test them; and, (c) the comparative data.

A. Terminology

Comparing the phylogeny, plasticity, and rate of differential trait evolution must include how concepts and different traits are defined. We use the following operational definitions:

1. Behavior, morphology, and life histories

Defining a trait as "behavioral," "morphological," or some other label is extremely difficult and is undoubtedly restricted to a specific problem or study. Operationally, we categorize a "behavior" as a characteristic that conveys some kind of "doing" in a species (Plotkin 1988), particularly in terms of a spatial relation (Martin & Bateson 1993). This is contrasted with morphological or life history traits that reflect more of a "having" quality. The two traits selected as behavioral have been previously treated as such in the literature (e.g., Brown 1975; Krebs & Davies 1993) and convey spatial movements in everyday activities (home range size) and spatial position with conspecifics (population group size). Body weight and brain weight are then used as morphological traits. Life history traits (herein including birth weight and gestation length) are somewhat more ambiguous in this usage. We therefore do not categorize them, a priori, but rather analyze patterns among traits in an investigative manner. That is, even though some traits cannot be characterized as behavioral, life history or morphological, all traits can be analytically compared with respect to relative phyletic change.

2. Phylogeny

The key element to examining patterns of trait variation and phylogeny is the nature of the phylogenetic information. For some questions in comparative biology it may be acceptable to use a phylogeny based on the same as traits as those being studied (Maddison 1990; Donoghue & Sanderson 1992). For present purposes we only use molecular phylogenies that are independent from the trait(s) in question; details of these phylogenies are given below.

3. Lability

We explicitly define evolutionary lability in phylogenetic terms: a trait has greater lability if it is not correlated or has a low correlation with phylogenetic distance, measured from node to node at speciation events. In a sense, lability is viewed as a release from phylogeny; our definition also resembles the concept as discussed by de Queiroz & Wimberger (1993). We recognize, however, that plasticity itself is subject to selection (Williams 1966) and therefore we do not consider it to be a nongenetic phenomenon.

B. Comparative Hypotheses and Methods

1. Phylogenetic correlation hypothesis

Quantitative comparative traits are significantly correlated with phylogeny. This hypothesis is motivated by the assertion, and some empirical observations, that traits evolve along phyletic lines and that trait variation should generally reflect evolutionary history. The following describes a statistical method that will be used to test this hypothesis.

Cheverud et al. (1985) introduced an autocorrelation approach to comparative studies with the concept that a species' attributes are in part a function of the relationships between species. That is, one may estimate the covariance among trait values of related species with a network autoregressive model that takes the form,

$$\mathbf{y} = \rho\, \mathbf{Wy} + \varepsilon$$

where **W** is an N x N weighting matrix, with N the number of species. The vector **y** of standardized trait values takes the linear combination $\rho\mathbf{Wy}$ as its phylogenetic component and the residual vector ε as its specific component. The autocorrelation coefficient, ρ, measures the correlation between the phenotypic trait vector **y** and the purely phylogenetic value **Wy**. The residual ε depicts the independent evolution of each species.

Critical to this model is the phylogenetic connectivity matrix **W** (Gittleman & Kot 1990). **W** is essentially a weighting matrix: w_{ij} is the weight assigned to species j in computing the value for species i. The weighting reflects the phylogenetic distances of all pairs of species. Specifically, these pairs reflect anticipated phenotypic similarities given these distances. In this fashion, the model views the qualities of an individual species as a weighted average of the qualities of all other species. Close relatives, because they share more recent ancestors, are expected to be most similar and are assigned the greatest weights; distant relatives are assigned low weights. It is important to recognize that the w_{ij} is a relative measure: even when all the phylogenetic distances are known, the comparative researcher must still set a relationship between phenotypic similarity and phylogenetic or patristic distance by using the maximum likelihood estimator. In our own work, we set the w_{ii} equal to 0 (a species is clearly its own best predictor) and, because we average over neighbors, we row normalize the w_{ij} and the summation of row is 1.

In Cheverud et al's (1985) original paper, trait similarity was arbitrarily set as the reciprocal of taxonomic distance. That is, all species within the same genus were assigned a value of 1.0, all species within the same family a value of 0.5, all species within the same superfamily 0.33, and so on up to the ordinal level. In order to lend greater flexibility to the model so that phylogenetic correlation can be detected with both conservative and plastic traits, we take either

$$w_{ij} = \frac{1}{d_{ij}^{\alpha}}$$

or

$$w_{ij} = \begin{cases} \dfrac{1}{d_{ij}^{\alpha}} & d_{ij} \leq c \\ 0 & d_{ij} > c \end{cases}$$

as the weights (prior to row normalization), where d_{ij} is the distance between the species i and j, and c a cut-off distance. A variable alpha is introduced to allow for a better fit of trait variation with phylogenetic distance. A cut-off distance is used when a negative correlation at larger distances counteracts a positive correlation at shorter distances and decreases the ability of the autoregressive model to fit the data.

The variable alpha exponent and the autocorrelation coefficient are determined by the method of maximum likelihood. Once the model is fit, we can calculate the residual variance and

$$R^2 = 1 - \frac{\hat{\sigma}^2(\varepsilon_i)}{\sigma^2(y_i)}$$

to estimate the proportion of total variance in each trait accounted for by phylogeny (Gittleman & Kot 1990).

2. Evolutionary lability hypotheses

Phenotypic trait variation, as measured by phylogenetic correlation (Fig. 1), follows four hypothetical patterns: (a) No relationship with phylogeny, as possibly observed under strong stabilizing selection. (b) A specific autoregressive (directional) form with correlation falling-off with phyletic distance, a pattern possibly observed from a simple Brownian motion random walk (i.e., pure random genetic drift). (c) A stepwise function with phylogenetic distance, representing stasis followed by change between nodes (note- this is not, in actuality, an alternative hypothesis to the second hypothesis because the overall pattern of change is in proportion to phyletic distance). (d) A piecewise function with phylogenetic distance, representing linear change of significant positive and negative (phylogenetic) correlation and, importantly, greater phenotypic lability than in the directional or stepwise patterns; this pattern might reflect changes in evolutionary rate (e.g., punctuational change) between speciation events. Our aim is to examine these patterns among traits, with the prediction that behavioral traits (home range size; population group size) will show either no phylogenetic pattern or less phylogenetic correlation than morphological and possibly life history traits. We now describe a method to examine the specific location of correlation in a phylogeny and, using graphical representation (correlogram), how traits are changing through time (i.e., lability).

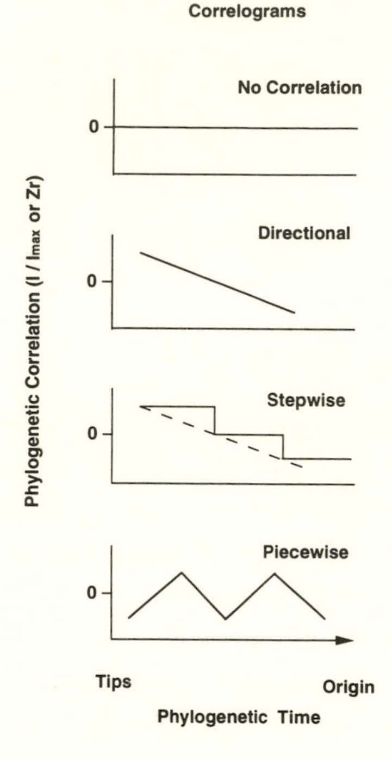

Figure 1. Hypothetical patterns of phylogenetic change in quantitative traits. Patterns are detected using a spatial autocorrelation statistic, the Moran's I (see text for details). Evolutionary lability of traits can be measured by relative differences in observed I 's (or associated z transformations) among traits, with no phylogenetic pattern expressed by zero and strong phylogenetic pattern represented by $\pm$ 1.

To measure effects that vary with phylogenetic distance (i.e., in a cladogram or phylogenetic tree) we borrow concepts and methods from the extensive literature on spatial autocorrelation (Cliff & Ord 1973, 1981; Upton & Fingleton 1985; Legendre 1993). Moran's (1950) I is used to examine the relationship between phylogenetic distance and phenotypic variation. We denote the observed phenotypic trait for species i by y_i and write $\overline{y}$ for the average of (y_i) over the N species of the data set. Using this notation, Moran's coefficient is given as

$$I = \frac{n}{S_0} \frac{\sum_{i=1}^{n} \sum_{j=1}^{n} w_{ij} (y_i - \bar{y})(y_j - \bar{y})}{\sum_{i=1}^{n} (y_i - \bar{y})^2}$$

where

$$S_0 = \sum_{i=1}^{n} \sum_{j=1}^{n} w_{ij}$$

Moran's I is, in essence, an (estimated) autocorrelation coefficient: the numerator is a measure of covariance among the (y_i) and the denominator is a measure of variance. At the heart of this statistic lies the weighting matrix **W**, **W**= [w_{ij}]. In searching out phylogenetic autocorrelation at some level, I will compare the phenotypic trait of a species with a weighted average of the trait over a set of neighbors. The ijth element of the **W** matrix, w_{ij}, is the weight assigned to the jth species in computing the weighted average for species i. $w_{ii}$is always set to 0 (as a species is clearly always its own best predictor), and because we average over neighbors, we will, for nonzero rows, transform the matrix such that:

$$\sum_{j=1}^{n} w_{ij} = 1, \qquad i = 1, 2, \ldots, n$$

The w_{ij} are determined by the correlation we are trying to ascertain. With molecular phylogenetic information (i.e., specific branch lengths and nodes), as in the present study, we average over all species within some fine interval of distance, typically set in accord with the frequency distribution of branch length intervals (see below). We use **W**, in effect, to flag the species that are to be averaged.

Because we refer to I as an autocorrelation, it is natural to assume that $-1 \leq I \leq 1$, with values $\pm$ 1 indicating a strong phylogenetic correlation. However, this is not always so. To standardize values of I to compare among traits, we scale by the maximum possible value. To interpret the values of I, we tabulate associated standard deviates, z, in which z scores greater than 1.96 may be used to reject the null hypothesis of no phylogenetic correlation at the 0.05 level. Further procedures of standardization of I and hypothesis testing with this statistic are given in Gittleman & Kot (1990).

To assess relative lability among traits, we employ a phylogenetic correlogram (Gittleman & Kot 1990). This is simply a graph showing how autocorrelation (observed I 's or associated z's) varies with phylogenetic distance, as hypothesized in Fig. 1. With detailed, complete phylogenetic information (i.e., branching patterns), we first calculate a distance set for each phylogeny (Purvis et al. 1994). The distance set is formed by arranging frequency distributions of the distances observed for each phylogeny and then selecting those distances with the highest frequency (D); this *maximizes* the chance of finding trait change at a given distance in a phylogeny (i.e., phylogenetic correlation). With this methodology, weighting matrices are then defined by applying the Moran's I as follows:

$$w_{ij} = 1 \text{ if the distance of species } i \text{ and } j \text{ belongs to distance set } D,$$
$$= 0, \text{ otherwise.}$$

and the weighting matrices of the autoregressive model

$$w_{ij} = \frac{1}{d_{ij}^{\alpha}} \text{ if the distance of species } i \text{ and } j \text{ belongs to distance set } D,$$
$$= 0, \text{ otherwise.}$$

We then use the correlogram to show where phylogenetic correlation occurs with divergence and branching patterns in the phylogeny.

It should be noted that we have presented the methodology of determining phylogenetic correlation first and then searched for the location of such correlation. In reality, the Moran's I is executed before the autoregressive model. Indeed, the Moran's I is a statistic that aids in diagnosing where the phylogeny/trait relations occur in order to properly apply the autoregressive model (Gittleman & Kot 1990).

3. Rates of evolution hypotheses

Using the measure darwins, we test whether different phenotypic traits (behavior, life histories, morphology) evolve at the same rate, behavioral traits evolve at faster rates than morphological/life history traits, or vice versa.

As described above, rates of evolution will be measured in darwins which is essentially a mathematical estimation of rate change. Similar to Haldane's (1949) definition and Gingerich's (1983, 1993) usage, darwins will be calculated as:

$$r = \frac{\ln(x_2) - \ln(x_1)}{\Delta t}$$

where r is the rate of change (in darwins), x_1 is the initial dimension of a character, x_2 is the final dimension of a character, and $\Delta t = t_2 - t_1$ is the amount of time elapsed between the ages of x_1 and x_2. Because rates of evolution change with temporal scale (i.e., evolutionary rate appears inversely proportional to time interval, Gingerich 1983, 1993), darwins will be plotted against time intervals, as measured from phylogenetic distances and calibrated from a molecular clock specific for each taxonomic group. Plots of darwins (y axis) and time intervals (x axis) will produce measures of evolutionary rate across taxa that can be statistically represented by intercepts and slopes (Gingerich 1983, 1993). These statistical measures reveal different kinds of rate information. Observed intercepts represent an estimate of instantaneous rate among traits; slopes (on a log-log plot) reflect how rapidly net evolutionary rates decay with time interval and whether rates of trait evolution occur in a random walk, directional, or stabilizing (stasis) manner. We are aware of the problem that statistical intercepts and slopes for representing rate change are potentially influenced by different units of measurement among variables (Gingerich 1993). For example, rates of change for areas and volumes cannot be compared directly with rates for linear measurements. To somewhat account for dimensionality, we display rates as a function of time interval on a (common) log-log plot. Because time calibration for molecular phylogenies can only be made within each group of independently-derived organisms (Hillis & Moritz 1990), patterns of rate are detected for each taxon and then compared.

Measurement of evolutionary rate in darwins may well be a biased estimate of rate due to statistical nonindependence in the data and lack of an explicit null model of phenotypic evolution. Following Lynch's (1990) metric for measuring evolutionary rate under neutral selection, Martins (1994) proposes a statistical approach for estimating evolutionary rate under neutral evolution or stabilizing selection while also controlling for statistical nonindependence of trait values. Future work will apply this method for estimating evolutionary rate.

C. The data

1. Molecular phylogenies of mammals

The limiting step in this study is clearly the availability of phylogenies. We found 97 molecular phylogenies of the Mammalia published since 1978. Of these, only 8 were usable because of the following reasons (Table 6.1): inappropriate genetic samples for present purposes (total 31: 12 included albumin/immunological samples; 19 were based on karyotypic G and C bands) because these techniques are not sensitive

Table 6.1. Molecular phylogenies of mammals used in comparative analyses

Taxon	Number of Species	Genetic data	Method[a]	Source
Primates	26	103 Beta-type sequences	P	Goodman et al. (1982)
Ceboidea	19	27 e-globin seqs.	P	Schneider et al. (1993)
Carnivora	39	DNA hybridization	D	Wayne et al. (1989)
Canidae	28	DNA hybridization	D	Wayne et al. (in prep.)
Bovidae	27	40 allozyme loci	D	Georgiadis et al. (1990)
Cervidae	11	mtDNA restriction sites	P	Cronin (1991)
Cetacea	13	cytochrome b (mtDNA)	P	Arnason and Gullberg (1994)
Arvicolinae	15	protein electrophoresis	D	Chaline and Graf (1988)

[a]Method used to infer phylogenetic relationships: P = parsimony; D = distance.

enough to detect changes over relatively short time scales, as in many of the examined taxa, as well as the phenetic nature of these types of studies; no time scale and/or divergence information (12), as necessary for analyzing evolutionary lability and rates; restricted to higher taxa (i.e., families and above), thus excluding species level study; of those phylogenies at the species level, 17 were of prohibitive sample sizes (<10 spp.) and 16 included species for which little, if any, behavioral, life history or morphological information is available.

There is no consensus as to which methods are best for constructing phylogenies from molecular data (Felsenstein 1988; Moritz & Hillis 1990; Hillis et al. 1994). Precedence was given to phylogenetic estimates based on DNA sequences rather than protein-based methods because of their inherently greater accuracy (Moritz & Hillis 1990). The selected molecular phylogenies are generally appropriate for the present study because phylogenetic inferences for mammals involve ranges of divergence times between 0 to 5 and 5 to 50 million years ago (Hillis & Moritz 1990). Nevertheless, given the disparate nature of the genetic material and reconstruction methods used in the phylogenies of this study, as indeed in most molecular phylogenies (Felsenstein 1988), we anticipate error is introduced when comparing differential trait evolution from different kinds of molecular information.

None of the published sources for the 8 identified phylogenies presented the original distance matrices for generating the phylogenies. Thus, to estimate distances and points of divergence, we measured with a standard ruler all pairwise distances between species in a given phylogeny.

2. Species traits

For species represented in the molecular phylogenies, trait values were calculated from published literature for the following — brain weight: average brain weight (g) of adult male and adult female calculated from volumetric measures of skull braincases; body weight: average body weight (kg or g in the case of Arvicolinae) of adult male and adult female; gestation length: average time from conception to birth (days), minus any possible period of delayed implantation (e.g., some mustelids and ursids); birth weight: average weight (g) of young at birth; home range size: average total area (km^2) used by an individual (or group in social species) during normal activities; group size: average number of individuals which regularly associate together and share a home range.

All analyses were performed on logarithmically transformed data because cross-species values of the examined trait variables approximate a lognormal distribution; logarithmic transformations are necessary to reduce skew. For the rates analysis, standardization procedures followed Gingerich (1983, 1993) in which logarithmic trait values were used to calculate darwins and then darwins along with time intervals were similarly transformed; this somewhat minimizes differences in measurement of the traits and in time scales among taxa.

The original data files, including both the molecular distance matrices and the species trait values, are available from the authors.

Results and Discussion

In the following, sections on empirical results and discussion are combined for each problem. We do this for brevity but also so we can more directly address the questions posed at the outset. Throughout, we emphasize comparisons of traits rather than comparisons among taxa. This reflects both the nature of the questions asked as well as a limitation in using, at present, molecular data: molecular phylogenies across independent monophyletic groups cannot be collated because of differences in rates of molecular evolution, types of characters and methods of analysis, and the complexity in inferring homology in molecular systematics (Moritz & Hillis 1990; Hillis 1994).

A. Phylogenetic correlation

The coefficients of phylogenetic determination for each trait across taxa are given in Table 2. All traits are correlated with phylogeny at the 0.05 or 0.01 level of significance except: group size (across all taxa), all traits in cervids, home range size in cetaceans, and birth weight in Arvicolinae. Lack of significance across the cervids is likely due to small sample sizes; across arvicolinae, as observed with most traits, birth weight reveals too little variation to detect any correlation. There is also a generally positive, but not significant ($r = 0.671$, $n = 8$; $p < 0.07$) relationship between number of terminal taxa (species) and average level of phylogenetic correlation within each monophyletic group.

Table 6.2. Proportion of total variance (R^2) accounted for by phylogeny among traits for different mammal taxa using a spatial autoregressive model

Taxon (No. Species)	Brain size	Body size	Gestation length	Birth weight	Home range size	Group size
Primates (26)	0.76	0.61	0.28	0.64	0.36	0.01[a]
Ceboidea (19)	0.66	0.66	0.49	0.48	0.60	0.14[a]
Carnivora (39)	0.72	0.76	0.53	0.51	0.53	0.01[a]
Canidae (28)	0.33	0.54	0.39	0.21	0.31	0.05[a]
Bovidae (27)	0.20	0.23	0.31	0.34	0.47	0.12[a]
Arvicolinae (15)	0.41	0.27	0.21	0.19[a]	0.29	0.10[a]
Cervidae (11)	0.13[a]	0.17[a]	0.22[a]	0.16[a]	0.17[a]	0.20[a]
Cetacea (13)	0.48	0.27	0.66	0.31	0.05[a]	0.01[a]

[a]Not significant; all remaining values are significant at 0.05 or 0.01 levels.

The proposed hypothesis in testing phylogenetic correlation is that all traits, behavioral, morphological, and life history, will relate to phylogenetic hierarchy. As expected, though, the behavioral trait of group size showed no relationship with phylogenies in any taxon. The observed correlations convey four general points: (a) Despite the fact that the molecular phylogenies are based on different genetic material and include disparate methods for analyzing this material, the trait variables of morphology, life history and most home range values are related to phylogeny. Generally, this gives some validity to the phylogenies and verifies an overall fit between trait variation and molecular systematic data. (b) Even though it was hypothesized, it is unexpected that home range size would significantly follow phyletic lines to a similar degree as morphology, especially given that home range size is quite malleable with ecological conditions (Gittleman & Harvey 1982; Gompper & Gittleman 1991). This could simply relate to the fact that home range size is, like many life histories, an allometric variable (Harvey & Pagel 1991); thus, certain traits may relate to phylogeny because of size-related effects. Allometric effects were not examined because we are mainly interested in relative phylogenetic pattern in single traits. Interestingly, failure to find correlation with home range size in the cetaceans may have a functional basis: whales essentially migrate annually rather than traverse a regular daily movement pattern. (c) Even though some taxa show considerable variation in group size, phylogenetic relations are not observed. For example, group size in carnivores ranges from 1 to 55 and in ceboid monkeys from 2 to 45. This result has a couple of implications. One is that there is now some evidence that this behavioral trait is more flexible phylogenetically than the other morphological and life history traits; this will be examined more directly in the next section. Two, comparative studies analyzing behavioral traits should not presume phylogenetic relations. Unless empirical study confirms phylogenetic correlation, data transformation with a comparative method should be adopted with caution and justified for reasons other than the assumption that sister taxa are statistically nonindependent. (d) Many of the highest correlations are across taxa with older origins of divergence including the primates (60 million years ago), ceboids (35 million years ago) and carnivores (55 million years ago). This phenomenon may relate to longer time scales increasing the quality of phylogenies, the resolution in detecting divergence patterns and/or stable selection of traits. Simulation studies (e.g., bootstrapping) are necessary to examine these possibilities with the phylogenies used here.

Measuring phylogenetic correlation among different types of traits parallels de Queiroz and Wimberger's (1993; Chap. 7 in this volume) analysis of comparing consistency indexes (*CI*s) between morphological and behavioral characters. (Interestingly, in terms of methodology, the phylogenetic correlation coefficient may serve as a useful analog (Sanderson & Donoghue 1989) of the *CI*. Even though *CI*s are used for discrete data and phylogenetic correlation for quantitative data, both are statistical measures of observed trait change divided by overall possible change throughout the tree; that is, each assesses the general relationship between phylogenetic tree topology and trait change.) In essence, de Queiroz and Wimberger found that, across 22 data sets for a wide diversity of taxa, the *CI*s for behavioral and morphological characters were not significantly different. Therefore, in assessing the usefulness of behavioral characters for estimation of phylogeny, the assumption of greater evolutionary lability of behavior than morphology is questioned. The overall motivation for this study was to question differential trait evolution and intuitive assumptions about how different traits evolve. Our study, however, differs in a few respects. First, we used phylogenies based on molecular information independent of the traits in question, whereas de Queiroz and Wimberger considered only behavioral and morphological data. Second, branch length information was incorporated into calculating phylogenetic correlation between the tree topology and trait change whereas their study used strictly cladistic techniques and set branch lengths equal. Last, trait evolution in our work is measured across broader phylogenetic distances and within different regions of the tree, in contrast to de Queiroz and Wimberger's mostly species and genus level analysis. Similar differences exist between our study and that of Irwin (Chap. 8 in this volume).

In addition to these methodological differences, our overall conclusions appear at odds because we found little, if any, phylogenetic pattern in the behavioral trait of group size whereas de Queiroz and Wimberger showed considerable phylogenetic relationships in many behavioral characters. Such different conclusions are, in actuality, complementary: using different phylogenetic information, traits, and comparative methods, both studies question deeply ingrained assumptions about behavioral versus morphological evolution. That is, patterns of trait evolution should be independently examined for various traits irrespective of intuitive biases about behavioral evolution being extremely labile or more homoplasic. All behavior patterns will not relate to phylogeny. The point is that as other traits such as morphology and life history are investigated in terms of phylogenetic pattern, so should behavioral traits. Cladistic type analyses of discrete traits have successfully carried out such study for a wide variety of evolutionary problems. As shown here, quantitative comparative studies should do likewise.

B. Evolutionary lability

To examine hypotheses of evolutionary lability, general trends will be discussed from phylogenetic correlograms in Fig. 2*a* and *b*. Immediate caution should be given from the fact that scale is not constant across correlograms: observed *z*-scores and time intervals among taxa are not drawn on the same scale so that actual comparative patterns are more easily observed. Moreover, comparisons of the correlograms among taxa are qualitative and do not necessarily indicate that observed differences are statistically significant.

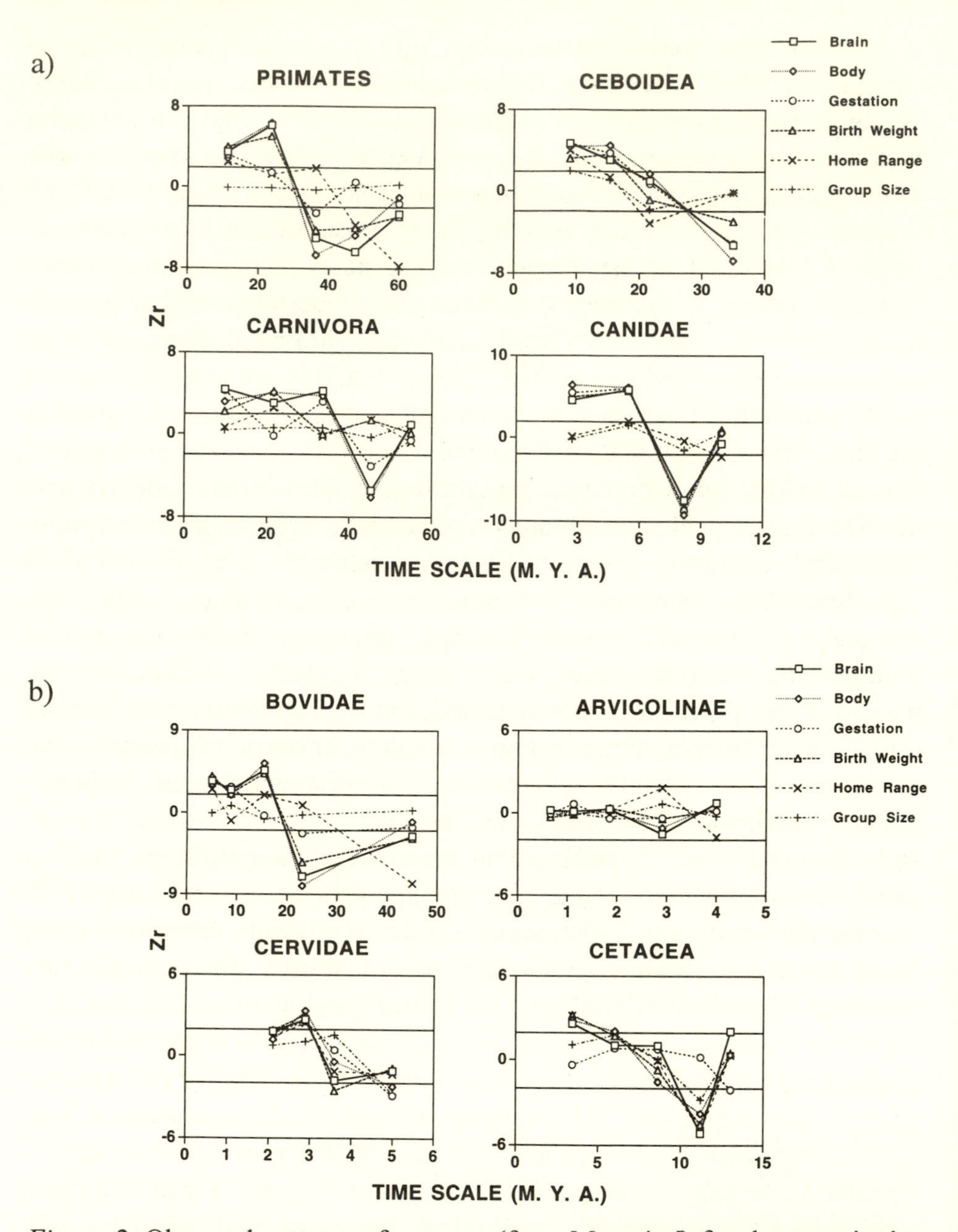

Figure 2. Observed patterns of *z* scores (from Moran's *I*) for the quantitative traits of: brain weight, body weight, gestation length, birth weight, home range size, and group size (see text for definitions). Phylogenetic patterns of each trait are shown across various mammal taxa including: (a) Primates, Carnivora, Ceboidea, Canidae, and (b) Bovidae, Cervidae, Arvicolinae, and Cetacea. Intervals of time scale and *z* scores are not standardized in order to accurately measure relative correlation among traits for each taxon.

Before discussing our results, we want to recognize two precursors that have addressed similar problems of phylogenetic pattern. Plotkin and Odling-Smee (1979) presented a conceptual model for analyzing how animal learning adjusts to "the limits of phylogenesis versus unlimited environmental change" (p. 21). Following Thoday (1964), the authors suggested that generation time of populations of different species produce resultant types of phenotypic change (i.e., in this case, learning ability) that reflect directional, singular (stepwise) or periodic (piecewise) patterns (Fig. 2 of their paper). In our study, we use phylogenetic time periods and speciation events rather than generational time. The types of change we investigated are, in essence, identical to Plotkin and Odling-Smee's except that we provide a statistical dimension for actually measuring and testing the degree of significance in (phylogenetic) change and pattern. In a more empirically based paper, Cherry et al. (1982) carried out a comparative analysis in which morphological differences (overall variability) among species were shown to be greater in more distantly related taxa; this observation parallels our null hypothesis. Nevertheless, our analysis differs from that of Cherry et al. because we considered traits other than morphological ones, used phylogenetic branch length information rather than taxonomic ranks, and directly measured correlation between phylogeny and trait variation rather than variability itself.

The main focus of this section is to ask whether different types of traits change phylogenetically at points in time in different ways. The simple answer is yes. As expected, behavioral traits, especially group size, show no relationship with phylogeny at any phylogenetic time interval. In contrast to searching for overall phylogenetic correlation, as in the previous section, this verifies that there is little if any behavioral change with cladogenesis. This is an important finding for two reasons. First, even in relatively old taxonomic groups (bovids; carnivores) there is little evidence of phylogenetic relations at different points (nodes) of speciation, at least in the variables examined. Second, a trait like group size which has important ecological functions and relations with morphological traits, may reveal an evolutionary pattern quite independent of other traits even across long (phylogenetic) time scales.

In terms of characterizing trait change in relation to our hypothesized patterns (Fig. 1), all of the traits generally follow a directional or piecewise pattern essentially revealing an autoregressive model. Closely related taxa, especially at the species level, reveal positive and significant correlation with a precipitous drop-off at greater phylogenetic distances. This general pattern emerges independent of different types of traits, molecular phylogenies, time scale and, except for Arvicolinae, sample size.

Interesting variation occurs, however, in relative correlation (*z* scores) among traits and in considering the time scale in which significant negative correlation occurs. The observed correlations for life history traits are generally lower than for morphological traits; this holds at most time intervals across the bovids, cetaceans, primates, carnivores, ceboids and canids (Fig. 2*a* and *b*). Such a result is consistent with the idea that life history traits are relatively more flexible than many morphological traits (Gittleman 1993). Following significant positive correlation at more recent time periods, significant negative correlation is observed at later times, specifically points of time occurring at two (cervids, primates, canids) or three (ceboids, carnivores, bovids, cetaceans) speciation events. Although this result could simply reflect tree topology, a possibility that warrants further study, this pattern may suggest some parallelism at lower taxonomic levels due to genetic and developmental similarities and, as consequence, similar phenotypic responses to selection. Sanderson and Donoghue (1989) argue differently, showing that consistency indexes for a wide variety of characters across plants and animals are not affected by taxonomic rank. Although it is difficult to draw direct comparisons between these studies because of different comparative methods, taxa, and characters, there is the possibility that the lack of change in the *CI*s in Sanderson and Donoghue's study relates to sample size; in one illustration, there appears to be a decline in *CI*s from lower to higher taxa and then an increase in *CI*s at the highest rank, but this is perhaps not statistically significant because of small samples.

Last, in relation to patterns of phylogenetic correlation, it is observed that in all cases but two, correlations generally converge around a value of 0 at the initial (basal) ancestor; this is due to there being little observed variation, positive or negative, at the split from the initial ancestor at the oldest time interval. The interesting exceptions are in most traits across the Ceboidea and home range size across the Bovidae. In the ceboids the phylogenetic tree is quite unbalanced between two major clades: one group, including the subfamily

Callitrichinae and the genera *Cebus, Saimiri* and *Aotus*, has different distance intervals and topologies than another group including the tribe Pitheciini, the subfamily Atelinae and the genus *Callicebus* (Schneider et al. 1993). In the bovids, there is a significant negative correlation with home range size. This may be because the oldest phylogenetic point appears to relate to the combination of a comparatively large value of home range size with the (presumed, or actually an outgroup) ancestral species of the bovids, the giraffe. Both of these patterns underscore the need to investigate the effects of tree topology on observed phylogenetic correlation, both throughout the entire tree and at different time intervals.

C. Evolutionary rate

We estimate evolutionary rate, as measured in plots of darwins against time intervals, among traits for each taxonomic group examined. Results are discussed in terms of "instantaneous rates" of evolution (i.e., observed intercepts) and the directedness of evolution (i.e., observed slopes). We insert a large dose of caution when interpreting the results given here because of potential problems in different measures of evolutionary rate, possible nonindependence of the data, lack of an expected (null) evolutionary model for trait change, and, perhaps most importantly, the literal usage of divergence time in the molecular phylogenies. Yet, we are encouraged by these results because they show that macroevolutionary measures of evolutionary rate (Gingerich 1983, 1993) can be useful for comparing rates of behavioral evolution with other traits across broad, independently evolved, taxa.

Initially, to get an overall impression of the kind of patterns that emerge, we present a plot of evolutionary rate on time across the primates for one morphological trait (body weight, Fig. 3*a*) and one behavioral trait (group size, Fig. 3*b*). Three patterns are evident in this one example. First, both variables show the expected inverse relationship between evolutionary rate and time scale, though the values of slope are shallow. Second, slope is more negative in group size than in body weight, which suggests that evolutionary change may reflect more of a random walk in the behavioral trait and (relatively) more directional evolution in the morphological trait (Gingerich 1993). Third, "instantaneous" rate is greater in group size than body weight, which indicates that evolutionary rate (over a million years) is greater in the behavioral variable. Due to problems of sample size and standardizing time scale, it is difficult to directly compare these results with studies of other taxonomic groups and traits or even to make taxonomic comparisons within this study. Nevertheless, it should be mentioned that all of the values of slope observed in this study are within a standard deviation of a value of slope -0.50 which is expected for a random walk model of evolutionary rate (Gingerich 1993).

It is instructive, however, to compare observed slopes and intercepts of rate change against time intervals while controlling for time scale within each taxonomic group. Comparisons of net evolutionary change can be made by plotting values of slope for each trait across taxa (Fig. 4). Observed slopes among traits are significantly different ($F_{4,70}$= 6.91, p <0.0001), with values for the two behavioral variables (home range size and group size) steeper than the morphological and life history variables. This generally indicates that, within taxa, net rate of evolution in the behavioral traits may be different, as hypothesized, and may change more randomly over time than the other traits. The slopes of the morphological and life history traits are all negative; however, in comparison to the behavioral traits, it is interesting that they approximate more of a directional evolutionary process, as expressed by a horizontal distribution of average slope zero (0.0) on a rate-versus-interval plot (Gingerich 1993).

Similarly, observed intercepts among traits are significantly different ($F_{4,70}$= 2.29, p < 0.05), with intercepts for the two behavioral traits consistently higher than for the other traits (Fig. 5). This reflects that evolutionary rate for short intervals of time are higher in the behavioral traits, as expected. As with the slopes, the intercepts for the two behavioral traits generally cluster together and the four other traits cluster together with the exception of gestation length. Gestation has an unusually low intercept relative to the other traits, which suggests a slower evolutionary rate. This may lie with the evolution of gestation length changing with female size (allometry) whereas selection will directly change the other traits. Similar arguments have been suggested for various life history traits which are composites of other morphological traits (Riska 1989).

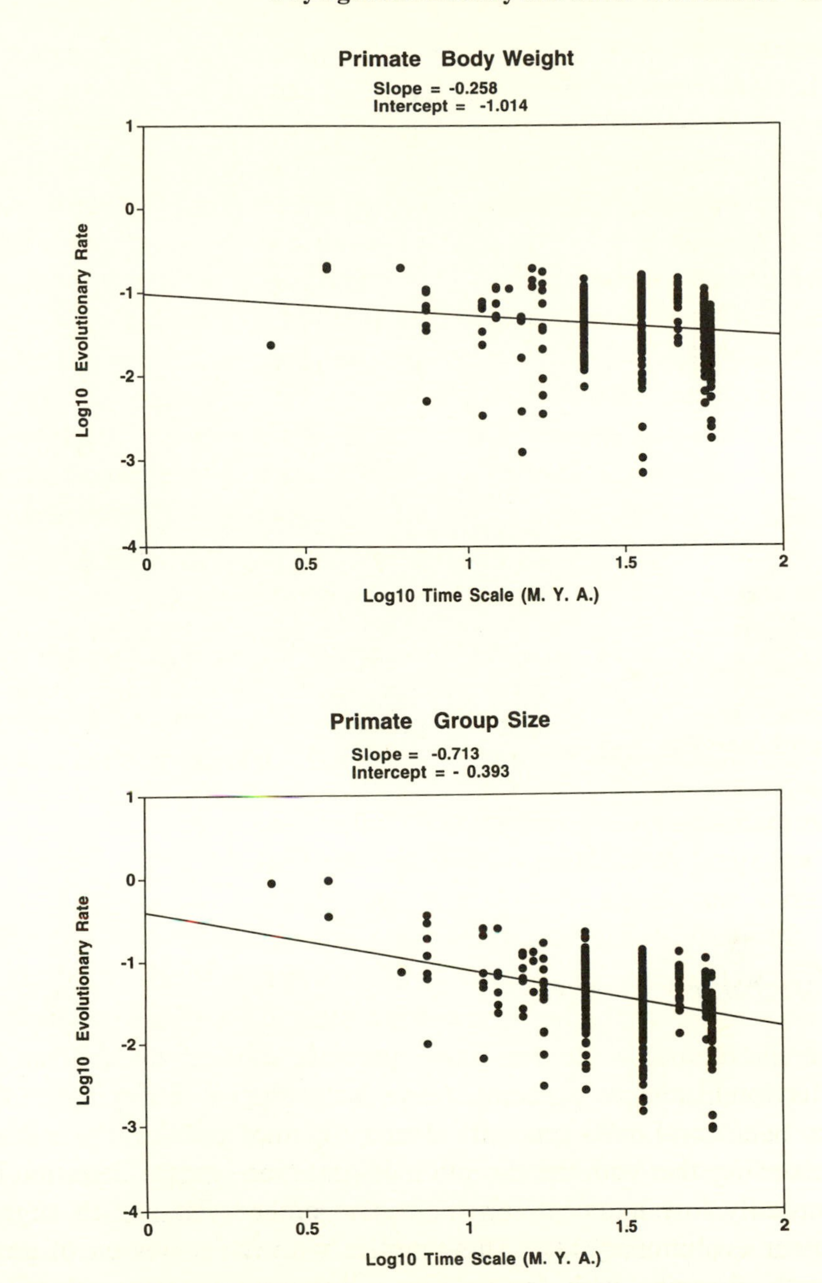

Figure 3. Log evolutionary rate (darwins) versus log time interval (million years ago) observed across Primates for (a) body weight and (b) group size. Values of slope represent directedness of evolution and values of intercept represent rates over a million years of evolution.

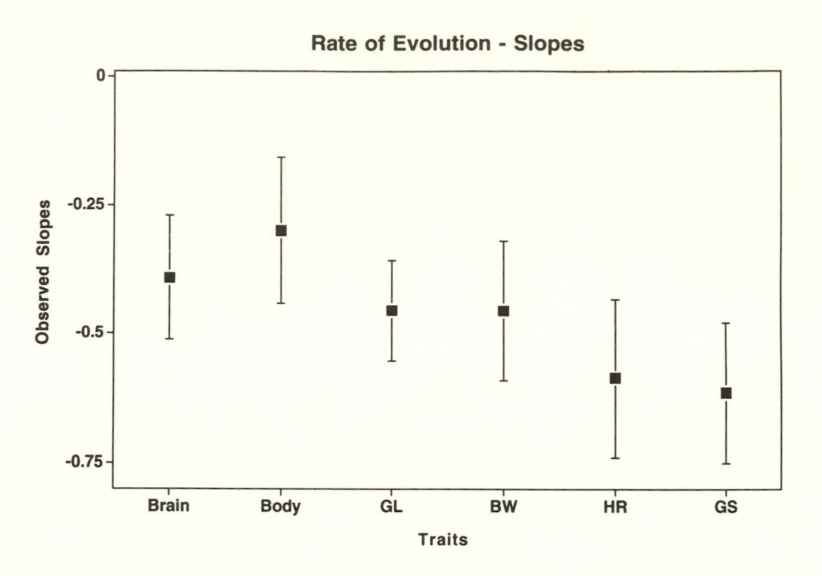

Figure 4. Combined values of slope across all mammal taxa observed for each quantitative trait. Traits are denoted by the following: Brain- brain weight; Body- body weight; GL- gestation length; BW- birth weight; HR- home range size; GS- group size. Means and standard errors are included for each trait.

Concluding Remarks

We end with some general thoughts about applications and future uses of molecular data in comparative biology: (a) Unifying molecular phylogenies and comparative data is currently frustrating. We were startled by the fact that, despite mammals receiving a disproportionate amount of attention in phylogenetic studies relative to other taxa (Sanderson et al. 1993), so few molecular phylogenies were usable for this comparative study. A primary reason for this is that most molecular phylogenies of the mammals are at higher levels or within small clades (Novacek 1992; Honeycutt & Adkins 1993); a prime example is the most intensively studied group, the Primates, in which molecular study mainly comprises familial groups (Miyamoto & Goodman 1990) or extant hominids (*Gorilla, Homo, Pan*: Patterson et al. 1993). There is evidently a clear lack of phylogenies for taxa which are the focus of study in animal behavior, ecology, and general natural history and this omission will continue to impede further work in analyzing the phylogeny and lability of trait evolution; as Winterbottom & McLennan (1993) wrote, "... interest in comparative biology has outstripped the production of robust trees, leaving many ecologists and ethologists understandably frustrated" (p. 1569). (b) The patterns detected in this study are clearly determined by the quality of the molecular phylogenies; quite literally, "garbage in, garbage out"! Although we assume the phylogenies to be accurate, we realize that tree topologies are significantly affected by the rate of genomic evolution which in turn varies among genes and among taxa. More careful analysis is required to evaluate the quality of the molecular information used in our analysis. As others have noted, it is unlikely that one gene will elucidate phylogenetic relationships across taxa (Hillis & Moritz 1990; Graybeal 1994). Thus, for comparative purposes, future work will have to match specific genes with particular types of divergences such as with long time frames or speciose regions of a tree. (c) The similar autoregressive relationships of the morphological and life history traits as well as the markedly different patterns between behavior and the other traits begs the question of what are the driving forces among these relationships. Critical to answering this question is a desperate need for multivariate comparative methods. We are presently working on multivariate techniques using autocorrelation (Luh, in prep).

Last, we would like to make a brief comment about how this study fits into the current state of comparative methodology. We deliberately have not touched on causal relations in the observed patterns of phylogenetic correlation, plasticity and rates. This is primarily because our goal was to redirect some attention away from hypothesis-testing, especially of adaptation, and toward more descriptive problems with quantitative comparative methods similar to those successfully carried out with cladistic comparative methods. Secondarily, we wanted to emphasize the value and application of using molecular phylogenetic information with quantitative comparative methods. Our intentions in no way should be interpreted as finding fundamental flaws in hypothesis testing with comparative study, as recently leveled by Leroi et al. (1994). Vast empirical work has, without question, shown that comparative study is successful in describing evolutionary pattern and

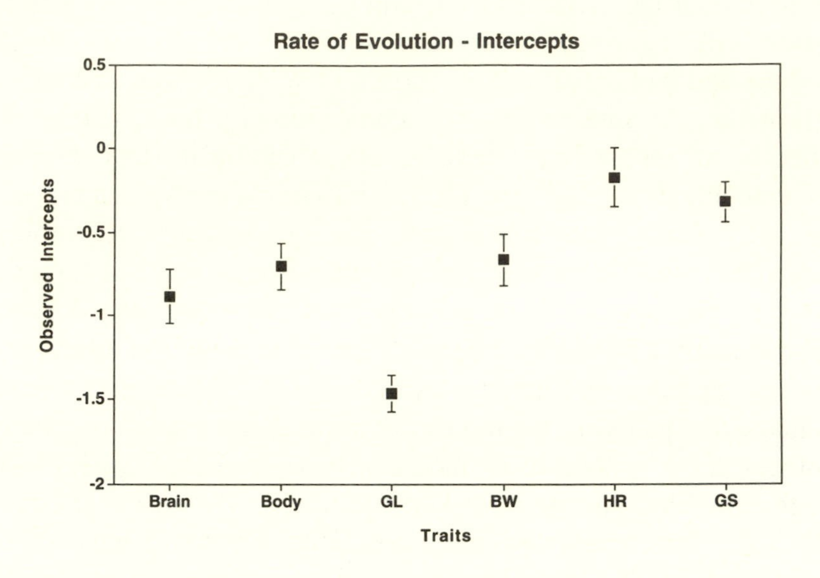

Figure 5. Combined values of intercept across all mammal taxa observed for each quantitative trait. Traits are denoted as in Figure 4. Means and standard errors are included for each trait.

adaptive hypothesis testing. Any limitations or flaws in the comparative method are, in the end, with the data at hand. The comparative method cannot be viewed as a black box that will turn inadequate or bad data into good science. On the other hand, to rephrase Harvey and Pagel's (1991) concluding line in their book on the comparative method in evolutionary biology, "Tell us how to reconstruct the past *and give us complete information on genetic variation and trait values*, and we shall perform the comparative analysis with precision."

Acknowledgments

We are most grateful to Robert Wayne for discussion on the availability and use of molecular phylogenies and Trevor Price, Harry Greene, Gordon Burghardt, two anonymous reviewers, and especially Emília Martins for helpful comments on the manuscript. Financial support was received from the recently deceased Department of Zoology and

Science Alliance of the University of Tennessee (JLG, CGA) and the National Science Foundation (H-KL).

References

Arnason, U., & A. Gullberg. 1994. Relationship of baleen whales established by cytochrome *b* gene sequence comparison. *Nature*, 367, 726–728.

Arnold, S. J. 1992. Constraints on phenotypic evolution. *Am. Nat.,* 140, S85–S107.

Basolo, A. L. 1990. Female preference predates the evolution of the sword in swordtail fish. *Science*, 250, 808–810.

Bateson, P. 1988. The active role of behaviour in evolution. In: *Evolutionary Processes and Metaphors* (Ed. by M.-W. Ho & S. W. Fox), pp. 191–207. London: Wiley & Sons.

Bateson, W. 1894. *Materials for the Study of Variation: Treated with special regard to Discontinuity in the Origin of Species.* London: Macmillan.

Bennett, P. M., & P. H. Harvey. 1985a. Relative brain size and ecology in birds. *J. Zool.,* 207, 151–169.

Bennett, P. M., & P. H. Harvey. 1985b. Brain size, development and metabolism in birds and mammals. *J. Zool.,* 207, 491–509.

Berger, J. 1988. Social systems, resources, and phylogenetic inertia: An experimental test and its limitations. In: *The Ecology of Social Behavior* (Ed by C. N. Slobodchikoff), pp. 157–186. New York: Academic Press.

Berrigan, D., E. L. Charnov, A. Purvis, & P. H. Harvey. 1993. Phylogenetic contrasts and the evolution of mammalian life histories. *Evol. Ecol.,* 7, 270–278.

Bradshaw, A. D. 1965. Evolutionary significance of phenotypic plasticity in plants. *Adv. Genet.*, 13, 115–155.

Brooks, D. R., & D. A. McLennan. 1991. *Phylogeny, Ecology, and Behavior.* Chicago: University of Chicago Press.

Brown, J. L. 1975. *The Evolution of Behavior*. New York: W. W. Norton.

Burghardt, G. M., & J. L. Gittleman. 1990. Comparative behavior and phylogenetic analyses: New wine, old bottles. In: *Interpretation and Explanation in the Study of Animal Behavior: Comparative Perspectives* (Ed. by M. Bekoff, & D. Jamieson), pp. 195–225. Boulder, Colorado: Westview.

Chaline, J., & J.-D. Graf. 1988. Phylogeny of the Arvicolidae (Rodentia): Biochemical and paleontological evidence. *J. Mamm.*, 69, 22–33.

Cherry, L. M., S. M. Case, J. G. Junkel, J. S. Wyles, & A. C. Wilson. 1982. Body shape metrics and organismal evolution. *Evolution,* 36, 914–933.

Cheverud, J. M., M. M. Dow, & W. Leutenegger. 1985. The quantitative assessment of phylogenetic constraints in comparative analyses: Sexual dimorphism in body weight among primates. *Evolution,* 39, 1335–1351.

Cliff, A. D., & Ord, J. K. 1973. *Spatial Autocorrelation.* London: Pion.

Cliff, A. D., & Ord, J. K. 1981. *Spatial Processes: Models and applications.* London: Pion.

Clutton-Brock, T. H., & P. H. Harvey. 1977. Primate ecology and social organization. *J. Zool.* 183, 1–39.

Coddington, J. A. 1988. Cladistic tests of adaptational hypotheses. *Cladistics*, 4, 3–22.
Colbert, E. H. 1958. Morphology and behavior. In: *Behavior and Evolution* (Ed. by A. Roe, & G. G. Simpson), pp. 27–47. New Haven, Connecticut: Yale University Press.
Cracraft, J. 1984. Conceptual and methodological aspects of the study of evolutionary rates, with some comments on bradytely in birds. In: *Living Fossils* (Ed. by N. Eldredge, & S. Stanley), pp. 95–104. New York: Springer-Verlag.
Cronin, M. A. 1991. Mitochondrial-DNA phylogeny of deer (Cervidae). *J. Mamm.*, 72, 533–566
Darwin, C. R. 1859. *The Origin of Species*. London: John Murray.
de Queiroz, A., & P. H. Wimberger. 1993. The usefulness of behavior for phylogeny estimation: Levels of homoplasy in behavioral and morphological characters. *Evolution,* 47, 46–60.
Donoghue, M. J., & M. J. Sanderson. 1992. The suitability of molecular and morphological evidence in reconstructing plant phylogeny. In: *Molecular Systematics of Plants* (Ed. by P. S. Soltis, D. E. Soltis, & J. J. Doyle), pp. 340–368. New York: Chapman & Hall.
Edwards, S. V., & S. Naeem. 1993. The phylogenetic component of cooperative breeding in perching birds. *Am. Nat.,* 141, 754–789.
Eldredge, N., & S. J. Gould. 1972. Punctuated equilibria: An alternative to phyletic gradualism. In: *Models in Paleobiology* (Ed. by T. J. M. Schopf), pp. 82–115. San Francisco: Freeman, Cooper & Co.
Elton, C. 1927. *Animal Ecology*. London: Macmillan.
Endler, J. A. 1986. *Natural Selection in the Wild.* Princeton: Princeton University Press.
Endler, J. A., & T. McLellan. 1988. The processes of evolution: toward a newer synthesis. *Ann. Rev. Ecol. Syst.,* 19, 395–421.
Felsenstein, J. 1985. Phylogenies and the comparative method. *Am. Nat.,* 125, 1–15.
Felsenstein, J. 1988. Phylogenies from molecular sequences: inference and reliability. *Annu. Rev. Genet.,* 22, 521–565.
Fenster, E. J., & U. Sorhannus. 1991. On the measurement of morphological rates of evolution. A review. *Evol. Biol.,* 25, 375–410.
Gause, G. F. 1947. Problems of evolution. *Trans. Conn. Acad. Sci.,* 37, 17–68.
Georgiadis, N. J., P. W. Kat, H. Oketch, & J. Patton. (1990). Allozyme divergence within the Bovidae. *Evolution*, 44, 2135–2149.
Gingerich, P. D. 1983. Rates of evolution: Effects of time and temporal scaling. *Science,* 222, 159–161.
Gingerich, P. D. 1993. Quantification and comparison of evolutionary rates. *Am. J. Sci.,* 293-A, 453–478.
Gittleman, J. L. 1991. Carnivore olfactory bulb size: Allometry, phylogeny, and ecology. *J. Zool.,* 225, 253–272.
Gittleman, J. L. 1993. Carnivore life histories: A re-analysis in the light of new models. In: *Mammals as Predators* (Ed. by N. Dunstone, & M. L. Gorman), pp. 65–86. Oxford, England: Oxford University Press.
Gittleman, J. L. 1994. Female brain size and parental care in carnivores. *Proc. Natl. Acad. Sci. USA,* 91, 5495–5497.

Gittleman, J. L., & D. Decker. 1994. The phylogeny of behaviour. In: *Behaviour and Evolution* (Ed. by P. J. B. Slater, & T. R. Halliday), pp. 80–105. Cambridge, England: Cambridge University Press.

Gittleman, J. L., & P. H. Harvey. 1982. Carnivore home-range size, metabolic needs and ecology. *Behav. Ecol. Sociobiol.*, 10, 57–63.

Gittleman, J. L., & M. Kot. 1990. Adaptation: Statistics and a null model for estimating phylogenetic effects. *Syst. Zool.*, 39, 227–241.

Gittleman, J. L., & H.-K. Luh. 1992. On comparing comparative methods. *Ann. Rev. Ecol. Syst.*, 23, 383–404.

Gittleman, J. L., & H.-K. Luh. 1994. Phylogeny, evolutionary models, and comparative methods: A simulation study. In: *Phylogenetics and Ecology* (Ed. by P. Eggleton, & D. Vane-Wright), pp. 103–122. London: Academic Press.

Gomendio, M., & E. R. S. Roldan. 1993. Coevolution between male ejaculates and female reproductive biology in eutherian mammals. *Proc. R. Soc. Lond.*, 252, 7–12.

Gompper, M. E., & J. L. Gittleman. 1991. Home range scaling: intraspecific and comparative trends. *Oecologia*, 87, 343–348.

Goodman, M., A. E. Romero-Herrara, H. Dene, J. Czelusniak, & R. E. Tashian. 1982. Amino acid sequence evidence on the phylogeny of primates and other eutherians. In: *Macromolecular Sequence in Systematic and Evolutionary Biology*. (Ed. by M. Goodman), pp. 115–191. New York: Plenum Press.

Gould, S. J. 1993. Fulfilling the spandrels of world and mind. In: *Understanding Scientific Prose* (Ed. by J. Selzer), pp. 310–336. Madison, Wisconsin: University of Wisconsin Press.

Gould, S. J., & R. C. Lewontin. 1979. The spandrels of San Marco and the Panglossian paradigm: A critique of the adaptationist program. *Proc. R. Soc. Lond.*, 205B, 581–598.

Grafen, A. 1989. The phylogenetic regression. *Phil. Trans. R. Soc. Lond.*, 326,119–157.

Graybeal, A. 1994. Evaluating the phylogenetic utility of genes: A search for genes informative about deep divergences among vertebrates. *Syst. Biol.*, 43, 174–193.

Greene, H. W. 1994. Homology and behavioral repertoires. In: *Homology: The Hierarchical Basis of Comparative Biology* (Ed. by B. K. Hall), pp. 369–391. San Diego, California: Academic Press.

Greene, H. W., & G. M. Burghardt, G. M. 1978. Behavior and phylogeny: Constriction in ancient and modern snakes. *Science*, 200, 74–77.

Haldane, J. B. S. 1949. Suggestions as to quantitative measurement of rates of evolution. *Evolution*, 3, 51–56.

Hall, B. K., editor. 1994. *Homology: The Hierarchical Basis of Comparative Biology*. San Diego: Academic Press.

Harvey, P. H., & T. H. Clutton-Brock, T. H. 1985. Life history variation in primates. *Evolution*, 39, 559–581.

Harvey, P. H., & M. D. Pagel. 1991. *The Comparative Method in Evolutionary Biology*. Oxford, England: Oxford University Press.

Harvey, P. H., & A. Purvis. 1991. Comparative methods for explaining adaptations. *Nature*, 351, 619–624.

Hecht, M. K. 1965. The role of natural selection and evolutionary rates in the origin of higher levels of organization. *Syst. Zool.*, 14, 301–317.

Hillis, D. M. 1987. Molecular versus morphological approaches to systematics. *Annu. Rev. Ecol. Syst.*, 18, 23–42.

Hillis, D. M. 1994. Homology in molecular biology. In: *Homology: The Hierarchical Basis of Comparative Biology* (Ed. by B. K. Hall), pp. 339-368. San Diego, California: Academic Press.

Hillis, D. M., & C. Moritz. 1990. An overview of applications of molecular systematics. In: *Molecular Systematics* (Ed. by D. M. Hillis & C. Moritz), pp. 502–515. Sunderland, Massachusetts: Sinauer and Associates.

Hillis, D. M., J. P. Helsenbeck, & C. W. Cunningham. 1994. Application and accuracy of molecular phylogenies. *Science* 264, 671–677.

Hinde, R. A. 1959. Behaviour and speciation in birds and lower vertebrates. *Biol. Rev.*, 34, 85–128.

Hinde, R. A., & N. Tinbergen. 1958. The comparative study of species-specific behavior. In: *Behavior and Evolution* (Eds., A. Roe, & G. G. Simpson), pp. 251–268. New Haven, Connecticut: Yale University Press.

Honeycutt, R. L., & R. M. Adkins. 1993. Higher level systematics of eutherian mammals: An assessment of molecular characters and phylogenetic hypotheses. *Annu. Rev. Ecol. Syst.,* 24, 279–305.

Huey, R. B. 1987. Phylogeny, history, and the comparative method. In: *New Directions in Ecological Physiology* (Ed. by M. E. Feder, A. F. Bennett, W. W. Burggren, & R. B. Huey), pp. 76–98. Cambridge, England: Cambridge University Press.

Huey, R. B. 1991. Physiological consequences of habitat selection. *Am. Nat.,* 137, S91–S115.

Huey, R. B., & A. F. Bennett. 1987. Phylogenetic studies of coadaptation: Preferred temperatures versus optimal performance temperatures of lizards. *Evolution,* 41, 1098–1115.

Huxley, J. S. 1942. *Evolution, The Modern Synthesis*. Oxford: Clarendon Press.

James, F. C. 1983. Environmental component of morphological differentiation in birds. *Science,* 221, 184–186.

Krebs, J. R., & N. B. Davies. 1993. *An Introduction to Behavioural Ecology.* Oxford, England: Blackwell.

Lauder, G. V. 1986. Homology, analogy, and the evolution of behavior. In: *Evolution of Animal Behavior* (Ed. by M. H. Nitecki, & J. A. Kitchell), pp. 9–40. New York: Oxford University Press.

Lauder, G. V. 1990. Functional morphology and systematics: Studying functional patterns in an historical context. *Annu. Rev. Ecol. Syst.,* 21, 317–340.

Lauder, G. V. 1994. Homology, form, and function. In: *Homology- The Hierarchical Basis of Comparative Biology* (Ed. by B. K. Hall), pp. 151–196. San Diego, California: Academic Press.

Legendre, P. 1993. Spatial autocorrelation: Trouble or new paradigm? *Ecology*, 74, 1659–1673.

Leroi, A. M., M. R. Rose, & G. V. Lauder. 1994. What does the comparative method reveal about adaptation? *Am. Nat.*, 143, 381–402.

Lorenz, K. 1965. *Evolution and Modification of Behavior*. Chicago: University of Chicago Press.

Lynch, M. 1990. The rate of morphological evolution in mammals from the standpoint of the neutral expectation. *Am. Nat.*, 136, 727–741.

Mace, G. M., P. H. Harvey, & T. H. Clutton-Brock. 1981. Brain size and ecology in small mammals. *J. Zool.*, 193, 333–354.

Maddison, W. P. 1990. A method for testing the correlated evolution of two binary characters: Are gains or losses concentrated on certain branches of a phylogenetic tree? *Evolution,* 44, 539–557.

Maddison, W. P., & D. R. Maddison. 1992. *MacClade: Analysis of Phylogeny and Character Evolution.* Sunderland, Massachusetts: Sinauer and Associates.

Martin, P., & P. Bateson. 1993. *Measuring Behaviour. An Introductory Guide.* 2nd ed. Cambridge, England: Cambridge University Press.

Martins, E. P. 1994. Estimating the rate of phenotypic evolution from comparative data. *Am. Nat.*, 144, 193–209.

Martins, E. P., & T. Garland, Jr. 1991. Phylogenetic analysis of the correlated evolution of continuous characters: A simulation study. *Evolution,* 45, 534–557.

Mayr, E. 1963. *Animal Species and Evolution.* Cambridge, Massachusetts: Harvard University Press.

Mayr, E. 1974. Behavior programs and evolutionary strategies. *Am. Sci.,* 62, 650–659.

McKitrick, M. C. 1993. Phylogenetic constraint in evolutionary theory: Has it any power? *Annu. Rev. Ecol. Syst.,* 24, 307–330.

McLennan, D. A. 1991. Integrating phylogeny and experimental ethology: from pattern to process. *Evolution,* 45, 1773–1789.

Miles, D. B., & A. E. Dunham. 1992. Comparative analyses of phylogenetic effects in the life-history patterns of iguanid reptiles. *Am. Nat.,* 139, 848–869.

Miles, D. B., & A. E. Dunham. 1993. Historical perspectives in ecology and evolutionary biology: the use of phylogenetic comparative analyses. *Annu. Rev. Ecol. Syst.,* 24, 587–619.

Miyamoto, M. M., & M. Goodman. 1990. DNA systematics and evolution of primates. *Annu. Rev. Ecol. Syst.*, 21, 197–220.

Moran, P. A. P. 1950. Notes on continuous stochastic phenomena. *Biometrika*, 37, 17–23.

Morgan, T. H. 1903. *Evolution and Adaptation.* New York: Macmillan.

Moritz, C. & D. M. Hillis. 1990. Molecular systematics: Context and controversies. In: *Molecular Systematics* (Ed. by D. M. Hillis, & C. Moritz), pp. 1–10. Sunderland, Massachusetts: Sinauer and Associates.

Novacek, M. J. 1992. Mammalian phylogeny: Shaking the tree. *Nature*, 356, 121–125.

Patterson, C., D. M. Williams, & C. J. Humphries. 1993. Congruence between molecular and morphological phylogenies. *Ann. Rev. Ecol. Syst.*, 24, 153–188.

Plotkin, H. C. 1988. Behavior and evolution. In: *The Role of Behavior in Evolution* (Ed. by H. C. Plotkin), pp. 1–17. Cambridge, Massachusetts: MIT Press.

Plotkin, H. C., & F. J. Odling-Smee. 1979. Learning, change, and evolution: An enquiry into the teleonomy of learning. *Adv. Study Behav.,* 10, 1–41.

Pontier, D., J.-M. Gaillard, & D. Allaine. 1993. Maternal investment per offspring and demographic tactics in placental mammals. *Oikos*, 66, 424–430.

Promislow, D. E. L. 1991. The evolution of mammalian blood parameters: Patterns and their interpretation. *Phys. Zool.,* 64, 393–431.

Promislow, D. E. L., R. Montgomerie, & T. E. Martin. 1992. Mortality costs of sexual dimorphism in birds. *Phil. Trans. R. Soc. Lond.,* 250, 143–150.

Provine, W. B. 1985. Adaptation and mechanisms of evolution after Darwin: A study in persistent controversies. In: *The Darwinian Heritage* (Ed. by D. Kohn), pp. 825–866. Princeton, New Jersey: Princeton University Press.

Purvis, A., J. L. Gittleman, & H.-K. Luh. 1994. Truth or consequences: Effects of phylogenetic accuracy on two comparative methods. *J. Theor. Biol.,* 167, 293–300.

Read, A. F., & D. M. Weary. 1992. The evolution of bird song: comparative analyses. *Phil. Trans. R. Soc. Lond.,* 338, 165–187.

Riska, B. 1989. Composite traits, selection response, and evolution. *Evolution,* 43, 1172–1191.

Robson, G. C. & O. W. Richards. 1936. *The Variations of Animals and Plants in Nature.* London: Longmans.

Rosenzweig, M. R. 1971. Effects of environment on development of brain and behavior. In: *The Biopsychology of Development* (Ed. by E. Tobach, L. R. Aronson, & E. Shaw), pp. 303–342. New York: Academic Press.

Rosenzweig, M. R., & E. L. Bennett. 1977. Experiential influences on brain anatomy and brain chemistry in rodents. In: *Studies on the Development of Behavior and the Nervous System* (Ed. by G. Gottlieb), pp. 289–327. New York: Academic Press.

Roth, V. L. 1988. On the biological basis of homology. In: *Ontogeny and Systematics* (Ed. by C. J. Humphries), pp. 1–26. New York: Columbia University Press.

Sanderson, M. J., & M. J. Donoghue. 1989. Patterns of variation in levels of homoplasy. *Evolution,* 43, 1781–1795.

Sanderson, M. J., B. G. Baldwin, G. Bharathan, C. S. Campbell, C. von Dohlen, D. Ferguson, J. M. Porter, M. F. Wojciechowski, & M. J. Donoghue. 1993. The growth of phylogenetic information and the need for a phylogenetic data base. *Syst. Biol.,* 42, 562–568.

Scheiner, S. M. 1993. Genetics and evolution of phenotypic plasticity. *Annu. Rev. Ecol. Syst.,* 24, 35–68.

Schmalhausen, I. I. 1949. *Factors of Evolution: The Theory of Stabilizing Selection.* Philadelphia: Blakiston.

Schneider, H., M. P. C. Schneider, I. Sampaio, M. L. Harada, M. Stanhope, J. Czelusniak, & M. Goodman. (1993). Molecular phylogeny of the New World monkeys (Platyrrhini, Primates). *Mol. Phylogenet. Evol.*, 2, 225–242.

Simpson, G. G. 1944. *Tempo and Mode in Evolution.* New York: Columbia University Press.

Simpson, G. G. 1953. *The Major Features of Evolution.* New York: Columbia University Press.

Simpson, G. G. 1958. Behavior and evolution. In: *Behavior and Evolution* (Ed. by A. Roe, & G. G. Simpson), pp. 507–535. New Haven, Connecticut: Yale Univ. Press.

Slater, P. J. B. 1986. Individual differences in animal behavior: A functional interpretation. *Accad. Nazion. Lincei,* 259, 159–170.

Thoday, J. M. 1964. Genetics and the integration of reproductive systems. *Symp. R. Entomol. Soc. Lond.,* 2, 108–119.

Tinbergen, N. 1951. *The Study of Instinct*. Oxford, England: Clarendon Press.

Upton, G. J. G., & G. Fingleton. 1985. *Spatial Data Analysis by Example*. Chichester, England: John Wiley & Sons.

Wayne, R. K., R. E. Benveniste, D. N. Janczewski, & S. J. O'Brien. (1989). Molecular and biochemical evolution of the Carnivora. In: *Carnivore Behavior, Ecology, and Evolution* (Ed. J. L. Gittleman), pp. 465–494. Ithaca, New York: Cornell University Press.

Wcislo, W. T. 1989. Behavioral environments and evolutionary change. *Annu. Rev. Ecol. Syst.,* 20, 137–169.

West-Eberhard, M. J. 1987. Flexible strategy and social evolution. In: *Animal Societies: Theories and Facts* (Ed. by Y. Ito, J. L. Brown, & J. Kikkawa), pp. 35–51. Tokyo: Japan Scientific Societies Press.

West-Eberhard, M. J. 1989. Phenotypic plasticity and the origins of diversity. *Annu. Rev. Ecol. Syst.,* 20, 249–278.

Williams, G. C. 1966. *Adaptation and Natural Selection*. Princeton, New Jersey: Princeton University Press.

Wilson, E. O. 1975. *Sociobiology: A Modern Synthesis*. Cambridge, Massachusetts: Harvard University Press.

Winterbottom, R., & D. A. McLennan. 1993. Cladogram versatility: evolution and biogeography of Acanthuroid fishes. *Evolution,* 47, 1557–1571.

CHAPTER 7

Comparing Behavioral and Morphological Characters as Indicators of Phylogeny

Peter H. Wimberger and Alan de Queiroz

(Writers share equal authorship)

> "...we can understand on the principle of inheritance, how it is that the thrush of South America lines its nest with mud, in the same peculiar manner as does our British thrush...."
>
> -- Darwin (1859, p. 243)

As the above quote indicates, Darwin recognized that behavioral traits, like morphology, could be inherited and thus reflect evolutionary affinities. Through the work of Whitman (1899) and Heinroth (1911) and, especially, early ethologists such as Lorenz (1941) and Tinbergen (1959), the idea that behavioral traits can and should be used as indicators of phylogeny has been maintained (McLennan et al. 1988; de Queiroz and Wimberger 1993). However, many biologists still view behavioral characters as inferior to morphology for estimating phylogenetic relationships. In the extreme, it has even been claimed that behavior is essentially worthless for reconstructing phylogenetic history (Atz 1970; Aronson 1981).

Initially we have to ask whether the dichotomy between behavioral and morphological characters is, in any sense, "real." The division of a systematic data set into behavioral and morphological subsets assumes at least the possibility that these classes of characters have different properties for phylogeny estimation. Recently Kluge and Wolf (1993) have questioned the existence of classes of characters as "mind-independent categories [p. 190]." They suggest that systematists must "question artificial subdivisions of evidence because there is no reason to believe those definitions have discoverable boundaries [p. 190]." Under this view, it would seem there is no reason even to consider comparing behavioral and morphological characters.

Without delving into what is meant by a "mind-independent category," we take the position that one can define classes of characters that have systematically different properties with respect to estimating phylogenetic relationships. For example, codons in pseudogenes generally evolve more rapidly than those in protein-coding genes (Li & Graur 1991). Similarly, within protein-coding genes, third base positions in codons tend to evolve more rapidly than first and second positions (Li et al. 1985). Different genes may even experience different branching histories (Avise et al. 1983; Tajima 1983). Whether such categorizations are mind-independent or not, their recognition seems to be useful for phylogenetic studies — for example, in refining methods of phylogeny estimation (Bull et al. 1993; Miyamoto et al. 1994).

Although we believe that recognizing classes of characters is useful, it is not clear that traditional categorizations have meaning for phylogenetic analyses. Behavior, morphology, molecules, and physiology may be definable categories, but do characters in these different classes have different properties for phylogeny estimation?

In this chapter, we are concerned with the question of whether behavior and morphology have such different properties. As argued above, we consider it valid to examine different classes of characters. However, at least in this case, we do not consider it valid to assume that one class of characters is inferior to the other without empirical proof. Both behavior and morphology subsume an enormous diversity of traits that defy easy generalization. Furthermore, behavior and morphology are often strongly associated with each other. For example, the distinction between substrate spawning and mouthbrooding in cichlid fishes involves much more than the mouthbrooders simply picking their eggs up and orally brooding them. A large suite of morphological changes accompanies the appearance of mouthbrooding. Mouthbrooders typically have wider heads, more and longer gill rakers, and more yolk in their eggs. In addition, the mouthbrooding species have non-adhesive eggs and a different developmental pattern than substrate spawners (Fishelson 1966; Chardon & Vandewalle 1971). The behavioral character, parental care mode, involves morphological change as well as behavioral change. Many behavioral traits show this dependence on morphology and, ultimately, since behavior is mediated by the nervous system, all behavior is dependent on morphology at some level. These considerations suggest there will be no absolute distinction between properties of behavioral and morphological characters.

Nonetheless, it could be that behavioral characters tend to be less reliable indicators of phylogeny. At least two reasons have been given for why this might be so. One is that behavioral characters are more evolutionarily labile than morphology (Atz 1970; Baroni Urbani 1989) and, therefore, the phylogenetic information in behavioral traits erodes more quickly through time than for morphology. Wilson (1975) states that, "behavior [is] the part of the phenotype most likely to change in response to long-term changes in the environment." In the extreme, this argument has been used to suggest that behavior is essentially uncorrelated with phylogeny. The second reason for dismissing behavioral characters is the claim that identifying homologous behavioral patterns is difficult, if not impossible (Atz 1970; Hodos 1976; Aronson 1981). Brooks and McLennan (1991) and de Queiroz and Wimberger (1993) discuss these ideas and their history more thoroughly.

Here we present several analyses bearing on the question of whether behavior and morphology should be considered different classes for phylogeny estimation. In an earlier study (de Queiroz & Wimberger 1993), we found no statistically significant difference in levels of homoplasy (evolutionary convergence and reversal) between behavioral and morphological characters from a wide variety of systematic studies. Systematic methodology provides a number of different measures of homoplasy. Here, we review our earlier results and present an additional analysis using a measure of homoplasy that may better reflect phylogenetic information content than the measure we originally used. Because differences in levels of homoplasy do not necessarily indicate differences between the estimated phylogenetic trees, we also make two kinds of direct comparisons of trees estimated from behavioral and morphological data. The first of these addresses the question of whether the behavioral and morphological trees are more similar than randomly selected pairs of trees (Penny et al. 1982; Hendy et al. 1984), the implication being that such similarity indicates that both kinds of data reflect the true phylogeny. The second tree comparison addresses the question of whether behavioral and morphological data disagree more in their estimates of phylogeny than one would expect from random partitions of the total data set. This last analysis, which uses a test recently proposed by Swofford (1994), is preliminary in nature but represents perhaps the most appropriate method for assessing whether a data set should be partitioned into subsets prior to phylogenetic analysis, a controversial implication of the recognition of character classes (e.g., Bull et al. 1993; Kluge & Wolf 1993).

Levels of homoplasy in behavioral and morphological characters

A. General procedure

To assess the generality of statements made about the relative utility of behavioral and morphological characters for estimating phylogenetic relationships, we have examined a large number of systematic studies that used behavioral and/or morphological data. We compared levels of homoplasy in the two types of characters, the idea being that levels of homoplasy are generally inversely related to utility for phylogeny estimation (but see Goloboff 1991). We performed two kinds of comparisons. In the first, we compared homoplasy of behavioral and morphological characters within data sets that contained both kinds of characters. This analysis controls for potential differences among clades in rates of character evolution, tree topology, or other factors that might influence the level and/or measurement of homoplasy. In the second comparison, we examined overall levels of homoplasy for data sets containing either behavioral or morphological characters. This second analysis lacks the control by study group of the first analysis but provides a larger sample of taxa and a more powerful statistical test. The methods are fully described in de Queiroz and Wimberger (1993). Here we describe only the most salient points.

We collected morphological and behavioral data sets from the literature (Tables 7.1 and 7.2). We defined a behavioral character as any character representing movement of the organism or its parts. Thus, behavioral characters ranged from simple, stereotyped movements to more complex characters such as mating system or nesting dispersion (solitary vs. colonial). Epiphenomena of behavior, such as nest architecture, were also classified as behavioral characters. Courtship and territorial behavior were, by far, the most common types of behavior used in systematic studies, followed by nest site and architecture which were primarily used in studies of social insects and birds (Fig. 1). Ideally, it would have been preferable to have a more uniform distribution of behavioral types represented in this study. Morphological characters included gross external characteristics, color pattern, osteological characters and features of the soft anatomy.

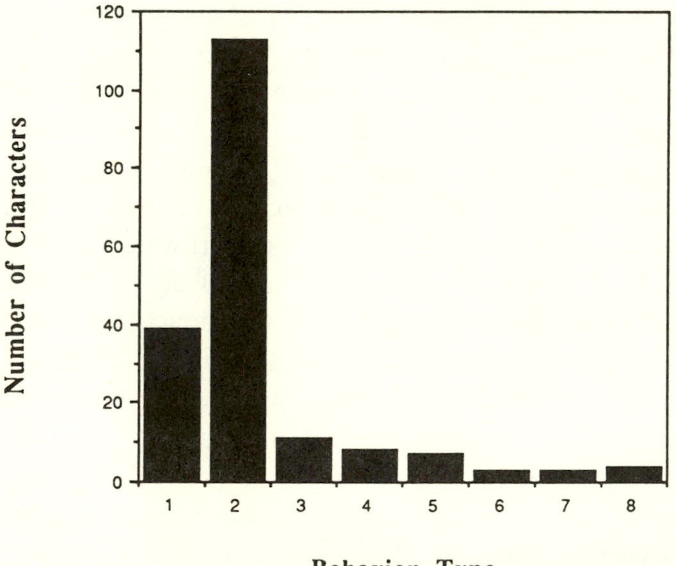

Figure 1. Behavioral characters used in the within-study comparison of *CI*s and *RI*s broken down into: (1) nest architecture and site; (2) courtship and territorial behavior; (3) parental care; (4) other reproductive behavior; (5) feeding and other maintenance (including locomotion); (6) general lifestyle; (7) social system; (8) other.

Table 7.1. Systematic data sets used for comparison of character *CI*s and *RI*s within data sets[a]

Taxon	Rank	# Taxa	MC	BC	*MCI*	*BCI*	*MRI*	*BRI*
Polistine wasps	Genera	28	34	18	0.60	0.59	0.65	0.81
Eumenine wasps	Genera	25	25	1	0.86	1.00	0.85	1.00
Vespine wasps	Species/Genera	7	17	8	0.88	0.71	0.79	0.38
Vespid wasps	Subfamilies	6	20	8	0.78	0.96	0.63	0.96
Apis bees	Species	6	9	3	0.94	0.83	0.94	0.83
Meloid beetles	Genera	9	21	2	0.87	1.00	0.90	1.00
Arachnids	Orders	11	63	1	0.71	1.00	0.64	1.00
Shrimps	Orders	5	3	1	0.83	1.00	0.67	1.00
Gasterosteiform fishes	Genera	6	18	17	0.86	0.96	0.84	0.91
North American Hylid frogs	Species	13	9	4	0.94	0.90	0.96	0.50
Pipid frogs	Species/Genera	7	52	2	0.88	0.75	0.79	0.50
Triturus newts	Species	9	2	11	0.52	0.81	0.62	0.78
Seaducks	Species/Genera	10	12	2	0.83	0.75	0.83	0.84
Alcid birds	Species	23	31	2	0.82	0.70	0.90	0.85
Pelecaniform birds	Families	7	26	5	0.92	1.00	0.87	1.00
Ochthoeca flycatchers	Genera	5	5	2	0.90	1.00	0.80	1.00
Tody-tyrant flycatchers	Genera	9	8	2	1.00	0.75	1.00	0.50
Myiobius flycatchers	Genera	5	8	2	0.85	1.00	0.81	1.00
Empidonax flycatchers	Genera	7	6	3	1.00	0.83	1.00	0.83
Manakin birds	Species	19	36	29	0.90	0.84	0.94	0.83
Squamate reptiles	Families	19	61	3	0.68	0.67	0.70	0.77
Sand lizards	Species	10	36	2	0.87	0.67	0.85	0.67
Francolin birds	Species	13	17	5	0.82	0.67	0.85	0.87

[a]Rank is the taxonomic rank of the terminal taxa. MC and BC are the number of morphological and behavioral characters, respectively. *MCI* and *BCI* are the mean *CI*s for morphological and behavioral characters; *MRI* and *BRI* are the mean *RI*s for morphological and behavioral characters. References are given in de Queiroz and Wimberger (1993), except for the francolin data set, which is from Crowe et al. 1992. See text and de Queiroz and Wimberger (1993) for further description.

When a single data set was used for a group we used the tree estimated by the author. For some groups we combined the available behavioral, morphological, and molecular data sets and, in these cases, we obtained trees using *PAUP 3.0* (Swofford 1989).

Table 7.2. Systematic data sets used for comparison of overall *CI*s for morphological and behavioral data sets[a]

Taxon	Character Type	Rank	#Taxa	#Chars	*CI*
Polistine wasps	Behavior	Genera	28	39	0.65
Vespine wasps	Behavior	Species/Genera	7	13	0.79
Vespid wasps	Behavior	Subfamilies	6	12	1.00
Zygothrica flies	Behavior	Species	7	14	0.75
Gasterosteiform fishes	Behavior	Genera	6	23	0.85
Triturus newts	Behavior	Species	9	13	0.73
Manakin birds	Behavior	Species	19	46	0.78
Cranes	Behavior	Species	13	39	0.64
Spilanthes composites	Morphology	Species	6	17	0.79
Pegolettia composites	Morphology	Species	9	19	0.79
Myrtaceae	Morphology	Species	14	31	0.72
Andropogoneae	Morphology	Species	20	28	0.44
Montanoa (Asteraceae)	Morphology	Species	25	51	0.63
Eucalyptus (Myrtaceae)	Morphology	Species	29	65	0.43
Pontederiaceae	Morphology	Species	37	65	0.39
Triticeae (Poaceae)	Morphology	Genera	29	126	0.31
Asteraceae	Morphology	Genera	29	81	0.56
Clusiaceae*	Morphology	Genera	68	94	0.32
Monocots	Morphology	Families	15	56	0.60
Centrospermae	Morphology	Families	20	71	0.35
Angiosperms*	Morphology	Families	47	61	0.26
Rhodophyta	Morphology	Orders	15	37	0.55
Green algae & bryophytes	Morphology	Classes	11	41	0.82
Seed plants	Morphology	Classes	20	31	0.62
Seed plants	Morphology	Classes	20	62	0.50
Steamer-ducks	Morphology	Species	4	23	1.00
Anarbylus geckos	Morphology	Species	6	34	0.84
Australian birds	Morphology	Species	9	33	0.81
Oligochaete annelids	Morphology	Species	11	83	0.64
Stomias teleosts	Morphology	Species	12	32	0.70
Mackerels	Morphology	Species	20	74	0.68
Anole lizards	Morphology	Species	24	31	0.60
Bledius beetles	Morphology	Species	35	72	0.43
Xantusiid lizards	Morphology	Genera	5	36	0.72
Leptopodomorph bugs	Morphology	Genera	10	49	0.80
Gomphaeshnine odonates	Morphology	Genera	16	15	0.50
Stomiid teleosts*	Morphology	Genera	27	323	0.49
Plusiine lepidoptera*	Morphology	Genera	57	307	0.45

Table 7.2. (cont.)

Taxon	Character Type	Rank	#Taxa	#Chars	*CI*
Plecoptera	Morphology	Families	22	113	0.63
Amniotes	Morphology	Classes	8	84	0.89
Cercomerian flatworms	Morphology	Classes	9	39	0.93
Collybia basidiomycetes	Morphology	Species	11	25	0.53
Rusts	Morphology	Genera	17	24	0.53
Eukaryotes	Morphology	Classes	36	105	0.34

[a]#Chars is the number of binary-character equivalents, i.e., the number of characters that would constitute the data set if all multistate characters were converted to binary (two-state) characters. References are given in de Queiroz and Wimberger (1993; for behavioral data sets) and Sanderson and Donoghue (1989; for morphological data sets). Asterisks indicate data sets not used in the analyses reported here. See the text and de Queiroz and Wimberger (1993) for further description.

We tested the assertion that behavioral characters are more prone to convergence and reversal (show more homoplasy) than morphological characters by examining two homoplasy measures, the consistency index (*CI*, Kluge & Farris 1969) and the retention index (*RI*, Farris 1989). The *CI* is the minimum number of steps (evolutionary character state changes) required by the character assuming no homoplasy divided by the number of steps required by the character on the specified tree. The minimum number of changes assuming no homoplasy is always just one less than the number of character states. Thus, if a two-state character changes once on the tree in question, it will have a *CI* of 1.0 indicating no homoplasy for that character on the given tree. If the character requires two changes on the tree (say from A→A' and then A'→A) the *CI* will equal 0.5 (minimum number of changes = 1, actual number of changes = 2; *CI* = 1/2 = 0.5).

The character retention index (Farris 1989) is:

$$RI = \frac{(g - s)}{(g - m)}$$

where g is the greatest possible number of steps for the character on any tree, s is the actual reconstructed number of steps on the specified tree, and *m* is the minimum possible number of steps on any tree. The *RI*, like the *CI*, compares the number of steps on the specified tree with the minimum number of steps. However, unlike the *CI*, the *RI* takes into account the maximum possible number of steps for the character given the number of taxa that have each state.

A character with a high *CI* is one that rarely changes and thus, it could be argued, should be given high weight (Farris 1969; Carpenter 1988b; Sundberg 1989); the same is not necessarily true of a character with a high *RI*. However, the *RI*, precisely because it does take into account the number of taxa that have each state, is probably a better measure of the phylogenetic information content of a character; as Farris (1989, p. 407) puts it, the *RI* "reflects the degree to which similarities apparent in the data can be retained as homologies on a tree." Because of the somewhat different implications of these two measures, we report results for both here.

We excluded phylogenetically uninformative characters (i.e., characters that were either invariant within the ingroup or in which derived character states were unique to single taxa) from our calculations because including these characters inflates the *CI*. (The *RI* cannot be calculated for such characters because the denominator is zero.) Ideally, because derived character states unique to single taxa (autapomorphies) represent evolutionary change, they should be included in the analysis (de Queiroz & Wimberger 1993). However, because some of the data sets we used excluded autapomorphies, we were unable to perform a separate analysis including them.

B. Relative homoplasy levels within data sets

To compare the average levels of homoplasy within data sets, we calculated the mean *CI* and *RI* for each character on the tree(s) given by the author(s) or that we estimated from data sets that included both behavioral and morphological characters (Table 7.1). The average morphological *CI* (*MCI*), behavioral *CI* (*BCI*), morphological retention index (*MRI*), and behavioral *RI* (*BRI*) were then calculated for each data set. Although the distribution of *CI*s is highly nonnormal, the distributions of the differences between *MCI*s and *BCI*s and *MRI*s and *BRI*s within data sets were not significantly different from normality (Lilliefors test for *CI*: $N = 23$, maximum distance = 0.114, $p > 0.6$; for

RI: $N = 23$, maximum distance = 0.134, $p > 0.3$). As a result, we used paired *t*-tests to examine the difference between *MCI*s and *BCI*s, and *MRI*s and *BRI*s; each data set provided an independent paired comparison for the analysis.

Behavioral characters were no more prone to convergence and reversal than the morphological characters used in systematic studies. Paired *t*-tests indicated no significant differences between morphological and behavioral character *CI*s or *RI*s (for *CI*: $X_{morph} - X_{behav} = -0.01 \pm .160$, $t = -0.195$, $p > 0.8$; for *RI*: $X_{morph} - X_{behav} = 0.00 \pm 0.247$, $t = 0.003$, $p > 0.9$; for both tests $N = 23$).

C. Comparison of morphological and behavioral data sets

The second approach we used to compare relative homoplasy levels was to examine the ensemble *CI* of data sets composed exclusively of morphological or behavioral characters. The ensemble *CI* is the sum of the numerators of the character *CI*s divided by the sum of their denominators. The morphological data sets were taken from Sanderson and Donoghue (1989) and the behavioral data sets were gleaned from the literature (Table 7.2).

The data were analyzed using analysis of covariance where number of characters and taxa were used as covariates because both variables affect the *CI* (Archie 1989; Sanderson & Donoghue 1989). The *CI*s were natural log transformed to make the variance independent of the means and because *CI* is curvilinearly related to number of taxa. We looked for differences in intercept of the log *CI* between morphological and behavioral characters at the mean number of taxa and characters for behavioral data sets because making the comparison at the mean for all data sets would have meant that the comparison was made outside the range of number of characters for the behavioral data sets.

The ANCOVAs indicated no significant difference between morphological and behavioral character *CI*s when either number of characters or number of taxa was used alone as a covariate, or when the two covariates were used together (Table 7.3, Fig. 2). Both covariates were significant when used alone; however, when used in the same analysis, only number of taxa had a significant effect. Interaction effects were not significant for any of the analyses.

Table 7.3. Analyses of covariance of behavioral and morphological data sets[a]

Covariate: Adjusted number of taxa
Multiple $R = 0.806$
Squared multiple $R = 0.650$

Source	SS	*df*	MS	*F* ratio	*p*
Adjusted number of taxa	1.859	1	1.859	56.915	<0.000
Character type	0.076	1	0.076	2.337	0.135
Error	1.208	37	0.033		

Covariate: Adjusted number of characters
Multiple $R = 0.563$
Squared Multiple $R = 0.317$

Source	SS	*df*	MS	*F* ratio	*p*
Adjusted number of characters	0.711	1	0.711	11.169	0.002
Character type	0.065	1	0.065	1.018	0.319
Error	2.356	37	0.064		

Covariates: Adjusted numbers of taxa and characters
Multiple $R = 0.807$
Squared multiple $R = 0.651$

Source	SS	*df*	MS	*F* ratio	*p*
Adjusted number of taxa	1.151	1	1.151	34.368	<0.000
Adjusted number of characters	0.003	1	0.003	0.091	0.764
Character type	0.061	1	0.061	1.819	0.186
Error	1.205	36	0.033		

[a]Analyses of covariance on log *CI*s of behavioral and morphological data sets with adjusted number of binary-character equivalents, adjusted number of taxa, or both entered as covariates. Character type refers to behavioral versus morphological data. $N = 40$ in all cases. See text and de Queiroz and Wimberger (1993) for further description.

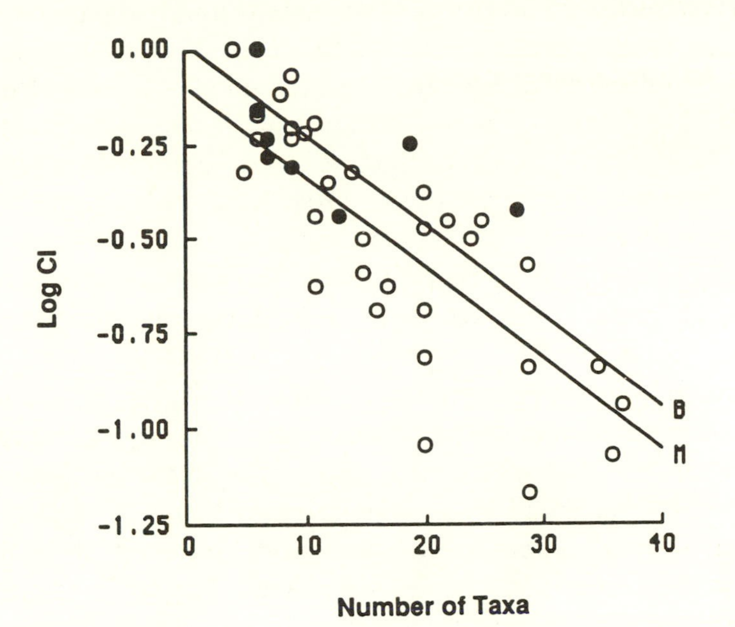

Figure 2. Overall log *CI*s for entire data sets plotted against number of taxa. Open circles are for morphological data sets, closed circles are for behavioral data sets. Lines labeled B and M are regressions of log *CI* on number of taxa for behavioral and morphological data sets, respectively, assuming a common slope for the two regressions. See text and de Queiroz and Wimberger (1993) for further description.

Are phylogenetic estimates derived from behavioral and morphological data similar?

The second question we examined is whether or not behavioral data sets and morphological data sets give similar estimates of relationships. A number of authors have remarked on the similarity of trees generated from behavioral and morphological characters (Lorenz 1941; McLennan et al. 1988; Prum 1990; Halliday & Arano 1991). However, as far as we know, there has been no quantitative test to show that such trees are significantly more similar than one would expect by chance.

Differences in tree shape and the arrangement of the taxa among the branches can be summarized with tree distance metrics (Robinson & Foulds 1979; Hendy et al. 1984; Steel & Penny 1993). One can compare the observed distance between trees generated from two data sets to the distribution of the tree distance metric generated from comparing random pairs of trees that have the observed number of taxa (Robinson & Foulds 1979; Hendy et al. 1984; Penny & Hendy 1985; Steel 1988; Steel & Penny 1993). This method tests for non-random similarity between the observed trees. Such tests have been performed for comparisons of morphology- and molecule-based trees (Penny et al. 1982; Rodrigo et al. 1993). Here we apply this method to comparisons of pairs of trees generated from behavioral and morphological characters.

The six pairs of trees we used are shown in Figure 3. These cases were chosen because they contain enough behavioral and morphological characters to at least potentially give fully resolved (completely dichotomously branching) trees when the two kinds of characters are used separately. The trees were taken from the original references, except for the vespid and vespine wasp trees. We obtained the latter using *PAUP 3.1.1*, employing the branch and bound option which guarantees finding the most parsimonious trees (Swofford 1993). We used a tree distance metric called the symmetric difference index (*SDI*, sometimes called the partition metric, Robinson & Foulds 1979; Hendy et al. 1984). The *SDI* can be defined as the sum of the number of clades found on the first tree (e.g., morphological) that are not found on the second tree (e.g. behavioral), and the number of clades found on the second tree that are not found on the first tree. In each case we calculated the *SDI* between the unrooted morphological and behavioral trees and compared this observed tree distance to a null distribution of *SDI* on random unrooted dichotomous trees (Penny et al. 1982; Hendy et al. 1984). Some of the observed trees were not fully resolved and thus not directly comparable to the null distribution. In these cases, we calculated the largest possible *SDI* between the trees (i.e., we resolved the trees so as to maximize their difference) and compared this value to the null distribution. This procedure makes our test conservative. Other tree metrics such as the path difference and quartet metric may ultimately provide more sensitive estimates of tree difference (Steel & Penny 1993), but their null distributions have not been calculated.

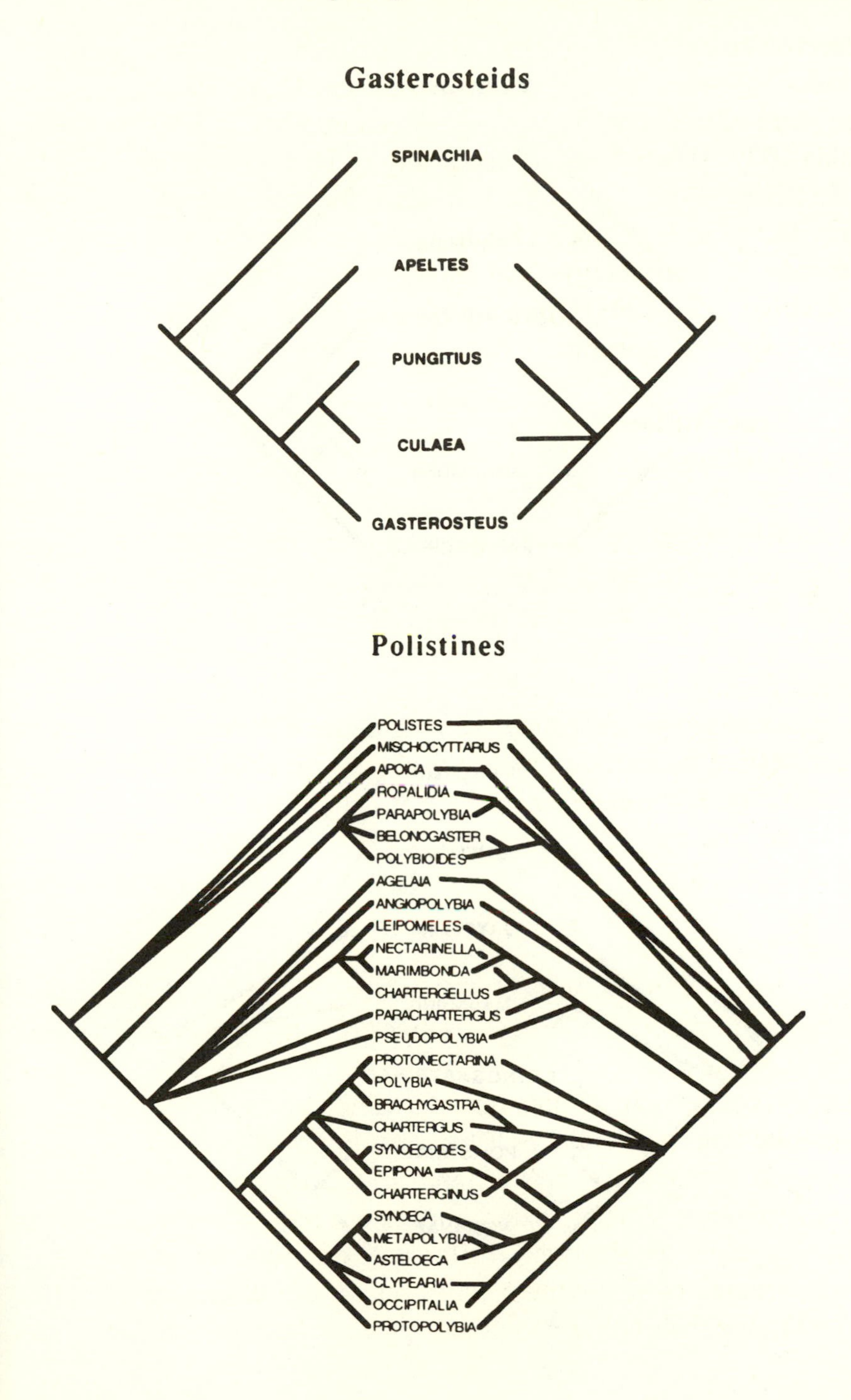

Figure 3. (Continued on following pages.)

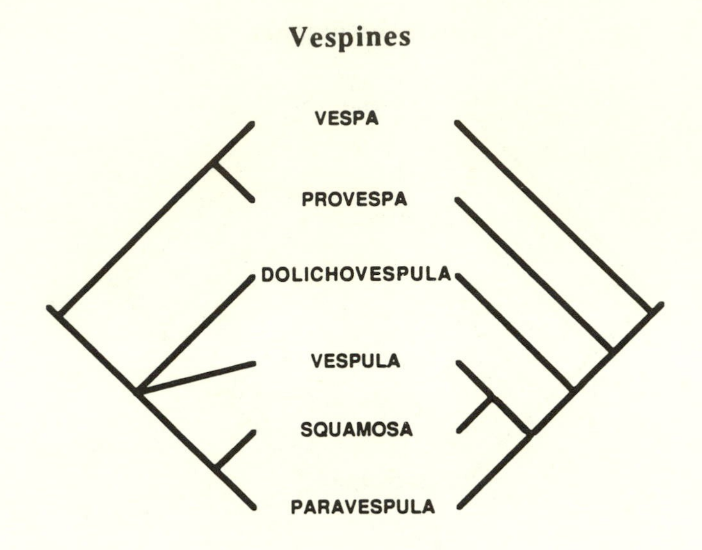
Vespines
VESPA
PROVESPA
DOLICHOVESPULA
VESPULA
SQUAMOSA
PARAVESPULA

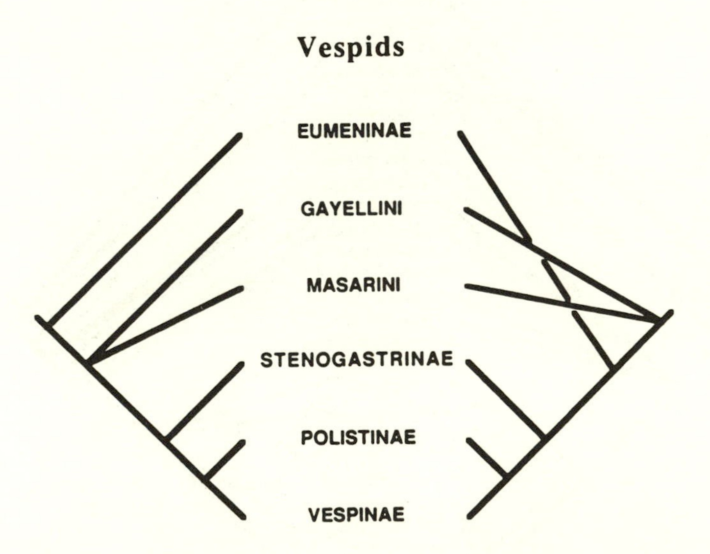
Vespids
EUMENINAE
GAYELLINI
MASARINI
STENOGASTRINAE
POLISTINAE
VESPINAE

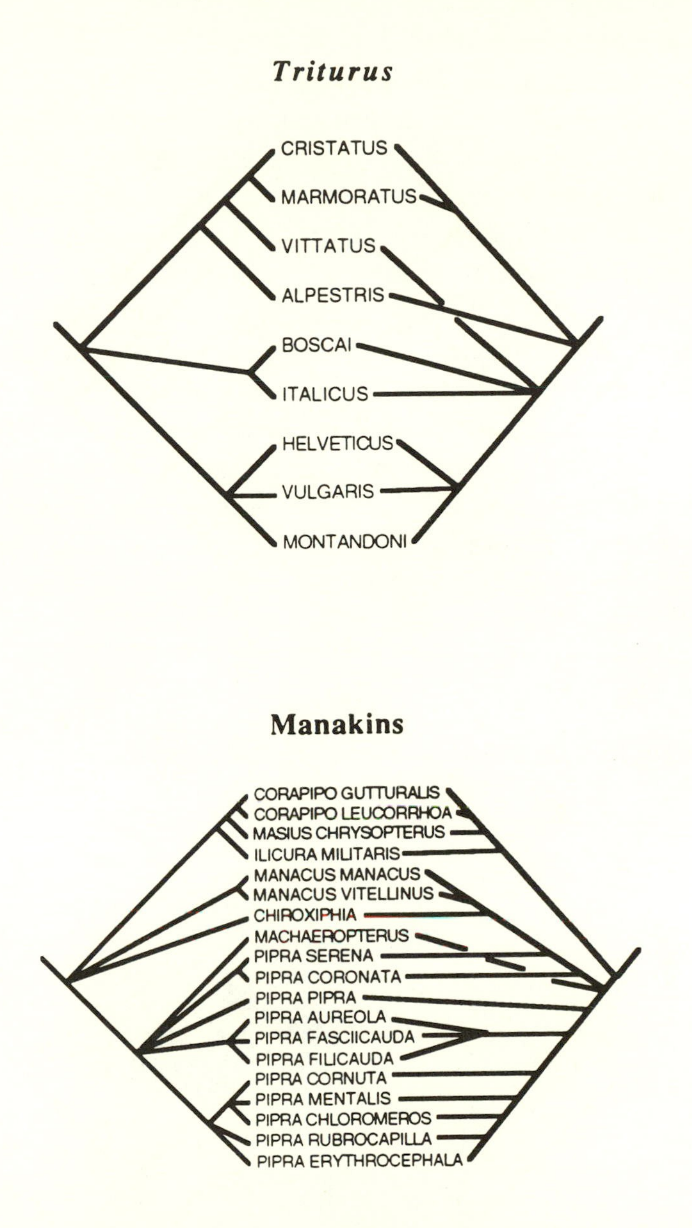

Figure 3. Paired behavioral and morphological phylogenies. In each case the behavioral tree is on the left and the morphological tree is on the right. References are as follows: gasterosteid fish (McLennan et al. 1988, McLennan 1993); polistine wasps (Wenzel 1993), vespine wasps (Carpenter 1987); vespid wasps (Carpenter 1982, 1987, 1988a,c); *Triturus* newts (Arntzen & Sparreboom 1989, Halliday & Arano 1993); manakin birds (Prum 1990, 1992).

Three of the six pairs of trees were significantly more similar than expected by chance (Fig. 3; Table 7.4). The three pairs of trees that were not significantly similar were at least more similar than the average for randomly selected pairs of trees. We suspect that the nonsignificant results are due to small numbers of taxa; it is the three groups with the fewest taxa that are not significantly similar. When the trees being compared have few taxa, the test is not very powerful (i.e., it has little power to reject the null hypothesis). For example, the unrooted behavioral and morphological trees for gasterosteid fishes and vespid wasps (when resolved to maximize their differences) are as similar as possibile without being identical, yet they are not significantly similar using the *SDI* test. In general then, the results suggest that morphological and behavioral trees are more similar than one would expect by chance. Irwin's comparison (Chap. 8 in this volume) of several other pairs of behavioral and morphological trees provides further support for this conclusion.

Table 7.4. SDIs for paired behavioral and morphological trees[a]

Group	Number of Taxa	*SDI*	p value
Gasterosteids	5	2	0.334
Vespines	6	4	0.304
Vespids	6	2	0.067
Triturus	9	8	0.037
Manakins	19	20	<<0.001
Polistines	28	38	<<0.001

[a]The p values refer to probabilities of obtaining an *SDI* as small or smaller by choosing two random trees with the observed number of taxa.

Do different data sets give significantly different estimates of phylogeny?

If one has shown that two data sets produce trees that are more similar than expected by chance, it is reasonable to move on to another, perhaps more interesting question: Do the data sets nonetheless produce systematically incongruent estimates of phylogeny? Answering this question is especially significant because it bears on whether or not one should combine data sets for phylogeny estimation, a controversial topic in systematics (Barrett et al. 1991; Swofford 1991; de Queiroz 1993; Kluge & Wolf 1993; Bull et al., 1993; Chippindale & Wiens 1994).

Swofford (1994) presented a resampling method to examine this question based in part on a procedure originally proposed by Mickevich and Farris (1981). Mickevich and Farris examined two data sets for the same taxa, and separated the incongruence (homoplasies) within each data set from the incongruence between the two data sets. They determined the incongruence within each data set by calculating the number of extra steps (homoplasious character state changes) required by the trees obtained through separate parsimony analyses of each data set. The number of extra steps required by the tree obtained from parsimony analysis of the two data sets combined was then calculated. The incongruence between data sets was defined as the difference between the extra steps for the combined analysis and the sum of the extra steps for the separate analyses. (The latter is always equal to or smaller than the former.) In other words, the incongruence between data sets is the number of extra steps the data sets force upon each other beyond what was required by each data set individually. This measure of conflict between data sets has an advantage over simply comparing the estimated trees (e.g., by the *SDI*) in that it reflects the amount of character support for different trees.

Swofford's test essentially takes the between-data-set incongruence as defined above and asks whether it is unexpectedly high. In Swofford's test, the original pair of data sets are combined into a single data set and each resampling replicate consists of splitting the combined data randomly into partitions of characters equal in size to the original data sets. Now, for each replicate and the observed data, the difference between the tree length (number of character state changes required) from analysis of the total (combined) data set and the sum of tree lengths from analysis of the two partitions is exactly the same as Mickevich and Farris' between-data-set incongruence. However, since the tree length from the combined analysis is always the same, we can ignore it and simply examine the sum of lengths from separate analysis of the two data partitions. Specifically, the sums of lengths of the random-partition pairs of trees form a distribution against which one compares the sum of lengths of the observed two trees. If there is significant incongruence between the original data sets, one expects their observed sum of lengths to fall in the lower tail of the distribution of sums of lengths for the random partitions. For example, consider the extreme case in which characters within each original partition are in complete agreement with each other but disagree with all characters in the other partition. In this case the observed sum of lengths of the two separately generated trees will be the minimum length possible for the data, whereas random partitions of the total data set (except those that exactly recreate the original data sets) will result in trees with some homoplasy and thus with a greater summed length.

We applied Swofford's test to Prum's (1990, 1992) morphological and behavioral data on manakin phylogeny. The two data sets were the 30 informative courtship display characters from Prum (1990) and the 40 informative syringeal characters from Prum (1992). All of the characters were binary.

We modified a Hypercard program created by W. Maddison to perform the test. The modified program randomly partitioned the total data set into subsets of 30 and 40 characters (the numbers of behavioral and morphological characters, respectively) and interfaced with PAUP 3.1.1 (Swofford 1993) to estimate most parsimonious trees from each of the subsets. For the tree estimations we used the heuristic search option with all default settings except that 10 replicates of random stepwise addition were performed for each search. Heuristic searches do not guarantee finding the most parsimonious trees. However, because the manakin data has relatively little homoplasy, it is highly likely that all minimum length trees were in fact found. The Hypercard program repeated the entire process 300 times. Sums of lengths were taken from the *PAUP* output.

The results of the test are shown in Figure 4. The sum of lengths of the morphological and behavioral trees is smaller than one would expect from random partitions of the total data set ($p < 0.04$). This result

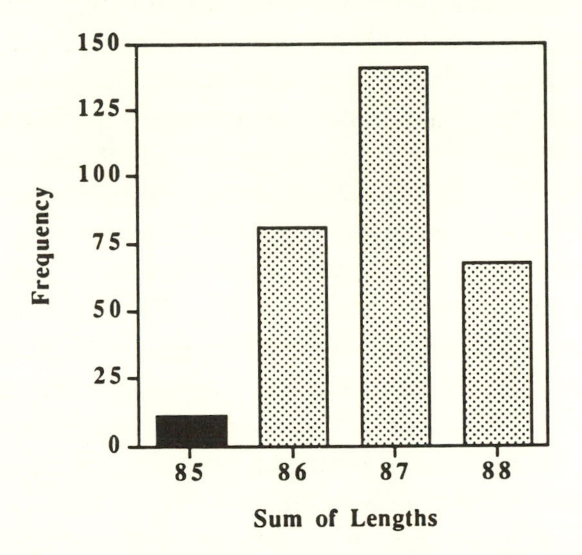

Figure 4. Test for incongruence between behavioral and morphological estimates of manakin phylogeny (data from Prum 1990, 1992). The histogram represents the distribution of sums of lengths for random partitions of the combined data set. The observed sum of the behavioral and morphological tree lengths is 85 steps, thus the black area indicates the portion of the distribution (0.037) as or more extreme than the observed value. See text for further description.

suggests that the morphological and behavioral characters are systematically incongruent (but see below).

Discussion

Our results provide quantitative validation of the use of behavioral characters for estimating phylogenetic relationships. We would like to point out that these results are relevant for characters used by systematists and do not necessarily reflect properties of behavior and morphology in general. The implications of each of the tests are somewhat different and worth considering separately.

The test comparing *SDI*s of behavioral and morphological trees with *SDI*s of random pairs of trees indicates that behavioral and morphological trees are significantly similar (Fig. 3; Table 7.4). The most obvious explanation for this result is that both kinds of characters reflect the true phylogenies of the groups in question (although other interpretations are possible, such as that the two kinds of characters tend to evolve in the same manner, but do not accurately reflect phylogeny). This is a very coarse validation for behavioral characters; it indicates only that they contain some phylogenetic information.

The comparison of consistency indices indicates there is no strong difference in levels of homoplasy between behavioral and morphological characters (Tables 7.1 to 7.3; Fig. 2). This suggests that any particular behavioral synapomorphy (shared derived character state) should be treated as seriously as any particular morphological synapomorphy; in other words, behavioral and morphological character state changes should not *a priori* be differentially weighted (Farris 1969; Carpenter 1988b; Sundberg 1989).

The retention index is the proportion of similarity that is retained as homology (Farris 1989) and therefore reflects the phylogenetic information content of a character. Thus, our comparison of retention indices suggests that the amount of phylogenetic information in a given behavioral character is not strongly different than that in a given morphological character. To the degree that these measures accurately reflect homoplasy and phylogenetic information, the tests comparing *CI*s and *RI*s show that behavioral characters are as good as morphological characters for estimating phylogenies.

Our comparisons of consistency and retention indices suggest how one should view particular characters in isolation. However, these tests do not address the possibility that behavioral and morphological data sets are more congruent within themselves than they are to each other. In other words, there may be a tendency for behavioral characters to estimate a different phylogeny than morphological characters. (Note that the test using random pairs of trees does not address this point; it asks whether trees are significantly similar, not whether they are significantly different.)

Swofford's (1994) test was designed to address this issue. Unfortunately, there are few data sets that contain enough behavioral and morphological data of a comparable type to make the test worth performing. We do not want to make too much of our one example, Prum's manakin data. The test shows significant incongruence between the behavioral and morphological data. However, since it is only one case, it is impossible to say whether the difference reflects a general dichotomy between behavior and morphology or factors specific to the particular sets of characters (in this case, courtship displays vs. syringeal characters). We think it likely that differences of similar magnitude will be found in comparisons within the two categories (e.g., courtship displays vs. nest architecture; osteology vs. plumage). Several other problems should also be considered. First, for the manakin data, the null distribution is very coarse; for example, if the observed sum of lengths was just one step longer (i.e., 86 rather than 85) the result would not even approach statistical significance. This suggests the *p*-value should be interpreted with caution. Second, a number of sets of the manakin characters logically could have been coded as multistate rather than binary; coding them as separate binary characters may have biased the test towards significance. Finallly, even aside from these issues, the general validity of Swofford's test needs to be investigated (e.g., through simulations).

Even if we assume the validity of Swofford's test, it does not indicate whether the behavioral data for manakins are better or worse than the morphological data, thus it neither strengthens nor weakens the conclusions from the comparisons of *CI*s and *RI*s. However, the significant between-data-set incongruence does suggest that combining these data sets prior to phylogenetic analysis may be problematic. The test indicates either that one or both of the data sets tend to give systematically misleading information about phylogeny or that one data set contains more phylogenetic "noise" than the other. In either case, simply combining the data and performing a standard parsimony analysis may be inappropriate because such an analysis implicitly assumes that the data are homogeneous with respect to estimating the phylogeny (see Bull et al. 1993 and Huelsenbeck et al. 1994 for a detailed discussion of this issue).

Our provisional answer to the question of whether behavior and morphology represent different classes of characters for phylogeny estimation is "perhaps." On a character by character basis, behavioral data used by systematists seem to be as good as morphological data. This is not meant to imply that all behavioral characters are good; undoubtedly some are good and some are bad for estimating relationships, just as some morphological characters are good and some are bad. Despite the similarity of behavioral and morphological *CI*s and

*RI*s, the incongruence between morphological and behavioral data for manakins hints at a systematic difference between the two character types. However, as indicated above, we suspect that the distinction between behavior and morphology may not be especially important compared to more specific characteristics of data sets. One preliminary conclusion from the analysis of the manakin data is that it may be reasonable to recognize data partitions for nonmolecular data in phylogenetic analyses (for molecular data, partitioning is a generally accepted practice). As pointed out by Chippindale and Wiens (1994), however, exactly how such partitioning should be done remains a problem.

Our results do not directly address the question of whether behavior, in general, evolves more rapidly than morphology. The ideal test of somehow identifying random sets of behavioral and morphological characters and then examining their relative rates of evolution would be difficult if not impossible. Barring this possibility, what conclusions regarding this point can we draw from our study? We might have expected some support for the idea that behavioral characters evolve more rapidly than morphological characters, even with the limited subset of characters used to construct systematic data sets. Given our results, if behavior does actually evolve more rapidly than morphology, then one would have to postulate that behavioral characters were subject to a more stringent selection process by systematists than the morphological characters. Most characters (with the occasional exception of DNA data) used in phylogenetic studies are subject to *a priori* culling depending on how much they vary in the group of interest. It is unlikely that researchers are more selective when choosing behavioral characters since behavioral data are, in general, more difficult and time consuming to collect than morphological data. The difficulty of collecting behavioral data probably explains the relative rarity of behavioral characters in our within-data-set comparison (see de Queiroz & Wimberger 1993 for further comments on this issue).

It is interesting to note the findings of Gittleman et al. (Chap. 6 in this volume) that the behavioral characters they examined tend to evolve more rapidly and in a qualitatively different manner than morphological characters. How do we explain the difference between their findings and ours? First, as mentioned above, our study examines a small subset of characters, i.e., characters used in systematic studies. The characters that

Gittleman et al. use, group size and home range size, were not chosen with their systematic potential in mind. Another notable difference between group size and home range size and the characters discussed in this paper are that the former are continuous characters, whereas the latter are all discrete. What would happen if group size were arbitrarily divided into a limited number of discrete states and a homoplasy index calculated? Would it still appear more homoplasious than morphological characters?

One possible explanation for the difference between these studies is that similarity of group or home range size among a set of taxa may be less indicative of underlying genetic and developmental similarity [one definition of the biological basis of homology (Wagner 1989)] than is the similarity of the traits used in our study (e.g., courtship displays). In the context of criteria for assessing homology prior to a phylogenetic analysis, we might say that similar group or home range sizes are not as similar in detail as, for example, the courtship displays of different stickleback species (see Edwards & Naeem 1994 and McLennan & Brooks 1994 for a recent discussion of a similar dichotomy). It is easier to imagine multiple genetic and developmental pathways for similar character states in characters such as group or home range size, which all organisms possess, than it is for more unique behavior patterns such as specific courtship or nest building behavior. This echoes one of the points made by Atz (1970) to argue against homologizing behavioral traits. When features coded as the same character state are not known to be similar in detail, one expects the character to show relatively rapid evolution and high homoplasy (assuming the number of character states is constrained). Although our study refutes Atz' argument as applied to behavior in general, we are suggesting that his argument may apply better to characters that are not unique to a group of taxa, such as group size or home range size.

Behavioral characters, like morphological and molecular characters include a broad array of different kinds of characters. The possible difference between the characters we examined and those used in the study by Gittleman et al. (Chap. 6 in this volume) underscores the diversity of behavioral trait evolution. We want to emphasize that neither extremely evolutionarily labile behavior nor more conservative characters, such as some of those used in our study, should be considered representative of behavioral characters, in general.

ACKNOWLEDGMENTS

We would like to thank Emília Martins for inviting us to contribute to this volume; Wayne Maddison for the PAUP cycler Hypercard program that we modified to obtain random partitions to perform Swofford's test, and for program troubleshooting; David Penny for advice on tree comparison metrics; and John Wiens, Emília Martins, and two anonymous reviewers for comments on the manuscript. AdQ was supported by an NSF Environmental Biology Postdoctoral Fellowship and an NSF Postdoctoral Fellowship through the Biological Diversification Research Training Grant at the University of Arizona. PHW was supported by an NSF Environmental Biology Postdoctoral Fellowship at the University of Michigan and a Murdock Foundation Faculty Summer Fellowship at the University of Puget Sound during the writing of this paper.

Note added in proof: After finishing this paper, we learned that the test devised by Swofford to assess conflict between data sets had been invented earlier by J. S. Farris. the latter presented the test at the annual meeting of the Willi Hennig Society, August 16-19, 1991, Toronto Canada.

References

Archie, J. W. 1989. Homoplasy excess ratios: New indices for measuring levels of homoplasy in phylogenetic systematics and a critique of the consistency index. *Syst. Zool.*, 38, 253–269.

Arntzen, J. W., & M. Sparreboom. 1989. A phylogeny for the Old World newts, genus *Triturus*: Biochemical and behavioural data. *J. Zool., Lond.*, 219, 645–664.

Aronson, L. R. 1981. Evolution of telencephalic function in lower vertebrates. In: *Brain Mechanisms of Behaviour in Lower Vertebrates* (Ed. by P. R. Laming), pp. 33–58. Cambridge, England: Cambridge University Press.

Atz., J. W. 1970. The application of the idea of homology to behavior. In: *Development and Evolution of Behavior: Essays in Honor of T. C. Schneirla* (Ed. by L. R. Aronson, E. Tobach, D. S. Lehrman & J. S. Rosenblatt), pp. 53–74. San Francisco: W. H. Freeman.

Avise, J. C., J. F. Shapira, S. W. Daniel, C. F. Aquadro, & R. A. Lansman. 1983. Mitochondrial DNA differentiation during the speciation process in *Peromyscus*. *Mol. Biol. Evol.*, 1, 38–56.

Baroni Urbani, C. 1989. Phylogeny and behavioural evolution in ants, with a discussion of the role of behaviour in evolutionary processes. *Ethol. Ecol. Evol.*, 1, 137–168.

Barrett, M., M. J. Donoghue, & E Sober. 1991. Against consensus. *Syst. Biol.*, 40, 486–493.

Brooks, D. R., & D. A. McLennan. 1991. *Phylogeny, Ecology, and Behavior.* Chicago: University of Chicago Press.

Bull, J. J., J. P. Huelsenbeck, C. W. Cunningham, D. L. Swofford, & P. J. Waddell. 1993. Partitioning and combining data in phylogenetic analysis. *Syst. Biol.*, 42, 384–397.

Carpenter, J. M. 1982. The phylogenetic relationships and natural classification of the Vespoidea (Hymenoptera). *Syst. Entomol.*, 7, 11–38.

Carpenter, J. M. 1987. Phylogenetic relationships and classification of the Vespinae (Hymenoptera: Vespidae). *Syst. Entomol.*, 12, 413–431.

Carpenter, J. M. 1988a. The phylogenetic system of the Stenogastrinae (Hymenoptera: Vespidae). *J. N. Y. Entomol. Soc.,* 96, 140–175.

Carpenter, J. M. 1988b. Choosing among equally parsimonious cladograms. *Cladistics*, 4, 291–296.

Carpenter, J. M. 1988c. The phylogenetic system of the Gayellini (Hymenoptera: Vespidae; Masarinae). *Psyche,* 95, 211–241.

Chardon, M., and P. Vandewalle. 1971. Comparison de la region cephalique chez cinq especes du genre *Tilapia*, dont trois incubateurs buccaux. *Ann. Soc. R. Zool. Belg.* 101, 3–24.

Chippindale, P. T., & J. J. Wiens. 1994. Weighting, partitioning, and combining characters in phylogenetic analysis. *Syst. Biol.*, 278–287.

Crowe, T. M., E. H. Harley, M. B. Jakutowicz, J. Komen, & A. A. Crowe. 1992. Phylogenetic, taxonomic and biogeographical implications of genetic, morphological, and behavioral variation in francolins (Phasianidae: *Francolinus*). *Auk*, 109, 24–42.

Darwin, C. 1859. *On the Origin of Species by Means of Natural Selection.* London: John Murray.

de Queiroz, A. 1993. For consensus (sometimes). *Syst. Biol.*, 42, 368–372.

de Queiroz, A.,& P. H. Wimberger. 1993. The usefulness of behavior for phylogeny estimation: levels of homoplasy in behavioral and morphological characters. *Evolution*, 47, 46–60.

Edwards, S. V., & S. Naeem. 1994. Homology and comparative methods in the study of avian cooperative breeding. *Am. Nat.*, 143, 723–733.

Farris, J. S. 1969. A successive approximations approach to character weighting. *Syst. Zool.*, 18, 374–385.

Farris, J. S. 1989. The retention index and homoplasy excess. *Syst. Zool.*, 38, 406–407.

Fishelson, L. 1966. Untersuchungen zur vergleichenden Entwicklungsgeschichte der Gattung *Tilapia* (Cichlidae, Teleostei). *Zool. Jb. Abt. Anat. Ontog. Tiere*, 83, 571–656.

Goloboff, P. A. 1991. Homoplasy and the choice among cladograms. *Cladistics*, 7, 215–232.

Halliday, T., & B. Arano. Resolving the phylogeny of the European newts. *Trends Ecol. Evol.*, 6, 113–117.

Heinroth, O. 1911. Beitrage zur Biologie, namentlich Ethologie und Psychologie der Anatiden. *Verh. Ver. Int. Ornithol. Kongr. (Berlin),* 1910, 589–702.

Hendy, M. D., H. C. Little, & D. Penny. 1984. Comparing trees with pendant vertices labeled. *SIAM J. Appl. Math.*, 44, 1054–1065.

Hodos, W. 1976. The concept of homology and the evolution of behavior. In: *Evolution, Brain, and Behavior: Persistent Problems* (Ed. by R. B. Masterton, W. Hodos, & H. Jerison), pp. 153–167. Hillsdale, New Jersey: Erlbaum.

Huelsenbeck, J. P., D. L. Swofford, C. W. Cunningham, J. J. Bull, & P. J. Waddell. 1994. Is character weighting a panacea for the problem of data heterogeneity in phylogenetic analysis? *Syst. Biol.,* 43, 288–291.

Kluge, A. G., & J. S. Farris. 1969. Quantitative phyletics and the evolution of anurans. *Syst. Zool.*, 18,.1–32.

Kluge, A. G., & A. J. Wolf. 1993. Cladistics: What's in a word? *Cladistics*, 9, 183–199.

Li, W.-H., & D. Graur, 1991. *Fundamentals of Molecular Evolution.* Sunderland, Massachusetts: Sinauer and Associates.

Li, W.-H., C. C. Luo, & C. I. Wu. 1985. Evolution of DNA sequences. In: *Molecular Evolutionary Genetics* (Ed. by R. J. MacIntyre), pp. 1–94. New York: Plenum Press.

Lorenz, K. 1941. Vergleichende Bewegungstudien an Anatien. *J. Ornithol.*, 89, 194–294.

McLennan, D. A. 1993. Phylogenetic relationships in the Gasterosteidae: An updated tree based on behavioral characters with a discussion of homoplasy. *Copeia*, 1993, 318–326.

McLennan, D. A., D. R. Brooks, & J. D. McPhail. 1988. The benefits of communication between comparative ethology and phylogenetic systematics: A case study using gasterosteid fishes. *Can. J. Zool.*, 66, 2177–2190.

McLennan, D. A., & D. R. Brooks. 1994. The phylogenetic component of cooperative breeding in perching birds: A commentary. *Am Nat.*, 141, 790–795.

Mickevich, M. F., & J. S. Farris. 1981. The implications of congruence in *Menidia. Syst. Zool.*, 30, 351–370.

Miyamoto, M. M., M. W. Allard, R. M. Adkins, L. L. Janecek, & R. L. Honeycutt. 1994. A congruence test of reliability using linked mitochondrial DNA sequences. *Syst. Biol.*, 43, 236–249.

Penny, D., L. R. Foulds, & M. D. Hendy. 1982. Testing the theory of evolution by comparing phylogenetic trees constructed from five different protein sequences. *Nature*, 297, 197–200.

Penny, D., & M. D. Hendy. 1985. The use of tree comparison metrics. *Syst. Zool.*, 34, 75–82.

Prum, R. O. 1990. Phylogenetic analysis of the evolution of display behavior in the neotropical manakins (Aves: Pipridae). *Ethology*, 84, 202–231.

Prum, R. O. 1992. Syringeal morphology, phylogeny, and the evolution of the neotropical manakins (Aves: Pipridae). *Am. Mus. Nov.*, 3043, 1–65.

Robinson, D. F., & L. R. Foulds. 1979. Comparison of weighted labeled trees. In: *Lecture Notes in Mathematics*, Vol. 748, pp. 119–126. Berlin: Springer-Verlag.

Rodrigo, A. G., M. Kelly-Borges, P. R. Bergquist, & P. L. Bergquist. 1993. A randomisation test of the null hypothesis that two cladograms are sample estimates of a parametric phylogenetic tree. *N. Z. J. Bot.*, 31, 257–268.

Sanderson, M. J., & M. J. Donoghue. 1989. Patterns of variation in levels of homoplasy. *Evolution*, 43, 1781–1795.

Steel, M. A. 1988. Distribution of the symmetric difference metric on phylogenetic trees. *SIAM J. Discr. Math.*, 1, 541–551.

Steel, M. A., & D. Penny. 1993. Distributions of tree comparison metrics — some new results. *Syst. Biol.*, 42, 126–141.

Sundberg, P. 1989. Phylogeny and cladistic classification of the paramonostiliferous family Plectonemertidae (Phylum Nemertea). *Cladistics*, 5, 87–100.

Swofford, D. L. 1989. *PAUP: Phylogenetic Analysis Using Parsimony, Version 3.0.* Computer program distributed by the Illinois Natural History Survey, Champaign, Illinois.

Swofford, D. L. 1991. When are phylogeny estimates from molecular and morphological data incongruent? In: *Phylogenetic Analysis of DNA Sequences* (Ed. by M. M. Miyamoto & J. Cracraft), pp. 295-333. Oxford, England: Oxford University Press.

Swofford, D. L. 1993. *PAUP: Phylogenetic Analysis Using Parsimony, Version 3.1.1.* Computer program distributed by the Smithsonian Institution, Washington, D. C.

Swofford, D. L. 1994. Inferring phylogenies from combined data. Look before you leap. Paper given at Annual Meeting of the Society of Systematic Biologists, June 15–19, 1994, Athens, Georgia.

Tajima, F. 1983. Evolutionary relationship of DNA sequences in finite populations. *Genetics*, 105, 437–460.

Tinbergen, N. 1959. Comparative studies of the behaviour of gulls (Laridae): a progress report. *Behaviour*, 15, 1–70.

Wagner, G. P. 1989. The origin of morphological characters and the biological meaning of homology. *Evolution*, 43, 1157–1171.

Wenzel, J. W. 1993. Application of the biogenetic law to behavioral ontogeny: A test using nest architecture in paper wasps. *J. Evol. Biol.*, 6, 229–247.

Whitman, C. O. 1899. Animal behavior. In: *Biological Lectures, Wood's Hole* (Ed. by C. O. Whitman), pp. 285–338. Boston: Ginn and Co.

Wilson, E. O. 1975. *Sociobiology*. Cambridge, Massachusetts: Belknap Press.

CHAPTER 8

The Phylogenetic Content of Avian Courtship Display and Song Evolution

Rebecca E. Irwin

The stereotyped movements and vocalizations given by birds during courtship and aggressive interactions are traits that historically were thought to reflect phylogeny (e.g., Lorenz 1941, van Tets 1965) but which are currently only rarely studied phylogenetically. In this chapter, my goal is to review those studies that allow evaluation of the phylogenetic content of visual displays and songs given during courtship and aggression. I review some of the more extensive studies of display phylogeny done by early ethologists, as well as the few more recent studies. I focus on studies in which phylogenies based on behavior can be compared to phylogenetic studies based on other traits, in these cases morphological traits, because well-supported molecular phylogenies for the relevant groups are not available, to determine whether behavioral phylogenies are congruent with phylogenies based on other data. If they are, it supports the contention that displays are phylogenetically informative. I review the kinds of information from display behavior that may reflect phylogeny and evaluate methods that have been used to identify possible behavioral homology. Since the early studies of display phylogeny were done without knowledge of modern cladistic methods of phylogenetic reconstruction, I reanalyze several data sets to determine the most parsimonious branching pattern based on the display data. Following de Queiroz and Wimberger (1993,

Chap. 7 in this volume), I use consistency indices as an indication of the level of convergent evolution of behavioral traits. I consider what the results of these studies imply about the processes of display evolution, the use of displays in phylogenetic reconstruction, and how phylogenetic studies of displays can be designed to evaluate adaptive hypotheses about display behavior, sexual selection, and possible parallels between cultural and genetic evolution.

Historical background

Because of their stereotypy and complexity, Lorenz (1941) considered courtship displays likely to be innate, and, as a result, better characteristics for studying phylogeny than most types of behavior. He and other ethologists studied the phylogenetic component of courtship displays, both using them as systematic characters (e.g., Lorenz 1941, van Tets 1965) and examining the evolutionary origins of displays (e.g., Tinbergen 1959). As discussed by Prum (1990), however, while systematic methods underwent a rapid increase in sophistication and standardization, interest in the phylogenetic component of courtship displays, as well as other types of behavior, waned. Studies of behavioral evolution conducted by most behavioral ecologists have, until relatively recently, taken a single species rather than comparative approach. Even the recent interest in the use of phylogenetic comparative methods (e.g., Harvey & Pagel 1991) has been focused more on the outcome of behavior, as addressed by sociobiological theory, than on actual behavioral acts such as displays. As a result, modern phylogenetic methods have been applied to few studies of courtship displays. In one of these few studies, Prum (1990) found that a phylogeny supported by displays was very similar to a phylogeny based on morphology in the manakins (Pipridae). As described by de Queiroz and Wimberger (1993), where behavioral acts have been used in systematic studies, their phylogenetic content does not differ from that of morphological traits. These results indicate that the phylogenetic patterns of courtship displays should be studied in more detail, both to evaluate Lorenz's contention that they are excellent characters for phylogenetic reconstruction, and to determine whether modern methods of studying traits in a phylogenetic context can be used to address other evolutionary questions of more general interest.

Lorenz's original phylogenetic work on display behavior in ducks inspired several other phylogenetic studies of courtship display, such as van Tets' (1965) study of the pelicans and their allies (Pelecaniformes). It also inspired argument as to whether behavior can be expected to reflect phylogeny. Two such arguments have been applied to behavior generally, and can be evaluated with regard to displays. Atz (1970) suggested that determining whether behavioral traits are homologous, that is, inherited from the same ancestral state, is problematic. Because there are a limited number of ways a behavioral act can be performed, he predicted that similar behavioral traits will arise repeatedly by chance and therefore show high levels of convergent evolution. Atz (1970) also predicted that behavior will evolve rapidly; rapid evolution would contribute further to the problem of chance convergence. The only way to evaluate these arguments is through studies of behavioral phylogeny. As discussed by de Queiroz and Wimberger (Chap. 7 in this volume), one way to evaluate levels of convergence is by using consistency indices; since the consistency index (*CI*) measures homoplasy, or independent evolution, its value will be low if there are high levels of convergence.

A second argument against a phylogenetic component to displays is that because of learning, behavioral traits may not be inherited and may not reflect phylogeny. The relative roles of learning and genetic inheritance in the development of most courtship displays are poorly understood. Lorenz' (1941) hypothesis that stereotypy in behavior indicated an innate basis to behavior has not been supported by more recent research. For example, the highly stereotyped sequences of vocalization given as courtship and agonistic song by oscine songbirds (Passeriformes: Passeres) are dependent on learning (reviewed in Kroodsma 1982). Songs and visual displays may reflect phylogeny despite the effects of learning. Even in the case of songbird song, for which the role of learning has been studied in detail in many species, the relative roles of learned and innate influences are not clear. While syllable sequences are frequently dependent on learning (reviewed by Kroodsma 1982), some aspects of microstructure (e.g., Marler & Pickert 1984), and song repertoire (Kroodsma & Canady 1985) in certain species are apparently innate. These aspects of song should evolve genetically and may thus reflect phylogeny. Further, as discussed by Payne (1986), learned songs, and most likely learned aspects of other courtship displays, are learned from conspecifics and inherited culturally within a species. As a result, they should be most similar within a species and gradually evolve culturally to differ after speciation; such traits should thus also reflect phylogeny.

Tinbergen (1959) proposed an additional argument against a phylogenetic component specifically of courtship displays. Based on his comparative work on gulls, Tinbergen (1959) described displays whose action reflected the display's functional context. For example, aggressive displays involve apparent size enhancement and displaying of "weapons", such as the bill. He argued that while displays may reflect phylogeny as a result of being homologous, they may also be likely to show convergent evolution because of the hypothesized relationship between structure of the display and its functional significance. As an example of convergence, he described the "bill-up" posture given by species as unrelated as the passerine great tit (*Parus major*), gannets (*Morus* spp.) and some terns (*Sterna* spp.). In these unrelated species, the "bill-up" is given in the context of appeasement during aggressive interactions. Tinbergen suggested that an appeasement display must involve pointing the bird's aggressive weapon, the bill, in a direction away from the interacting bird, and that there are a limited number of ways to do this; this, he hypothesized, could explain the observed convergent evolution of displays. Adaptive arguments such as these also require phylogenetic studies to be tested, since it is through phylogenetic studies that convergence of displays in relation to functional context, suggesting similar adaptation to similar circumstances, can be identified (Harvey & Pagel 1991).

Avian displays as systematic characters

A. Display characteristics used in phylogenetic analysis

Courtship displays in birds include various highly stereotyped actions, exaggerated movements of certain body parts as in the head-bobbing displays of some ducks (Anatidae), and raising of specific feathers, such as the crest of the hooded merganser (*Lophodytes cucullatus*). Some displays, such as the "song-spread" of many New World blackbirds (Icterinae), include both vocalizations and movement of the body and feathers (Nero 1964, Orians & Christman 1968). Long, complex displays involving a stereotyped sequence of movements and sometimes including sound production occur in some species, such as many manakins (Pipridae, reviewed in Prum 1990). Many oscine songbirds lack the more extreme physical displays found in some other birds but have complex songs involving stereotyped sequences of syllables that are also used as displays. Well-developed displays occur in species with diverse mating systems, including both socially monogamous species such as the gulls (Laridae, reviewed in Tinbergen 1959) and highly polygynous, lekking species such the manakins (reviewed in Prum 1990).

Similar displays are often associated with both courtship and aggression (e.g., Tinbergen 1959). The predicted pattern of evolution of displays differs depending on their function in courtship or aggression. For example, courtship displays are expected to be important in species recognition and should be distinctive, while aggressive acts are more likely to be subject to convergence based on their function (Tinbergen 1959). Since studies in which these separate functions can be distinguished are rare, I here consider displays associated with both of these functions. Further behavioral studies are needed to determine whether the phylogenetic pattern of display is related to its aggressive or courtship function.

The physical sequences of behavior that make up a display, and the acoustic structure of songs, are the characters used most commonly in studies of display phylogeny. Other display characters that have been used in phylogenetic studies include display site (Prum 1990), presence of displays given in social groups as opposed to pairs (Lorenz 1941), size or structure of repertoire of songs given by individual males (Irwin 1988, 1990; Spector 1992), and whether displays are given commonly or rarely (Wood 1984). Because of the stereotypy of displays, most of these are most easily coded as discrete characters and character states in phylogenetic analyses, although song repertoire size and display intensity require a ranking system if they are used as discrete characters and may be more appropriately analyzed as continuous traits. I have chosen to analyze all traits here as discrete, because current methods for using continuous traits are typically based on a "Brownian motion" model of evolution (see in this volume, Martins & Hansen, Chap. 2). This Brownian motion assumption is problematic with regard to display traits because displays are likely to be subject to sexual selection, which is predicted to result in highly variable rates of evolution in different lineages (e.g., Fisher 1954; Lande 1981), and at different times evolutionarily within a lineage (e.g., Lande & Arnold 1985). If continuous aspects of displays are studied phylogenetically, using such models, care should be taken to test for possible evolutionary rate variation. I use a parsimony analysis of discrete traits; parsimony analysis also depends on assumptions about evolution, but using this analysis I can compare results to similar analysis of morphology, using methods such as those used by de Queiroz and Wimberger (Chap. 7 in this volume).

As is standard in phylogenetic studies, and discussed elsewhere in this volume, I identified display traits that are apparently homologous, that is, have apparently evolved from the same ancestor, by examining similarity in "structure" — in the sequence of actions or vocalizations making up the behavioral act — either directly, through development, or through common similarity to intermediate forms (Atz 1970, Prum 1990). These similarities, or putative homologies, provide the basis for studying phylogeny. An additional criterion for determining putative homology that has been used implicitly or recommended (Payne 1986) is consideration of the context or function of the display. Prum (1990) argued against considering display context or function as a criterion for homology, since displays that appear structurally homologous may evolve to have different functions in different species. A possible example of this transference of function in a display is seen in many gull species, in which displays used in aggressive contexts throughout many gull species are also given, in a particular sequence, as part of the "meeting ceremony" involved in courtship within particular species (Tinbergen 1959). Transference of function also apparently occurs in vocalizations. Many New World blackbirds give a pure tone, down-slurred whistle as an alarm call (e.g., red-winged blackbirds *Agelaius phoeniceus*, Orians & Christman 1968; eastern and western meadowlarks, *Sturnella magna* and *S. neglecta*, Lanyon 1957). This distinctive vocalization occurs in most members of the blackbird subfamily, but not in apparently related avian groups. In three species in different blackbird clades, the brown-headed cowbird (*Molothrus ater*; Rothstein et al. 1986), yellow-winged blackbird (*Agelaius thilius*; Irwin 1989), and scarlet-rumped cacique (*Cacicus uropygialis*; Irwin 1989), a pure tone, down-slurred whistle, with acoustic structure similar to that used as an alarm call of other blackbirds, is used as a courtship vocalization. Similarly, in chickadees and their allies (Paridae), structurally similar vocalizations are used as calls, without a courtship or aggressive function, in some species, but as songs, with a courtship or aggressive function, in others (Hailman 1989). These examples indicate that the function of displays should not be considered as a criterion for determining putative homology, but rather should be considered separately as another display characteristic that may or may not reflect phylogeny.

Results of phylogenetic studies

A. Phylogenetic Analyses of Visual Displays

Phylogenies based on display behavior and on morphology have been compared explicitly for the manakins (Prum 1990) and the storks (Ciconiidae; Wood 1984). In addition to these, I was able to compare display based phylogenies with morphological phylogenies for the pelicans and allies [displays studied by van Tets (1965), morphological phylogeny developed by Cracraft (1986)], the pelecaniform family containing the cormorants [Phalacrocoracidae; displays studied by van Tets (1965), morphological phylogeny developed by Siegel-Causey (1988)], and the ducks and geese [Anseriformes; displays studied by Lorenz (1941), morphological phylogeny developed by Livezey (1986)]. Prum's study of manakins is done cladistically for both display traits and morphological traits, and should reflect phylogeny. Prum's behavioral phylogeny of manakins is the only extensive study of displays using explicitly cladistic methods. Wood's analysis of storks is phenetic, based on multivariate size and shape variables. He did conduct one morphological analysis in which he approximated outgroup comparison by mathematically "removing" information similar to related groups; he considered this his best analysis of morphological data, and I have used this as his morphological phylogeny. I base my phylogenetic analyses of display traits on the data tables from van Tets, Lorenz, and Wood. These data may contain errors; one particular problem is variation in intensity of sampling. Species described as lacking displays may in fact have been inadequately sampled, since the behavioral repertoire described for a species depends on intensity of sampling, and may give these displays. While this is a problem, incorrect behavioral data should not bias behavioral trees to make them more similar to morphological trees; even if errors are present, high levels of similarity between morphological and behavioral trees should still reflect phylogeny, not sampling bias.

I determine the most parsimonious distribution of the described display traits using the program *PAUP 3.0s* (Swofford 1991). Since Lorenz provided information on displays of geese (*Anser* and *Branta*) as well as ducks, I used the geese as outgroups to determine polarity of characters and root the phylogeny, that is, to identify the likely point of ancestry for the group. For the other groups, I was unable to root the trees based on an outgroup. To compare the nodes shared between the display trees and morphology based trees, I rooted the display trees at the same point that the morphology based trees were rooted.

A moderate to large number of distinct behavioral characteristics, most involving the presence versus absence of a particular stereotyped display, were described for most groups (Table 8.1). This suggests that displays are highly diverse, and unlikely to be subject to chance convergence caused by a limited possible number of ways to perform a display, as was suggested by Atz (1970). This conclusion is also supported by the fact that the consistency indices of display based trees

Table 8.1. Number of display characters and consistency indices (*CI*) for groups in which display based phylogenies could be compared to morphological phylogenies[a]

Group	Number of characters	*CI*	Source
Pelecaniformes	8	1.0	van Tets (1965)
Phalacrocoracidae	10	0.786	van Tets (1965)
Ciconiidae	42	0.65	Wood (1984)
Anseriformes	29	0.694	Lorenz (1941)
Pipridae	35	0.85	Prum (1990)

[a]Only display characters shared by more than one species are counted.

is relatively high, being at least 0.6 for the groups studied (Table 8.1). This is comparable to some, and higher than many, consistency indices of morphology based trees given in de Queiroz and Wimberger's (1993) comparison of morphological and behavioral trees. While *CI* values cannot be directly compared because they tend to decrease with the number of characters used, the overall high *CI* values even for groups with many described display traits, the ducks, storks, and manakins, suggest that display traits are evolving with relatively low levels of convergent evolution, that is, little independent evolution of similar traits in different groups.

There is general agreement between trees based on displays and trees based on morphology in the nodes identified, with perfect congruence between the two trees for the pelecaniform genera (Figs. 1 to 5). In all cases most nodes agree between the morphological and behavioral trees. The least congruence is seen in the ducks and the storks; in both cases, while most nodes agree, there is a group that differs strongly in placement. In neither case is the morphological tree highly consistent [*CI* for ducks = 0.59, Livezey 1986; Wood's (1984) analysis of storks was not cladistic, and his different phenetic analyses gave somewhat different results], so it is unclear whether the morphologically based tree or the display tree is a better representation of the true phylogeny.

Many of the display characters described in these studies occur in a relatively small number of species. Prum (1990) found only three behavioral characters to be broadly distributed in manakins; most traits were shared by only two or three species; many of the displays he examined were apparently derived in only a single species. The other data sets also show many display traits shared among only small numbers of species. van Tets (1965) found more display traits shared between subsets of species within the families he studied, the cormorants, and the boobies and gannets (Sulidae) than were shared by all members of a family. In both Wood's (1984) study of storks and Lorenz's (1941) study of ducks, a larger number of traits were shared between two, three, or four species than between larger numbers of species. This pattern could result from either recent derivation of these displays, or retention of displays in only a small number of species while others have evolved different displays. Either interpretation

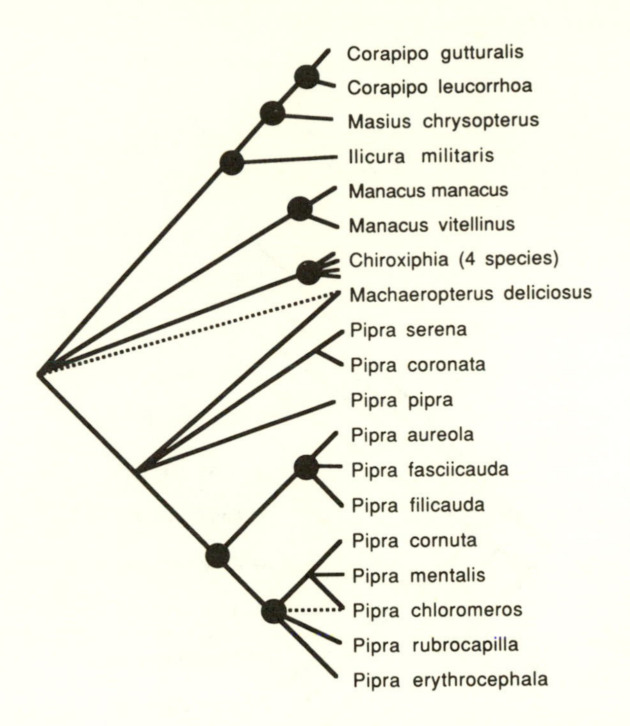

Figure 1. Phylogeny of the manakins based on behavioral traits as analyzed by Prum (1990). Solid circles indicate nodes that are also supported by Prum's (1990) morphology-based phylogeny; dashed lines indicate clades that are placed differently in Prum's morphology-based phylogeny.

suggests that many aspects of display behavior have changed relatively rapidly during evolution.

B. The Phylogeny of Vocal Displays

Several lines of evidence suggest that various aspects of courtship song behavior also reflect phylogeny. In his review of bird song and systematics, Payne (1986) developed two phylogenetic hypotheses based on acoustic structure of vocalizations, one for honeyguides and indicatorbirds (Indicatoridae) and the other for wood-warblers (Parulinae) of the genus *Dendroica*, in the *Dendroica virens* species complex. These phylogenies were based on relatively few vocal character states (three in the Indicatoridae, four in *Dendroica*), but these showed no independent evolution. As described by Payne, the

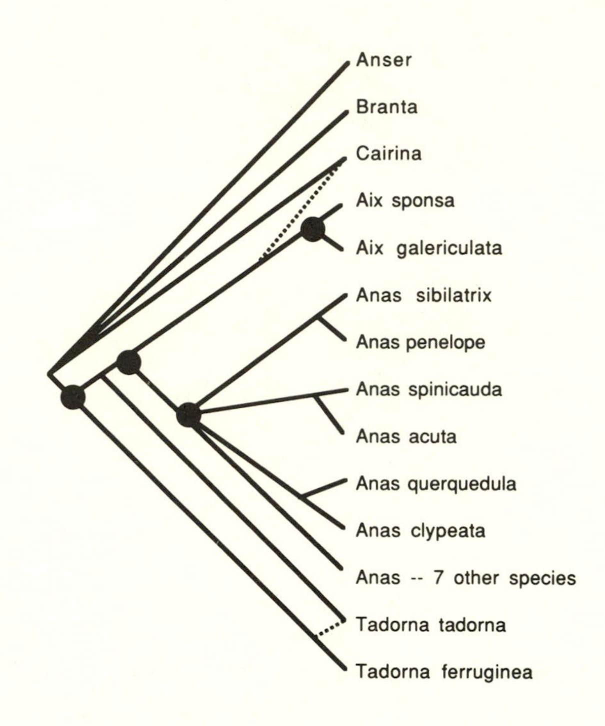

Figure 2. Phylogeny of the ducks based on behavioral traits presented in Lorenz (1941) and re-analyzed using PAUP (Swofford 1991). Tree rooted using *Branta* and *Anser* as the outgroup. Solid circles indicate nodes that are also supported by Livezey's (1986) morphology-based phylogeny; dashed lines indicate the position of differently placed clades in Livezey's phylogeny.

phylogeny of the Indicatoridae is consistent with morphological characteristics given by Friedmann (1955, in Payne 1986), and the phylogeny of the *Dendroica virens* group is consistent with Mengel's (1964, in Payne 1986) proposed pattern of speciation in the group based on biogeography, plumage, and glaciation events. These studies suggest that song structure reflects phylogeny.

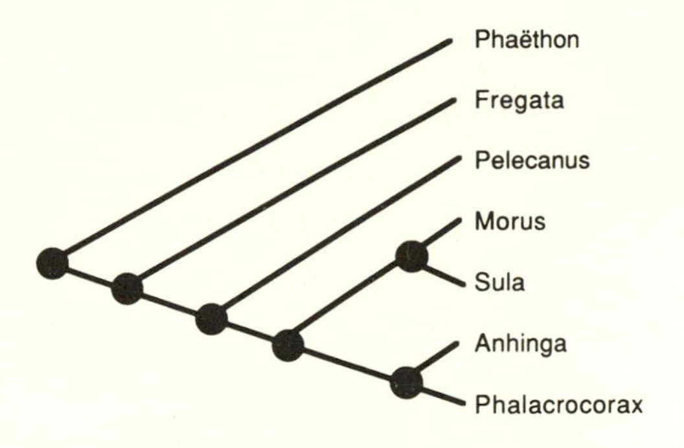

Figure 3. Phylogeny of the Pelecaniformes based on behavioral traits presented in van Tets (1965) and reanalyzed using *PAUP* (Swofford 1991). Tree is rooted to be consistent with Cracraft's (1985) morphological analysis so that nodes other than the root can be compared. Solid circles indicate nodes that are also supported by Cracraft's morphology-based phylogeny; dashed lines indicate the position of differently placed clades in Cracraft's phylogeny.

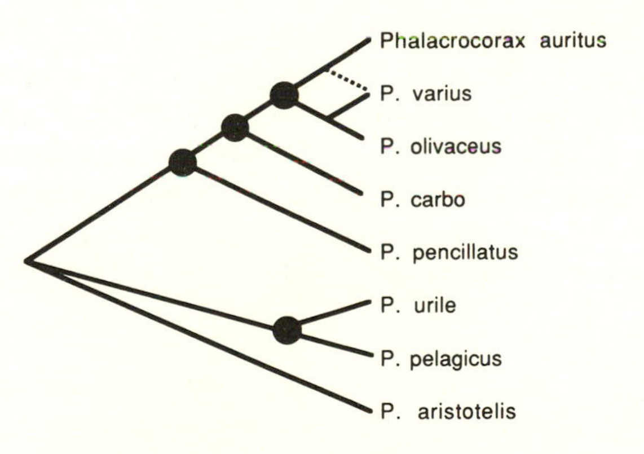

Figure 4. Phylogeny of the cormorants (Phalacrocoracidae) based on behavioral traits presented in van Tets (1965) and reanalyzed using *PAUP* (Swofford 1991). Tree is rooted to be consistent with Siegel-Causey's (1988) morphological analysis so that nodes other than the root can be compared. Solid circles indicate nodes that are also supported by Siegel-Causey's morphology-based phylogeny; dashed lines indicate the position of differently placed clades in Siegel-Causey's phylogeny.

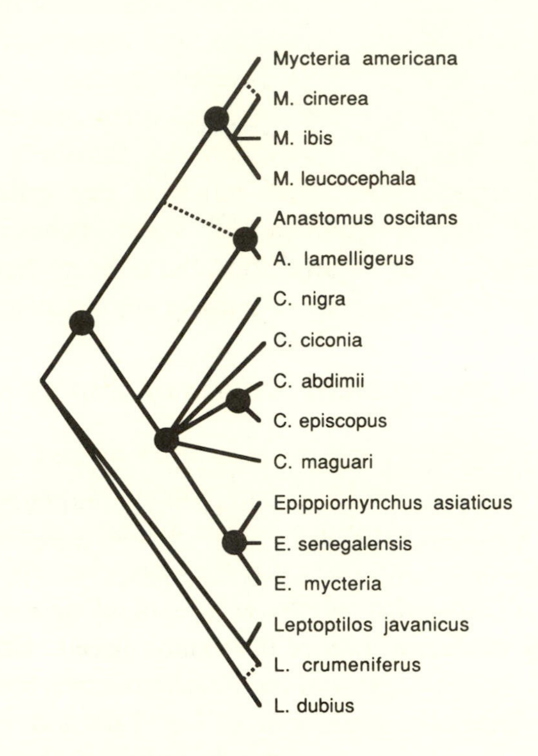

Figure 5. Phylogeny of the storks (Ciconiidae) based on behavioral traits presented in Wood (1984) and reanalyzed using *PAUP* (Swofford 1991). Tree is rooted to be consistent with Wood's (1984) morphological analysis so that nodes other than the root can be compared. Solid circles indicate nodes that are also supported by Wood's morphology-based phylogeny; dashed lines indicate the position of differently placed clades in Wood's phylogeny.

I found evidence that a repertoire of a single structural song type was a derived character shared by all five members of the sparrow genus *Zonotrichia* when I examined the evolution of sparrow song characteristic traits using a phylogeny developed by Zink (1982) and reanalyzed by Lanyon (1985) based on allozyme variation (Irwin 1988). The song structure of having only a single repeated syllable type in a song, however, was derived twice, independently. Song repertoire size is highly variable among related species in the New World blackbirds, however, suggesting that it does not strongly reflect phylogeny (Irwin 1990). Spector (1992) suggested several other traits that may reflect phylogeny in wood-warblers, including the presence of two functionally and structurally distinct song categories, and aspects of the repertoire within each song category. Finally, certain aspects of song structure, such as the presence of long, variable, and often mimetic songs in all members of the family of mockingbirds, thrashers, and catbirds (Mimidae), also suggest that vocal structure can reflect phylogeny. Taken together, the descriptions of vocal behavior suggest a phylogenetic component to vocalizations, but indicate that the aspects of song that are related to phylogeny may differ among avian groups.

Implications of phylogenetic studies of display behavior

Since both visual displays and some aspects of bird song apparently reflect phylogenetic relationships, they should be valuable systematic characters. The fact that many distinctive displays are found in only a small number of species suggests that they may be especially valuable in determining relationships within groups of closely related species. Their value may decline at higher taxonomic levels, although the fact that the ordinal level phylogeny of the Pelecaniformes based on displays is both highly internally consistent (CI = 1.0) and congruent with Cracraft's (1986) morphology-based tree indicates that some displays have been retained over long periods of evolutionary time. The phylogenetic component to display behavior also emphasizes the importance of considering phylogeny when conducting comparative studies of aspects of displays to test hypotheses about evolutionary process.

Atz' (1970) claim that behavior evolves rapidly is supported to some extent by the result that many display traits are found in only a relatively small number of species. This rapid evolution did not, however, cause high levels of convergence, as indicated by the relatively high consistency indices, and the congruence between morphological and display-based trees.

It is interesting to consider the implications of apparent high levels of evolutionary lability, but little convergent evolution. One possible explanation of this pattern is that the particular form of a display is relatively arbitrary; that is, there are diverse ways of giving equally effective displays for at least some contexts, so that convergence does not occur. With regard to courtship, a reason for this could be that a major function of courtship is species identification. Traits important in species identification might be initially considered unlikely to reflect phylogenetic similarities among species, but if they evolve such that specific aspects of ancestral displays are modified to be species-specific, leaving much of the display unchanged, they would show a pattern very much like the pattern seen here. This is highly speculative, but several lines of evidence suggest that it is worth further testing. The examples of transfer of function of gull "meeting ceremonies" and blackbird "whistle" songs discussed above both involve modification of apparently previously existing displays to form a distinctive courtship display; in both gulls and in the cowbirds that use the "whistle" as a song, the display is known to function specifically in courtship, rather than aggression (Tinbergen 1959, Rothstein et al. 1986). Further, while songs of mimids (mockingbirds and their allies) are, as discussed above, superficially similar, there are differences in details of structure, such as the number of times syllables are repeated, and it is these differences that function in species recognition (Boughey & Thompson 1976). These examples suggest that displays involved in courtship or species recognition are made up of many aspects that are more generally distributed among related species, with a few species specific details being adequate for species recognition.

The apparent rapid evolution of the aspects of display discussed here also has implications for interpreting results of other studies of the phylogenetic component of behavior. de Queiroz and Wimberger (Chap. 7 in this volume) found that behavioral traits that have been used in phylogenetic reconstruction have a phylogenetic component comparable to that of morphological traits, but Gittleman et al. (Chap. 6 in this volume) found that behavioral traits evolve more rapidly than do morphological traits, and suggested that behavioral traits show little dependence on phylogeny. The phylogenetic pattern that I have found for avian displays suggests that behavioral traits evolve rapidly, as found by Gittleman et al., but that they may still reflect phylogeny, as found by de Queiroz and Wimberger, if rapid evolution does not lead to a high degree of convergence. The traits discussed by Gittleman et al. (home range size and group size in mammals) seem likely to show convergence if they evolve rapidly. There are probably only a limited number of biologically reasonable home range or group sizes, so rapid evolution is expected to result in independent evolution of these traits by chance. As I have discussed, displays apparently are not limited strongly in the form they can take; the diverse behavioral traits studied by de Queiroz and Wimberger may also be relatively unlimited in possible form and may be less likely to evolve independently even if they do change rapidly.

Other aspects of display behavior may be more subject to convergence than those I have examined. The display traits that have been used for phylogenetic analysis so far are not sufficiently varied to test whether certain aspects, or categories, of displays are more likely to show convergence than others. Some of the early ethological studies involved description of more aspects of displays than were ultimately used to construct phylogenies and may show more convergent evolution, suggesting adaptation. For example, in his monograph on the Pelecaniformes, van Tets (1965) discussed display intensities, and whether displays were given by both members of a pair or only one sex. These data were not included in his phylogenetic analysis, probably because they were too variable among both individuals and groups of species other traits show to be close relatives. It is common practice for systematists not to use such "noisy" data, but studying this variation phylogenetically to test for adaptation could prove very interesting. Other display traits have been explicitly hypothesized to be adapted to particular ecological or social situations, and should be studied phylogenetically to test for convergence. For example, Morton (1975) suggested that certain aspects of acoustic structure of bird songs are adapted to the sound carrying properties of the habitat. The acoustic features he discussed, such as pitch, are generally not those details of syllable sequence that I have suggested reflect phylogeny. Since his comparative study was done before methods for distinguishing convergence from possible phylogenetic effects were developed, it should be reevaluated to determine whether these acoustic features do evolve independently in association with certain habitat types. Similar studies could be carried out for aspects of visual displays that are hypothesized to differ in signaling effectiveness in different habitats. Another area where phylogenetically based comparative studies should be applied to displays is in the study of sexual selection. For example, it might be that the exact form of a display is not subject to convergence, but that the energetics of a display, which could reflect male quality, is.

One additional area in which visual displays and songs may be used to study questions hard to address with other behavior is in the comparison of genetic and cultural evolution. Payne (1986) found indications of phylogenetic information in songs of two groups of bird, the Indicatoridae, in which songs are thought to be innate, and *Dendroica* warblers, the songs of which are influenced by learning, and suggested that cultural evolution of learned songs might in fact reflect phylogeny. Since bird song in songbirds does often have genetic as well as learned influences, studies of the specific song traits that reflect phylogeny in different groups need to be related to these different influences on the development of those traits. In addition, studies of visual display development are needed to determine whether they, too, have learned and innate influences and could provide a model for the relatively poorly understood processes of cultural evolution.

ACKNOWLEDGMENTS

I thank P. Kukuk, S. A. Sloane, D. A. Spector, and two anonymous reviewers for making constructive comments on the manuscript. I am very grateful to Emília Martins for doing an excellent job of organizing this symposium, and for inviting me to participate in it.

References

Atz, J. W. 1970. The application of the idea of homology to behavior. In: *Development and Evolution of Behavior* (Ed. by L. R. Aronson, & E. Tobach), pp. 53–74. San Francisco: W. H. Freeman.

Boughey, M. J., & N. S. Thompson. 1976. Species specificity and individual variation in the songs of the brown thrasher (*Toxostoma rufum*) and catbird (*Dumetella carolinensis*). *Behaviour*, 57, 64–90.

Cracraft, J. 1986. Monophyly and phylogenetic relationships of the Pelecaniformes: a numerical cladistic analysis. Auk, 102, 834–853.

de Queiroz, A., & P. H. Wimberger. 1993. The usefulness of behavior for phylogeny estimation: Levels of homoplasy in behavioral and morphological characters. *Evolution*, 47, 46–60.

Fisher, R. A. 1954. *The Genetical Theory of Natural Selection*, 2nd ed. New York: Dover.

Hailman, J. P. 1989. The organization of major vocalizations in the Paridae. *Wilson Bull.,* 101, 305–343.

Harvey, P. H., & M. D. Pagel. 1991. *The Comparative Method in Evolutionary Biology*. Oxford: Oxford University Press.

Irwin, R. E. 1988. The evolutionary significance of behavioural development: the ontogeny and phylogeny of bird song. *Anim. Behav.*, 36, 814–824.

Irwin, R. E. 1989. A comparative study of sexual selection on song repertoire size in the avian subfamily Icterinae (Passeriformes: Emberizidae). Ph.D. dissertation. University of Michigan.

Irwin, R. E. 1990. Directional sexual selection cannot explain variation in song repertoire size in the New World Blackbirds (Icterinae). *Ethology*, 85, 212–224.

Kroodsma, D. E. 1982. Learning and the ontogeny of sound signals in birds. In: *Acoustic Communication in Birds* (Ed. by D. E. Kroodsma, & E. J. Miller), pp. 1–23. New York: Academic Press.

Kroodsma, D. E., & R. A. Canady. 1985. Differences in repertoire size, singing behavior, and associated neuroanatomy among marsh wren populations have a genetic basis. *Auk*, 102, 439–446.

Lande, R. 1981. Models of speciation by sexual selection of polygenic traits. *Proc. Natl. Acad. Sci. USA*, 78, 3721–3725.

Lande, R., & S. J. Arnold. 1985. Evolution of mating preference and sexual dimorphism. *J. Theor. Biol.*, 117, 651–664.

Lanyon, W. E. 1957. The comparative biology of the meadowlarks (*Sturnella*) in Wisconsin. *Publ. Nuttall Ornithol. Club*, 1, 1–67.

Lanyon, S. M. 1985. Detecting internal inconsistencies in distance data. *Syst. Zool.*, 34, 397–403.

Livezey, B. C. 1986. A phylogenetic analysis of recent anseriform genera using morphological characters. *Auk*, 103, 737–754.

Lorenz, K. 1941. Vergleichende Bewegungsstudien an Anatinen. *J. Ornithol. Suppl.*, 89, 194–293.

Marler, P., & R. Pickert. 1984. Species-universal microstructure in the learned song of the swamp sparrow (*Melospiza georgiana*). *Anim. Behav.*, 32, 673–689.

Morton, E. S. 1975. Ecological sources of selection on avian sounds. *Am. Nat.*, 109, 17–34.

Nero, R. W. 1964. Comparative behavior of the yellow-headed blackbird, red-winged blackbird, and other icterids. *Wilson Bull.*, 75, 376–413.

Orians, G. H., & G. M. Christman. 1968. A comparative study of the behavior of red-winged, tricolored, and yellow-headed blackbirds. *Univ. Calif. Publ. Zool.*, 84, 1–81

Payne, R. B. 1986. Bird songs and avian systematics. *Curr. Ornithol.*, 3, 87–126.

Prum, R. O. 1990. Phylogenetic analysis of the evolution of display behavior in Neotropical manakins (Aves: Pipridae). *Ethology*, 84, 202–231.

Rothstein, S. I., D. A. Yokel, & R. C. Fleischer. 1986. Social dominance, mating and spacing systems, female fecundity, and vocal dialects in captive and free-ranging brown-headed cowbirds. *Curr. Ornithol.*, 3, 127–185.

Siegel-Causey, D. 1988. Phylogeny of the Phalacrocoracidae. *Condor*, 90, 885–905.

Spector, D. A. 1992. Wood-warbler song systems: a review of paruline singing behaviors. *Curr. Ornithol.*, 9, 199–238.

Swofford, D. L. 1991. *Phylogenetic Analysis Using Parsimony (PAUP), version 3.0s.* Champaign, Illinois: Illinois Natural History Survey.

Tinbergen, N. 1959. Comparative studies of the behaviour of gulls (Laridae): A progress report. *Behaviour*, 15, 1–70.

van Tets, G. F. 1965. A comparative study of some social communication patterns in the Pelecaniformes. *Ornithol. Monogr.*, 2, 1–88.

Wood, D. S. 1984. Concordance between classifications of the Ciconiidae based on behavioral and morphological data. *J. Ornithol.*, 125, 25–37.

Zink, R. M. 1982. Patterns of genic and morphologic variation among sparrows in the genera *Zonotrichia*, *Melospiza*, *Junco*, and *Passerella*. *Auk*, 99, 632–649.

CHAPTER 9

Comparative Analysis of the Origins and Losses of Eusociality: Causal Mosaics and Historical Uniqueness

Bernard J. Crespi

Since Darwin (1859) described the caste dimorphism of social insects as one special difficulty that appeared fatal to his whole theory, analysis of the causes of insect sociality has been in the forefront of research in evolutionary biology. Much of this research has been comparative, in that selective pressures common to diverse eusocial taxa have been sought as unitary explanations for the evolution of alloparental care and sterile castes (e.g., Hamilton 1964; Andersson 1984; Alexander et al. 1991). Although many such studies have implicitly or explicitly recognized the importance of analyzing independent origins of eusociality as data points (e.g., Michener 1974; Ross & Carpenter 1991; Edwards & Naeem 1993 on cooperative breeding), students of eusociality have yet to use any of the large suite of recent explicit comparative methods (reviewed in Maddison 1994; Losos & Miles 1994) to search for the causes of eusociality within or between clades comprising eusocial and noneusocial forms.

The purpose of this chapter is to discuss and apply comparative methods in the analysis of the causes of the origins and losses of eusociality in insects. First, I introduce the main questions to be addressed and the specific problems to be solved. Second, I briefly categorize the comparative methods that have burgeoned in the literature and discuss their advantages and disadvantages with respect to these questions and problems. Third, I apply what I consider to be the most appropriate methods to each clade that contains eusocial forms, with special emphasis on analysis of the evolution of soldier castes in Australian gall thrips (Crespi 1992a,b; Mound & Crespi 1994; Crespi & Mound in press). Finally, I evaluate the results of these analyses, and attempt to map out a productive course for future work on this endlessly fascinating difficulty for evolutionary theory.

What is eusociality?

Analyzing the evolution of eusociality requires consideration of just what trait has evolved. The term "eusociality" has been traditionally defined by the presence of generation overlap, cooperative brood care, and reproductive division of labor (Michener 1969, 1974; Wilson 1971). Crespi and Yanega (1994), Sherman et al. (1994), and Gadagkar (1994) have recently pointed out that the criterion of reproductive division of labor is sufficiently vague as to be interpreted in quite different ways by different authors. Crespi and Yanega (1994) suggest a more precise definition of eusociality, based on the presence or absence of discrete and permanent differences in life history after reproductive maturity. Individuals with distinct life histories with respect to helping and reproduction are considered to be of different castes, and societies with such individuals are eusocial; by contrast, where such differences are temporary, as in all birds with alloparental care, the resultant societies are referred to as cooperatively breeding. Sherman et al. (1994) propose instead that eusociality should be viewed as a continuum of degrees of reproductive skew, comprising all species with alloparental care. By their scheme, pooled values of reproductive skew for each species form a range along the skew axis and the search for convergent causes focuses on specific areas along the continuum. Gadagkar (1994) takes a similar view that the term "eusociality" should be extended to all forms with alloparental care, though he proposes no quantitative measure.

To test for the causes of variation in social behavior using all of the diverse arthropods and vertebrates that exhibit alloparental care, a definition of eusociality based on degree of reproductive skew is problematic because quite different societies may have similar values for skew for different reasons (Crespi & Yanega 1994). Thus, although analyses using reproductive skew as a dependent variable should be useful within ecologically homogeneous clades of social animals, such as polistine wasps (Reeve 1991), the use of skew as a universal indicator of social level or complexity may lead to more difficulties than it solves.

Wcislo (in press a) noted that all criteria for distinguishing biological phenomena that are not based in phylogenies are artificial (i.e., unnatural mental constructs), and hence are arbitrary with respect to their importance and priority for organizing and explaining diversity. He suggested that, following West-Eberhard (1978), the phenomenon under study should be defined only with respect to the specific test being conducted in the specific clade. I agree with Wcislo (in press a) that usefulness of definitions is the main criterion for adopting them, and believe that the Crespi and Yanega (1994) conception of castes, and division of societies into eusocial (with castes) and cooperatively breeding (without castes) provides a useful tool for analyzing the most important and taxonomically broad questions in the study of sociality. Thus, in our comparative search for the causes of eusociality, we seek causes of the origin of permanent differences in life history, whereby one caste of individuals, the workers or soldiers, engages in more alloparental care than the other, more reproductive individuals. We recognize, however, that different clades could have reached this more-or-less adaptive peak by different routes, such that a single explanation for eusociality is worth seeking but certainly may not exist.

The main questions in the study of the evolution of eusociality are why some taxa exhibit eusociality whereas others do not, and what selective pressures explain behavioral and morphological variation within eusocial colonies and taxa? Students of sociality are usually careful to distinguish between the origin of eusociality, its maintenance, and the explanation of variation within eusocial societies. The origin and loss of eusociality are historical questions, accessible using phylogenetic inference when information about past events has not been obscured by rapid anagenetic change. By contrast, the methods of functional design (*sensu* Williams 1966), and measurement of selection (e.g., Lande & Arnold 1983; Crespi 1990) can be used to analyze the causes of the maintenance and current forms of eusocial societies.

Hypotheses for explaining the origin and maintenance of eusociality

Three main forms and sources of selective pressure have been proposed to explain how and why alloparental care and eusociality evolve. First, ecological costs and benefits may vary between taxa in such a way as to favor helping in some lineages but not others. Ecological benefits of alloparental care may include better protection against predators and parasites (Lin 1964; Lin & Michener 1972), and habitation of an already constructed nest, proven as a good site for producing offspring. These ecological benefits are intimately associated with aspects of demography (Queller 1989, 1994; Gadagkar 1990) that may favor staying at home and helping. Data from temporal variation in ecological factors shows that within some bird species, higher costs of independent reproduction, relative to remaining at home, are associated with increased rates of helping (Emlen 1992). However, only one among-species comparative test of demographic and ecological causes of helping has so far been conducted (Edwards & Naeem 1993).

Second, variation among lineages in genetic factors, such as relatedness, relatedness asymmetry, biased sex ratios, inbreeding, and likelihood of fixation of alleles for helping (e.g., Hamilton 1964; Grafen 1986; Reeve 1993; Mueller et al. 1994) may favor the evolution of eusociality in some taxa. Although high relatedness, relatedness asymmetries, and female-biased sex ratios exist in many populations showing eusociality, genetic variables have not yet been shown to differ systematically between taxa in such a way as to implicate them in the origin or loss of eusociality.

Third, behavioral dynamics within societies may lead to eusociality if some individuals have control over others and adjust the costs and benefits of their behavior so as to favor helping over dispersal (Alexander 1974; Michener & Brothers 1974). This "parental manipulation" hypothesis predicts that mothers reduce the fertility or dispersal ability of offspring in such a way as to increase the manipulator's personal reproduction (e.g., Maeta et al. 1992). This hypothesis would be supported by the presence or higher intensity of manipulation in lineages with alloparental care than other lineages; no such data exists or has yet been sought in this context.

These three hypotheses for the evolution of eusociality are not mutually exclusive (Evans 1977), and treating them as independent, competing hypotheses probably hinders rather than helps progress in understanding. A useful unifying framework for considering the three hypotheses together is to view the evolution of alloparental care and eusociality as arising from selection on behavioral ontogenies. Thus, developing individuals are selected, over evolutionary time, either to stay in their natal nest or leave and attempt to found their own nest. If they stay, then they may either help or not help; if they help, then they may help either temporarily or permanently (Brockmann in press). Each route incurs various costs (c) and benefits (b), and each decision by an individual is expected to satisfy Hamilton's rule, $r\,b - c > 0$, where r is the coefficient of relatedness. Using this scheme, the search for the causes of eusociality becomes an analysis of selection of alternative ontogenetic decisions and their timing, by identifying and comparing the agents of selection, and the effects of selection, between lineages. Moreover, eusociality can be precluded by any factors that block the behavioral ontogeny at any stage, or facilitated or caused to varying degrees by factors at any of the stages. To analyze the causes of variation in ontogenetic trajectories, we must compare lineages with respect to morphology, ecology, demography, genetics, and the manipulation of individual's behavioral decisions (Table 9.1).

Comparative methods for testing hypotheses

Given a definition of the phenomenon under study, hypotheses to analyze, and a conceptual approach to considering the hypotheses jointly, analysis of the causes of alloparental care and eusociality next requires (a) choice of populations, clade or clades; (b) choice of ecological, morphological, demographic and genetic variables to measure; and (c) choice of philosophical and statistical methods for combining historical information with other information to judge the validity of the hypotheses. Each of these three decisions should be guided, at least in part, by the distribution of variation in traits related to social behavior at different genealogical levels, from individuals to higher taxa. Variance and covariance may thus be present among individuals, among nests, among populations within species, and among lineages recent or ancient.

Table 9.1. Genetic, morphological, developmental, demographic, and ecological traits that may influence the evolution of eusociality[a]

Type of trait	Trait	Taxon	Influence	References
Genetic	Haplodiploidy	Hymenoptera, Thysanoptera	Higher relatedness to siblings than to offspring	Hamilton (1964)
	Inbreeding	Isoptera, Naked mole rats	High within-colony relatedness	Bartz (1979), Myles and Nutting (1988), Shellmann-Reeve (in press), Reeve et al. (1990)
	Protected invasion	Hymenoptera, Thysanoptera	Easier fixation of alloparental care alleles	Reeve (1993)
	Biased sex ratios	Hymenoptera, Thysanoptera	Higher relatedness to helped offspring	Hamilton (1972), Trivers & Hare (1976)
Morphology	Ability to defend	All taxa with weapons	Increased gains from helping	Kukuk et al. (1989), Starr (1989)
	Ability to feed offspring	Hymenoptera, Isoptera, Naked mole rats	Increased gains from helping	—
	Ability to build	Hymenoptera, Isoptera	Increased gains from helping	Hamilton (1972)

Type of trait	Trait	Taxon	Influence	References
Morphology	Wing polymorphism	Isoptera	Higher fecundity of winged forms	Taylor (1978)
Behavior	Produce small subfertile offspring	Hymenoptera	Increase likeiihood of helping	Alexander (1974)
Symbioses	Gut symbionts	Isoptera	Group-living obligatory, permanent helping favored	Higashi et al. (1992), Shellman-Reeve (in press)
Natural variation in fertility	Higher variation	Hymenoptera: Polistinae	Favors reproductive division of labor	West-Eberhard (1978), Gadagkar (1991)
Natural variation in size	Higher variation	Hymenoptera: Allodapini	Promotes coexistence in nest	Sakagami and Maeta (1984)
Development	Hemimetaboly	Isoptera	Can work when juveniles	Alexander et al. (1991), Queller (1989)
	Longer life span of queens	Hymenoptera: Allodapini	Leads to generation overlap	Schwarz (in press)
	Shorter life span of workers	All	Favors staying and helping, loss of totipotency	Queller (1994)

(continued)

Table 9.1 (cont.)

Type of trait	Trait	Taxon	Influence	References
Development	Inheritance of the nest	Isoptera, others?	Favors temporary helping in hopes of inheritance	Myles (1988)
Breeding seasonality	Seasonal	Hymenoptera, Thysanoptera?, Homoptera?	Difficult to mate and start own colony some times of year	Yanega (in press)
Nature of food	Transportable	Hymenoptera, some Isoptera, Naked mole rats	More helping possible	—
	Can be shared	Hymenoptera, Isoptera	Helping more effective	—
	In tiny units	Hymenoptera: Sphecidae	Joint work on cells favoreed	Matthews (1991)
	In large, rare units	Naked mole rats	Foraging by helpers more important to survival	Jarvis et al. (1994)
Nature of nest	Structure allows interactions	Hymenoptera: Allodapini, Sphecidae	Creates conditions for dominance interactions	Matthews (1991), Schwarz (in press)

Type of trait	Trait	Taxon	Influence	References
Nature of nest	Persists long time	Hymenoptera, Isoptera, Coleoptera, Homoptera, Thysanoptera	Helping a better long-term strategy	Alexander et al. (1991)
	Defensible	Hymenoptera, Isoptera, Coleoptera, Homoptera, Thysanoptera	Helping increases nest success	Alexander et al. (1991)
	Expandable	Hymenoptera, Isoptera,	Increases benefits of helping	Alexander et al. (1991)
Predator or parasite pressure	High pressure from attackers	Hymenoptera	Selects for alloparental defense	Lin (1964), Lin and Michener (1972)
	High host-specificity of attackers	All?	Stronger selection for defense, via coevolution	Crespi (1994)
	Ability of attackers to destroy entire nest	All?	Stronger selection for defense	Alexander et al. (1991)

[a]Comparative support for the influence of the traits is shown in Table 9.4.

Analyses using individuals or nests as units are not usually considered comparative per se, since the anastomosis of pedigrees through sex obfuscates patterns of descent with modification. Thus, although comparative methods could certainly be applied to asexual forms, few eusocial taxa have taken this route, and comparisons of individual decisions to help or not help, or differences among nests that favor helping in some but not others, are best addressed using the methods of functional design or measurement of selection.

Most comparative studies of adaptation use species-level or genus-level phylogenies, and compare values of traits for extant taxa or inferred ancestral states. These analyses can be divided into two types, according to the breadth, depth, and form of their usage of phylogenies (Pagel 1994; in this volume, Martins & Hansen, Chap. 2). First, character state changes can be inferred along single clades, to analyze transformations of either single characters or sets of characters for which hypotheses of causal relationship have been proposed or are being sought (Coddington 1988; Baum & Larson 1991). This transformational approach to analyzing adaptation may also involve attempts to study the fitness consequences of macroevolutionary changes. However, the presumption that more recently evolved traits engender enhanced performance at the same selective task, rather than performance of somewhat different tasks more or less equally well, may be met in only some situations (Frumhoff & Reeve 1994). Moreover, the combined uncertainties in phylogeny reconstruction and character optimization make statistical tests, assessing the likelihood of one historical reconstruction relative to others (e.g., Templeton 1983; Kishino & Hasegawa 1989), necessary before much confidence can be placed in any specific result. Transformational studies have been conducted in the analysis of social behavior to infer how many origins or losses of different social systems have occurred (e.g., Winston & Michener 1977; Stern 1994), and which social systems tend to evolve into which others (e.g., West-Eberhard 1978; Carpenter 1989; Packer 1991; Richards 1994; Danforth & Eickwort 1995). Such studies are essential for in-depth analysis of the causes of specific transitions, as single data points for comparisons of sister-taxa that have diverged in sociality.

Second, hypotheses of common selective causes or facilitation for multiple parallel or convergent changes in traits can be tested (e.g., Felsenstein 1985; Martins & Garland 1991; Garland et al. 1993; Maddison 1994). The convergence approach has long been used informally in the analysis of insect eusociality, through comparisons of disparate lineages such as Hymenoptera with Isoptera (Andersson 1984), different families of Hymenoptera with one another (Michener 1974), and even social insects with vertebrates (Strassmann & Queller 1989; Alexander et al. 1991). Such comparisons suffer considerably from failure of the assumption of *ceteris paribus* in postulating specific causes, but they have the virtue of potentially yielding results of general significance for the study of sociality.

Determining which of these two approaches is best for analyzing eusociality requires considering just what we mean by the phrase "causes of eusociality." Various ecological, demographic, genetic, and behavioral conditions may influence the origin or loss of eusociality (Table 9.1); some conditions may be necessary, and one or more sets of these conditions may be sufficient. Thus, we are seeking one or more sets of variables whose values can accurately predict the presence or absence of eusociality in a particular monophyletic group (Hölldobler & Wilson 1990). The taxa comprising the dependent variable span a vast range of developmental, morphological, and behavioral bauplans, including Hymenoptera, Isoptera, Coleoptera, Thysanoptera, and Homoptera, and each species with eusocial forms within these clades should be viewed as a unique, historical individual. Given four considerations: (a) the present lack of well-resolved phylogenies for most clades containing eusocial forms; (b) the uncertainties concerning whether or not all cases of eusociality have the same set of causes; (c) the likelihood that one or two variables alone cannot accurately predict the origin or loss of eusociality, due to the absence of ceteris paribus among morphologically and ecologically-divergent lineages; and (d) the rarity of transitions to and from eusociality, I consider application of any of the current battery of statistical tests either premature or misleading for almost all groups of social insects. Instead, I will analyze the causes of eusociality using two approaches that are comparative but take account of the phylogenetic uniqueness of each clade and the possibility of diverse selective causes for eusociality.

First, for each insect taxon with eusocial taxa, I will use taxonomic and phylogenetic information to determine which transitions, of all possible transitions, have taken place between different social systems. This approach, which was developed by Gittleman (1981), allows assessment of how selection moves social systems between adaptive peaks, which provides clues as to the agents of selection that cause the transitions.

Second, I will compare the patterns of similarities and differences, in traits that are predicted from behavioral-ecological data to be important in the evolution of eusociality, between eusocial lineages and their noneusocial sister taxa and between clades that contain at least one eusocial species and their sister clades that do not. The latter comparisons may allow recognition of which character states are sufficient for the evolution of eusociality, and the former comparisons may provide information about which character states are necessary, sufficient, or preventative (e.g., Table 9.2). Characters are chosen for inclusion on the basis of behavioral-ecological studies indicating that their functional design is relevant to the evolution of the behavioral ontogenies expressed as eusociality. To the extent that sister taxa in diverse, phylogenetically distinct lineages differ in one or more variables from the same small set of hypothesized factors necessary for the origin or loss of eusociality, common causes of eusociality exist. By contrast, to the extent that taxonomic groups differ in the variables whose changes coincide with changes in the presence of eusociality, the causes of eusociality must be sought in the idiosyncrasies of history within each lineage. Thus, I use the transformation approach, at each origin or loss of eusociality, to infer which variables have precipitated the change, and to infer which variables and states of these variables are necessary or preventative for the transitions or their absence. Once the set or sets of relevant causal variables are inferred, the convergence approach can be used for testing the generality of hypotheses of these sets of necessary, sufficient, and preventative conditions. I focus on taxa for which sufficient taxonomic or phylogenetic information is available for meaningful discussion. For other taxa, I suggest clades most productive for future work, and variables that I consider most useful to measure.

Table 9.2. An example of the necessity and sufficiency approach to analyzing the causes of the evolution of eusociality[a].

Taxa	Characters				
	A	B	C	D	E
Eusocial species 1	2	2	2	2	1
Eusocial species 2	2	2	1	3	1
Not eusocial, sister-taxon	2	1	2	1	1
Not eusocial, other clade	2	2	2	3	2
Not eusocial, other clade	1	1	2	1	1

[a]Character states A2, B2 and E1 may be necessary for the origin of eusociality, A2+B2+C2+E1, or A2+B2+D3+E1 may be sufficient for the origin of eusociality, and E2 may prevent the origin of eusociality regardless of the states of the other characters. Choice of characters requires behavioral-ecological and demographic information, and determining which characters are necessary, sufficient, and preventative requires comparisons of eusocial and noneusocial taxa within diverse clades.

Analysis

A. Thysanoptera

Among Thysanoptera, eusociality is found only among species that form galls, and only among the Australian gall-formers. These species live on *Acacia*, usually in arid or semiarid habitats, and make galls that are more securely closed to the outside than those of gall-making thrips on other types of plant.

Morphological and mitochondrial DNA sequence data are most consistent with two origins and no losses of soldier castes in Australian gall-forming thrips, once each in the genera *Oncothrips* and *Kladothrips* (Crespi & Mound in press)(Fig. 1). The differences between the species of *Oncothrips* with soldiers, and their sister-group without soldiers (*O. rodwayi*), include (a) the presence of enlarged forelegs, used for fighting over incipient galls, in foundresses of the taxa with soldiers among their offspring, but not in *O. rodwayi*; (b) habitation of arid or semiarid regions in the species with soldiers, but a mesic habitat for *O. rodwayi*; and (c) high rates of kleptoparasitism (theft of galls) by *Koptothrips* in the species with soldiers, but low rates in *O. rodwayi*. The differences between the two *Kladothrips* species with soldiers, and their apparent sister group, a clade containing *K. ellobus, K. rugosus*, and two undescribed species, include (a) probable longer life-cycle duration in the clade with soldiers; (b) the lack of pupae in galls of the species without soldiers, which suggests that second-instar larvae of these other species leave the gall to pupate, and (c) high levels of *Koptothrips* invasion in the two species with soldiers, and low levels of invasion in *K. rugosus* and its sister taxa, but similarly high levels in *K. ellobus* and *K. acaciae*. All *Kladothrips* species exhibit enlarged forelegs and observed or inferred fighting among foundresses.

From these data on phylogeny, behavior, demography, and ecology, I hypothesize that three conditions — (a) high kleptoparasitism pressure, (b) the presence of enlarged forelegs and fighting among foundresses, and (c) pupation inside of the galls — are necessary and sufficient for the origin of soldiers (Fig. 1). Enlarged forelegs and fighting in foundresses may be a preadaptation for their evolution in soldiers; a lack of pupation in the gall may either preclude selection for defense by adult offspring or make defense less beneficial, since there are fewer siblings to defend; and the presence of high kleptoparasitism pressure may be the main selective force for defense, which only leads to soldiers given this particular phenotypic variation upon which to act.

We must still determine whether or not relatedness is higher, or sex ratios more female-biased, in clades with soldiers than without soldiers, and analyze the phylogenetics, ecology, and behavior of the remaining Australian gall thrips before these ideas can be tested further.

Among the thrips on *Acacia* that do not form galls but live in small cavities or glue phyllodes (petioles modified to serve as leaves) together to form a living space, at least several species form communal (nest-sharing without alloparental care) groups of adults (Crespi & Mound in press). Mitochondrial DNA evidence indicates that these phyllode-gluing taxa are phylogenetically distinct from the gall formers (Crespi, unpublished data). Thus, the only observed transitions are between eusocial and solitary forms and between solitary and communal forms.

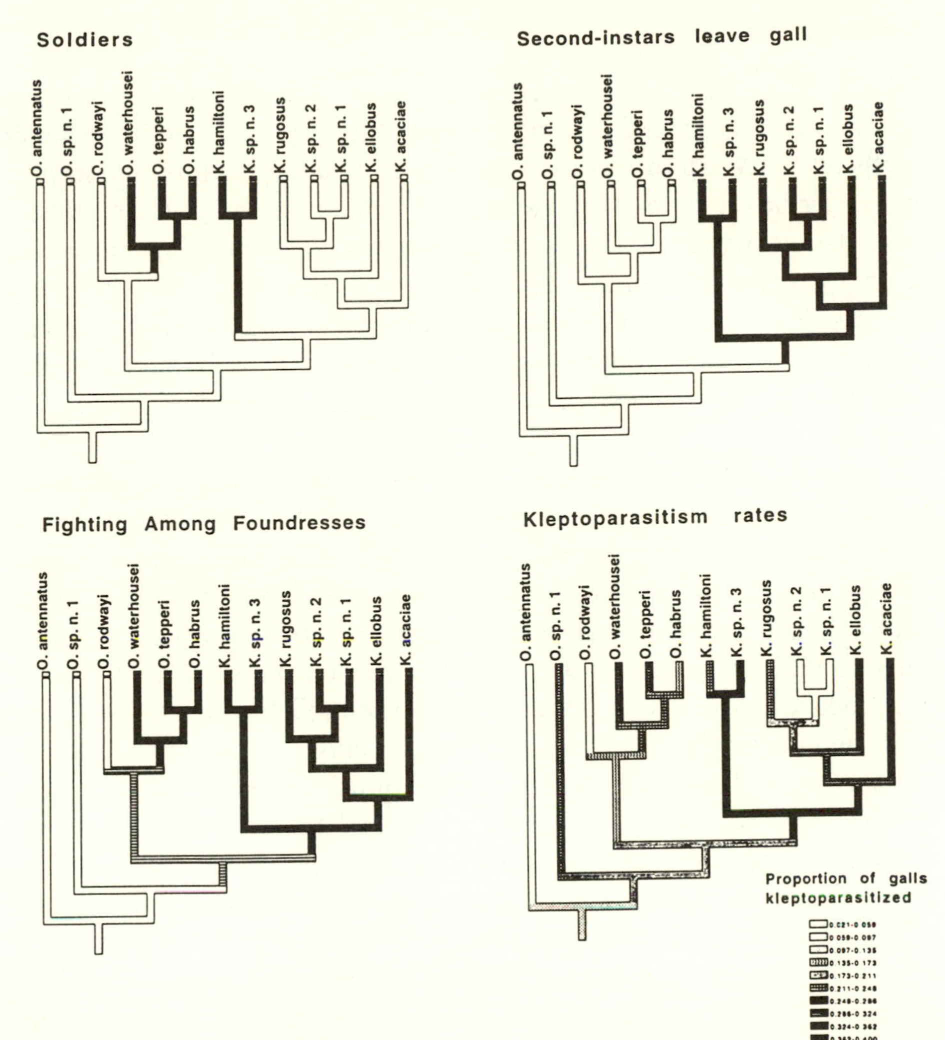

Figure 1. The evolution of soldier castes in Australian gall thrips, and three conditions hypothesized to be necessary and sufficient for the origin of soldiers. The phylogeny was inferred using data from cytochrome oxidase 1 and morphology. Some nodes (e.g., among *Oncothrips tepperi, habrus, rodwayi,* and *waterhousei*) are weakly supported and subject to revision. Kleptoparasitism rates were measured by dissecting galls, and traits were optimized on the phylogeny using ACCTRAN in *MacClade* (Maddison & Maddison 1992). For all traits except kleptoparasitism rates, black bars represent the inferred presence of the character state, white bars represent its inferred absence, and striped bars represent ambiguity in the optimization. Several taxa for which kleptoparasitism rate data are unavailable are not included in this phylogeny.

B. Homoptera

Soldiers have now been found in 30 to 40 aphid species from two families, the Hormaphididae and the Pemphigidae (Aoki 1987; Foster & Northcutt 1994; Stern & Foster in press). In some taxa, monomorphic first-instar larvae defend but usually molt to adulthood given sufficient time; these taxa may be regarded as cooperatively breeding, in having labor by juveniles similar to that of the termites with temporary workers. By contrast, other taxa have dimorphic first-instar or second-instar larvae, with soldiers highly specialized morphologically and behaviorally for defense and apparently incapable of molting to adulthood.

The main difference between aphid taxa with and without soldiers is that all taxa with soldiers form galls at some point in their life cycle (Itô 1989; Stern & Foster in press). Despite this association between gall-forming and the presence of soldiers, many gall-forming species do not have soldiers, and some gall-forming species with soldiers do not form galls on their secondary hosts (i. e., they live on the surface of plants), but produce soldiers there nonetheless. Foster and Northcutt (1994) and Stern and Foster (in press) have noted that, among gall-forming species, the species with soldiers appear to have much longer-lived galls than species without soldiers; indeed, some gall-forming species (e. g., *Pemphigus obesi*) do not migrate to secondary hosts at all, but spend the entire favorable season within their gall. But did longer-lived galls select for soldiers, or did the presence of soldiers, which arose for some other reason, create conditions favoring and allowing longer-lived galls (Stern & Foster in press)? Whichever scenario proves correct, the specificity of the function of soldiers, defense against natural enemies, suggests that the ultimate cause of the origins of soldiers was higher pressure from enemies in these lineages than others. Such higher selective pressure should persist in extant taxa, or else soldiers should have been lost, but whether or not higher pressure from enemies results in higher actual predation depends upon the efficacy of defense. Stern's (1994) recent phylogeny of Hormaphididae provides a framework for analyzing the importance of gall duration, enemy pressure, and other traits such as gall hygiene and fighting in foundresses, in the evolution of aphid soldiers.

C. Hymenoptera

1. Halictine bees

Halictine bees exhibit a large number of origins and losses of eusociality (Michener 1990a; Packer 1991, Danforth & Eickwort in press), as well as a bewildering and, as yet, little explored range of variation in social systems within and between populations (summarized in Wcislo in press b, Yanega in press). The main generalizations that can be made about phylogenetics of halictine sociality are threefold. First, large clades of bees comprise a mixture of either solitary and communal forms, or solitary and eusocial forms, and there are no cases of possible transitions between eusocial and communal life. Second, eusociality appears to be ancient in some lineages, and transitions to solitary life or an intraspecific mix of eusocial and solitary forms appear to be common (Packer 1991; Richards 1994; Danforth & Eickwort in press). Third, almost no species that is active year-round is known to be eusocial (Wcislo et al. 1993; Yanega in press).

In only a few cases is it possible to combine information on phylogeny and social systems to develop hypotheses for the causes of transitions between social systems. For example, Danforth and Eickwort (in press) noted that a transition from eusociality to solitary life in *Augochlora* bees was accompanied by a shift from living in open to woodland habitat, which may have led to a reduction in predation and parasitism pressure (Michener 1990a). Given the apparently ancient origins of eusociality in sweat bees, and the possibility of shifts between social levels occurring rapidly with respect to the timing of speciation events, analysis of the causes of such losses of eusociality, which are not due simply to a shortened season at high altitudes or latitudes, should be more productive than analysis of origins. But in organisms as socially flexible as halictines, even this level of comparative approach may be less useful than focusing on species such as *Halictus ligatus*, which show pronounced geographical variation in sociality (Michener & Bennett 1977). Indeed, halictine species might best be viewed as exhibiting not species-typical, or even population-typical behavior, but environmentally cued reaction norms of sociality, whereby different social systems can be expressed in different ecological situations

(Wcislo in press b). To the extent that this view is valid, the study of halictine sociality should comprise two levels: (a) elucidating the norms of reaction for different populations and what environmental cues trigger switches between social systems and (b) for lineages that differ in their reaction norms, determining what aspects of ecology, especially parasite and predator pressure and demography, covary with sociality.

2. Allodapine bees

Although some species of Allodapine bees are solitary, most show some form of social organization, and the few species that have been studied in detail show remarkable intraspecific variation in social systems, between both colonies and populations (Michener 1990b; Schwarz et al. in press). Factors that may be necessary for the evolution of eusociality in Allodapines include a lack of cell partitions, making defense of the nest more important, and progressive provisioning, which requires that adults be present throughout offspring development and thus selects for high adult longevity (Schwarz et al. in press). Variation in forms of sociality among populations and species of these bees may be due largely to variation in nest durability, with longer-lived nests providing more opportunities for inclusive fitness benefits via helping, and predator and parasite pressure, with higher attack rates favoring higher investment in defense of the nest.

3. Polistine wasps

The Polistine wasps comprise both eusocial species, such as some polybiines with pronounced queen-worker dimorphism, preimaginal caste determination, and task specialization (Richards and Richards 1951) and cooperatively breeding species, such as *Miscocyttarus drewseni* (Litte 1977) and some *Ropalidia* (Sinha et al. 1993), in which all individuals appear to be totipotent (Strassmann & Queller 1989; Gadagkar 1991). However, as Gadagkar (1991) notes, for most species, whether or not all individuals are totipotent is unknown. In highly seasonal environments, such as temperate North America, offspring of foundresses apparently do not overwinter and found new colonies in the next year; thus, these individuals are not totipotent (Crespi & Yanega 1994). However, too little is known at present about the social status and systematics of Polistinae to usefully compare cooperatively breeding and eusocial lineages.

4. Stenogastrine wasps

The Stenogastrinae, or hover wasps, comprise both cooperatively breeding species, some with linear dominance hierarchies, and, apparently, species with castes and permanent sterility (Carpenter 1991; Turillazzi 1991). Ecological, demographic, and phylogenetic data are currently insufficient to conduct comparative analysis with this clade, but it should provide excellent opportunities to analyze the causes of the transition from cooperative breeding to eusociality.

The main differences between Stenogastrinae and its sister-group, the Polistinae, which may be related to sociality include (a) less strong and durable nest materials in the Stenogastrinae, which may prevent colonies from growing large (Hansell, 1985, 1987) and (b) use of a unique form of abdominal secretion to feed offspring in Stenogastrinae, which may limit the rate of offspring production (Turillazzi 1991). These traits may tend to prevent the evolution of eusociality in Stenogastrines, if castes are more likely to evolve in large colonies where the odds of any individual becoming the primary reproductive are vanishingly small.

The clade (Stenogastrinae + Polistinae) differs from its sister-group, the Eumeninae (Carpenter 1988, 1991) in that Eumenines are trap-nesting (nesting in cavities of fixed size) with partitions between linearly arranged cells; thus, nests cannot be expanded, offspring cannot be reared as a group (as in Allodapines), and there is little or no contact between mothers and offspring. The Eumenines are a clear case where the physical nature of the nest may be the primary factor precluding the evolution of group living, cooperative breeding, and eusociality.

5. Sphecid wasps

Matthews (1968) has shown that the sphecid wasp *Microstigmus comes* is eusocial, and he notes that the 50-odd additional species in this genus may also be eusocial (Matthews 1991). Moreover, species in the related genera *Arpactophilus* and *Spilomena* may also contain eusocial species, although the range of sociality in these genera is little known as yet

(Matthews 1991). Matthews (1991) notes that these three genera differ from other sphecids in at least two respects, which may be related to their origin or origins of eusociality. First, these taxa mass provision with large numbers of tiny prey, which may facilitate cooperative provisioning of cells because single females provision prohibitively slowly; selection for specialized foragers could allow the evolution of reproductive dominance by mothers or other individuals with a competitive advantage. These tiny prey may also represent a seasonally constant food source (Matthews 1991), which could make helping a more predictably profitable enterprise, and independent founding a predictably non-profitable alternative. Second, species in all three genera use silk in nest construction, which could favor eusociality if silk production is costly and selects for joint work on nests or specialization. Although these ecological differences between sphecid taxa appear sufficiently large to result in different social systems, it is not yet clear why *Microstigmus*, and perhaps *Arpactophilus* and *Spilomena*, have evolved eusociality rather than communal life or cooperative breeding. For this remarkable group, the first necessary step in analyzing social evolution is study of the natural history of more species, to better understand the range of social variation and optimally choose clades for phylogenetic and in-depth behavioral-ecological analysis.

6. Formicidae

Most discussions of insect sociality treat the entire family Formicidae as being eusocial (e.g., Seger 1992). However, at least one species, *Pristomyrmex pungens*, has no queens, and all workers reproduce after laboring when young (Tsuji 1988, 1990). Moreover, in some lineages of Ponerinae, the queen caste has been lost, and behaviorally queen like workers, referred to as gamergates (reviewed in Peeters 1993), tend to monopolize reproduction; when a gamergate is removed, a former worker can mate and becomes the next gamergate (e.g., Ito & Higashi 1991). Both *P. pungens* and such Ponerines can be regarded as cooperatively breeding rather than eusocial, because all individuals appear totipotent except in cases where mating soon after eclosion apparently controls ability to become a gamergate (Peeters & Crewe 1984; Peeters 1987; Ito 1993). Among Ponerinae, which are considered among the ants most similar to ancestral formicids, it is unclear whether such cooperative breeding is ancestral or derived with respect to eusociality; in *Nothomyrmex macrops*, the ant species which by morphological criteria appears most similar to presumed ant ancestors, lack of insemination of workers suggests a caste system similar to that of most other ants (Hölldobler & Wilson 1983)

Although comparisons of the family Formicidae with any of its possible sister groups, such as Scoliidae or Bradynobaenidae (Hölldobler & Wilson, 1990, p. 25), are unlikely to yield clear insights, due to the ancientness of the split and current difficulty of reconstructing phylogenies and reconstructing character states, comparisons of ecological and demographic traits between ant groups in which helping is temporary or permanent may prove useful in analyzing gains or losses of eusociality. Conducting such analyses requires phylogenies for determining the groups to compare and detailed ecological and demographic field studies of the ants involved.

D. Isoptera

Because soldier castes are present in almost all termites and permanent workers are present in the few taxa without soldiers (reviewed in Shellman-Reeve in press), this order has traditionally been regarded as comprising only eusocial forms. However, because working is only temporary in many termite taxa, with juvenile workers dispersing upon adulthood, the order Isoptera may usefully be considered to comprise two more-or-less separate forms of sociality: the presence of soldiers is indicative of eusociality per se, but where workers are temporary, the species can be considered as a cooperative breeder with respect to working, and only where workers are permanent can the species be considered eusocial with respect to working. Thus, the causes of the evolution of eusociality in termites and their relatives can be analyzed in two ways: by comparing clades with and without soldiers, and by comparing clades with temporary and permanent workers. Gains of soldiers can be analyzed by comparing Isoptera with their nearest cockroach relatives, but a lack of phylogenetic data makes these comparisons difficult at present.

Are there any clear differences between termite clades with temporary and permanent workers? Abe (1990, 1991) found that termite taxa with temporary workers tend to inhabit "single-site" nests; that is, they stay and forage within a single piece of wood (and the ground below it) for the life of the colony (see also Higashi et al. 1991, 1992). Under these circumstances, selection has apparently favored worker alleles for retaining the option of leaving, since the habitat is temporary and working may become a poor means of increasing inclusive fitness. By contrast, termites with permanent workers tend to inhabit "multiple-site" nests; that is, they forage outside of the nest, and therefore do not irrevocably convert their home and nursery into termite biomass. The presence of multiple-site nests appears to be a sufficient condition to explain permanent workers in termites, but this hypothesis requires further analysis using a robust phylogeny for the termite taxa involved, to determine the closeness of the association between habitat and eusociality. If the association is not perfect, we may find that additional factors are necessary for the origin or loss of termite worker castes.

E. Coleoptera

Eusociality has been reported for only one species of beetle, *Austroplatypus incompertus* (Kent & Simpson 1992; Kirkendall et al. in press; this genus now transferred to *Platypus*). *Austroplatypus incompertus* is an ambrosia beetle that differs from nearly all other such beetles in several ways. First, this species breeds in the heartwood of living rather than dead trees; as a result, their habitat remains stable for long periods, although tree defenses, especially production by trees of "kino" resin into the beetle burrow, must continually be fought. Second, *A. incompertus* is the only platypodid in which males do not help with gallery establishment; as a result, founding of new colonies tends to be extremely risky and time-consuming.

The sister-group to *A. incompertus* is not yet known (although *Dendroplatypus* is a good candidate), but it probably comprises taxa with colony establishment aided by males. Breeding in living trees is found in a handful of ambrosia beetles (Kirkendall et al. in press), some of which may be closely related to *A. incompertus*, but details of social biology for these species are scarce. In *Trachyostus ghanensis*, multiple adult females are sometimes found breeding in the same gallery system of living trees (Roberts 1961); this species may therefore have communal societies, at least in some burrows.

Analysis of transitions between social systems

Analyzing the causes of the origin or loss of eusociality requires determining what transitions have taken place between social systems, because transitions between different pairs of social systems involve selection of different traits or sets of traits. These transitions were determined simply by inspecting phylogenies and taxonomic data, to infer changes from the pattern of differences present within monophyletic groups. Table 9.3 shows that transitions between solitary life, cooperative breeding, and eusociality are known for most higher taxa that exhibit variation in social systems. However, no transitions are known to have occurred between communal life and either cooperative breeding or eusociality, despite the presence of the former social system, and one or both of the latter systems, in Halictinae, Sphecidae, and Thysanoptera.

Among Halictinae, one of the main differences between communal and eusocial societies is that in communal societies the nest-sharing bees are non-relatives, whereas in eusocial societies high relatedness has been shown for all taxa investigated (Kukuk & Sage 1994). Among the Sphecidae, the one eusocial species for which relatedness has been estimated, *Microstigmus comes*, also shows high relatedness, and among the Thysanoptera, gall founding by single females may engender high relatedness among offspring. These data suggest that whenever nest-sharing occurs in situations where relatedness is high, some degree of reproductive dominance or asymmetry will be present among nestmates. Thus, the initial composition of groups of adults, which is probably influenced by demography and the nature and duration of the nest, may tip the balance between fundamentally different social systems.

Similarities and differences between clades and sister taxa

The differences between clades or populations of eusocial forms and their closest relatives without eusociality vary substantially among higher taxa, despite a remarkable paucity of data (Table 9.4). If longer-lived habitats or galls engender higher predator and parasite pressure, then higher rates of such attack in more social lineages may be emerging as a common theme. However, longer-lived nests in termites probably favor permanent helping because of more predictable benefits to long-term alloparental care (Shellman-Reeve in press), and in aphids longer-lived galls may be a consequence, rather than a cause, of the evolution of soldiers (Stern & Foster in press). These data can thus be safely interpreted as showing that no single difference, or small set of differences, is sufficient to explain all cases of the evolution of eusociality. I have proposed instead that a small set of traits, a habitat that combines food and shelter, strong selection for defense, and ability

Table 9.3. Inferred transitions between social systems[a]

	Communal	Cooperatively breeding	Eusocial
Solitary	Halictines Sphecids Thrips	Halictines? Polistines Xylocopines Aphids	Halictines Sphecids Thrips Ants? Termites? Aphids? Beetle?
Communal		None	None
Cooperatively breeding			Ants Termites Aphids? Polistines

[a]See text for explanation of social terminology.

Table 9.4. Differences between eusocial and non-eusocial sister groups

Taxon	Reference	Difference
Thrips	Crespi (1994)	Higher kleptoparasitism rates?; phenology; fighting in foundresses
Allodapine	Schwarz et al. (in press)	Higher parasitism rates (intraspecific comparison)
Termites	Abe (1987, 1990)	Longer-duration nests
Aphids	Stern and Foster (in press)	Longer-lived galls
Sphecids	Matthews (1991)	Use of silk; provisioning with tiny prey
Beetles	Kent and Simpson (1992)	Female not helped by male; nest in living trees

to defend, may be sufficient to explain the evolution of soldiers in Thysanoptera, Homoptera, and Isoptera, and the evolution of other forms of eusociality in naked mole rats and, perhaps, *Austroplatypus* beetles (Table 9.5; Crespi 1994). Testing this hypothesis requires measurement of the strength of selection for defense in closely related clades with and without eusociality, and searching for eusocial forms in unstudied taxa which appear to exhibit this unusual set of traits.

Table 9.5. Nest or domicile locations and food location of eusocial forms

Taxon	Nest location	Food location
Hymenoptera	Soil or dead wood, exposed with strong defense, or cryptic	Outside of nest
Thysanoptera	Galls	Inside domicile
Isoptera	Wood or soil	Inside nest in many
Coleoptera	Heartwood, living trees	Inside nest
Naked mole rats	Soil	Inside nest
Homoptera	Galls	Inside domicile

Discussion

The two main questions posed at the start of this chapter, how best to categorize and define social systems and how comparative methods should best be used to analyze these systems, turn out to be intimately related. The traditional definition of Michener and Wilson, though ensconced in the literature, is sufficiently vague as to obfuscate comparisons, and the definitions of Crespi and Yanega (1994), though more precise, are useful primarily to help identify convergences across broad taxonomic groups with respect to when helping is permanent or facultative and when totipotency is lost, which they see as the most important questions in the analysis of sociality. As Wcislo (in press a) has pointed out, all definitions not based on historical, phylogenetic criteria must be arbitrary with respect to the biology of the animals concerned. Our primary criteria for choosing words to describe social systems thus become the nature of the specific comparative question being addressed, and the likelihood that a given definition will lead to new insights.

The analysis of features common to diverse lineages showing permanent alloparental care by some individuals, and the comparison of differences between sister groups with and without this trait, demonstrate that no single morphological, ecological, or demographic variable, or set of variables, can predict all cases of the presence of permanent alloparental care or eusociality defined in any other way. Each lineage is unique as a historical individual, and each will require its own peculiar explanation for why a particular social system has evolved or not. These considerations direct our efforts to understand the evolution of sociality away from questions such as the relative importance of mutualism, relatedness, and parental manipulation, and toward asking specific questions about specific lineages, such as why certain species of gall thrips have soldiers, or why some ants have temporary rather than permanent helping. Among gall thrips, the states of three variables, fighting in foundresses, eclosing within the gall, and kleptoparasitism pressure, can accurately predict the presence and absence of soldiers, and these variables are rather idiosyncratic to this clade of insects. Traits that predict social systems in other clades, such as beetles and sphecids, are likely to be similarly idiosyncratic, and will be revealed only through combined phylogenetic and behavioral-ecological analysis.

The search for common features between clades that comprise eusocial forms suggests that (a) one small set of conditions, food-shelter coincidence, selection for defense, and ability to defend, appear sufficient to explain the origin of many cases of eusociality (Table 9.5) and (b) among the eusocial taxa that forage outside of the nest, a diverse array of ecological and genetic factors, including predator and parasite pressure, relatedness, nest structure and duration, prey type, and demography, are each important to different degrees in different lineages (Tables 9.1 and 9.4). Among the clades containing eusocial forms, any of multiple states of multiple characters could prevent eusociality from evolving, and different sets of states of multiple characters may lead to eusociality in different lineages. For example, various different combinations of relatedness levels and parasitism pressure may have led to eusociality in one species or clade of bees, but its sister clade may lack eusociality only because its life-history phenology, or nest-site requirements, are slightly different. For such a complex behavioural trait, we should not expect that the selective pressures that differ between eusocial and noneusocial species in one clade will be the same as in a different clade, and statistically based comparative tests must gloss over such historical uniqueness and hence miss patterns that might otherwise be obvious.

The comparative analysis of eusociality conducted in this chapter has several implications for future work. First, the vast majority of studies of the evolution of insect sociality conducted thus far have focused on eusocial species lacking morphologically distinct queens and workers. These species have been used as models for forms near the threshold of eusocial life, despite the general lack of phylogenetic evidence that such forms are close in time or behavior to noneusocial relatives. To study the evolution per se of eusociality, analysis of noneusocial lineages closely related to eusocial forms becomes essential. Choosing the most informative noneusocial lineages to study requires accurate phylogenies and a willingness to conduct basic natural history. However, only through such fine-scale studies can the necessary and sufficient conditions for eusociality or cooperative breeding in each clade be recognized.

Second, given a choice of clades to analyze, comparative analysis is likely to be most rewarding at the lowest available genealogical level, for several reasons: (a) unmeasured causal factors, which may differ between units of analysis and lead to spurious results (Ridley 1989), are less likely to differ among units at lower levels, because ecological and genetic traits converge as one approaches the level of individuals; (b) units of analysis are likely to differ in fewer traits at lower levels, such that isolation of possible causes is considerably simplified; and (c) the assumptions of many comparative methods, especially those that involve character optimization (e.g., Maddison 1994), are more likely to be satisfied at lower than higher levels. For example, some lineages of halictine bees may cross the eusociality threshold back and forth rapidly with respect to the timing of speciation events (Michener 1985), such that inference of ancestral states from extant taxa is prone to errors. Thus, given a choice between analyzing a species showing geographic variation in sociality, a genus showing variation in sociality among species, and a family with variation in sociality only among genera, the former levels are more likely to yield clear-cut results. Statistical tests for divergence or convergence are least likely to yield misleading results when important selective pressures have been identified and can be presumed to act against a more or less homogeneous ecological, demographic, and genetic background.

Third, although most studies of the causes of the evolution of eusociality have focused on its origins, analysis of losses of eusociality may be at least as informative, for two reasons: (a) multiple specific character states may be necessary for the origin of eusociality, but single changes may trigger its loss; thus, isolation of particular necessary conditions for the presence of eusociality may be easier by comparing noneusocial species with their sister groups in clades of eusocial taxa; and (b) analysis of the causes of losses of eusociality may be more straightforward than analysis of gains because secondary effects of eusocial life, which obscure analysis of origins, may be less likely to complicate analysis of losses. Thus, to the extent that known or unknown secondary effects reduce the likelihood of the evolution of noneusociality, they are less likely to obscure causes of the transitions. Recent phylogenetic analyses of taxa comprising eusocial and noneusocial species show that losses of eusociality are by no means rare

(Danforth & Eickwort in press), although in some cases they are a simple consequence of a shorter breeding season (e.g., Sakagami & Munakata 1972).

Finally, our measure of success in the comparative analysis of eusociality should be the ability to predict the social systems of species in particular clades from some more or less small set of genetic, morphological, demographic or behavioral characters. This set of characters may overlap broadly between ecologically-divergent clades with eusocial forms, but given the contingencies of history, a search for universal explanations may be much less rewarding than a search for the subtle differences between lineages that lead to differences in their social systems. Comparative biologists are historians, and history may be no less quirky for bees, wasps, and aphids than for ourselves. For the analysis of traits such as eusociality, each idiosyncrasy, in each lineage at the highest to lowest taxonomic levels, will frustrate and mislead the use of any but the most robust statistical tests.

ACKNOWLEDGMENTS

I am grateful to E. Martins, M. Winston, and two anonymous reviewers for helpful comments, and to the Natural Sciences and Engineering Research Council of Canada and the National Geographic Society.

References

Abe, T. 1990. Evolution of worker caste in termites. *Proceedings of the 11th International Congress of the International Union for the Study of Social Insects*, Bangalore, pp. 29–30.

Abe, T. 1991. Ecological factors associated with the evolution of worker and soldier castes in termites. *Ann. Entomol.*, 9, 101–107.

Alexander, R. D. 1974. The evolution of social behavior. *Annu. Rev. Ecol. Syst.*, 5, 325–383.

Alexander, R. D., K. M. Noonan, & B. J. Crespi. 1991. The evolution of eusociality. In: *The Biology of the Naked Mole Rat* (Ed. by P. W. Sherman, J. U. M. Jarvis, & R. D. Alexander), pp. 3–44. Princeton, New Jersey: Princeton University Press.

Andersson, M. 1984. The evolution of eusociality. *Annu. Rev. Ecol. Syst.*, 15, 165–189.

Aoki, S. 1987. Evolution of sterile soldiers in aphids. In: *Animal Societies: Theories and Facts* (Ed. By Y. Itô, J. L. Brown, & J. Kikkawa), pp. 53–65. Tokyo: Japan Science Society Press.

Bartz, S. H. 1979. Evolution of eusociality in termites. *Proc. Natl. Acad. Sci. USA*, 76, 5764–5768 (correction 77, 3070).

Baum, D. A., & A. Larson. 1991. Adaptation reviewed: A phylogenetic methodology for studying character macroevolution. *Syst. Zool.*, 40, 1–18.

Brockmann, H. J. (in press). "Primitive" eusociality, cooperative breeding, and the role of ecological constraints. In: *Sexual and Social Competition in Insects and Arachnids*. Vol. 2. *Evolution of Social Behavior* (Ed. By J. Choe, & B. J. Crespi). Princeton, New Jersey: Princeton University Press.

Carpenter, J. M. 1988. The phylogenetic system of the Stenogastrinae (Hymenoptera: Vespidae) *J. N. Y. Entomol. Soc.*, 96, 140–175.

Carpenter, J. 1989. Testing scenarios: wasp social behavior. *Cladistics*, 5, 131–144.

Carpenter, J. 1991. Phylogenetic relationships and the origin of social behavior in the Vespidae. In: *The Social Biology of Wasps*. (Ed. by K. G. Ross, & R. W. Matthews), pp. 7–32. Ithaca, New York: Cornell University Press.

Coddington, J. A. 1988. Cladistic tests of adaptational hypotheses. *Cladistics*, 4, 3–22.

Crespi, B. J. 1990. Measuring the effect of selection on phenotypic interaction systems. *Am. Nat.*, 135, 32–47.

Crespi, B. J. 1992a. Eusociality in Australian gall thrips. *Nature*, 359, 724–726.

Crespi, B. J. 1992b. The behavioural ecology of Australian gall thrips. *J. Nat. Hist.*, 26, 769–809.

Crespi, B. J. 1994. Three conditions for the evolution of eusociality: Are they sufficient? *Insectes Soc.*, 41, 395–400.

Crespi, B. J., & L. A. Mound. (in press). Ecology and evolution of social behavior among Australian gall thrips and their allies. In: *Sexual and Social Competition in Insects and Arachnids*. Volume 2: *Evolution of Social Behavior* (Ed. By J. Choe, & B. J. Crespi). Princeton, New Jersey: Princeton University Press.

Crespi, B. J., & D. Yanega. 1994. The definition of eusociality. *Behav. Ecol..*, 6, 109–115.

Danforth, B. N., & G. C. Eickwort. (in press). The evolution of social behavior in the Augochlorine sweat bees (Hymenoptera: Halictidae). In: *Sexual and Social Competition in Insects and Arachnids*. Volume 2: *Evolution of Social Behavior* (Ed. By J. Choe, & B. J. Crespi). Princeton, New Jersey: Princeton University Press.

Darwin, C. D. 1859. *On the Origin of Species*. London: Murray.

Edwards, S. V., & S. Naeem. 1993. The phylogenetic component of cooperative breeding in perching birds. *Am. Nat.*, 141, 754–789.

Emlen, S. 1992. Evolution of cooperative breeding in birds and mammals. In: *Behavioural Ecology. An Evolutionary Approach* (Ed. By J. R. Krebs, & N. B. Davies), pp. 301–337. Oxford, England: Blackwell Scientific Publications.

Evans, H. E. 1977. Extrinsic versus intrinsic factors in the evolution of insect eusociality. *BioScience*, 27, 613–617.

Felsenstein, J. 1985. Phylogenies and the comparative method. Am. Nat., 125, 1–15.

Foster, W. A. & P. A. Northcott. 1994. Galls and the evolution of social behaviour in aphids. In: *Plant Galls: organisms, interactions, populations*. (Ed. M. A. J. Williams), Oxford: Oxford University Press.

Frumhoff, P. & H. K. Reeve. 1994. Using phylogenies to test hypotheses of adaptation: a critique of some current proposals. *Evolution*, 48, 172–180.

Gadagkar, R. 1990. Evolution of eusociality: the advantage of assured fitness returns. *Phil. Trans. R. Soc. Lond. B.*, 329, 17–25.

Gadagkar, R. 1994. Why the definition of eusociality is not helpful to understand its evolution and what we should do about it. *Oikos*, 70, 485–487.

Garland, T. Jr., A. W. Dickerman, C. M. Janis & J. A. Jones. 1993. Phylogenetic analysis of covariance by computer simulation. *Syst. Biol.*, 42, 265–292.

Gittleman, J. L. 1981. The phylogeny of parental care in fishes. *Anim. Behav.*, 29, 926–941.

Grafen, A. 1986. Split sex ratios and the evolutionary origins of eusociality. *J. Theor. Biol.*, 122, 95–121.

Hamilton, W. D. 1964. The genetical evolution of social behaviour. *J. Theor. Biol.*, 7, 1–52.

Hamilton, W. D. 1972. Altruism and related phenomena, mainly in social insects. *Annu. Rev. Ecol. Syst.*, 3, 193–232.

Hansell, M. H. 1985. The nest material of Stenogastrinae (Hymenoptera, Vespidae) and its effect on the evolution of social behaviour and nest design. *Actes Coll. Insectes Soc.*, 2, 57–63.

Hansell, M. H. 1987. Nest building as a facilitating and limiting factor in the evolution of eusociality in the Hymenoptera. In: *Oxford Surveys in Evolutionary Biology*, Vol. 4. (Ed. by P. H. Harvey, & L. Partridge), pp. 155–181. Oxford, England: Oxford University Press.

Higashi, M., N. Yamamura, T. Abe, & T. P. Burns. 1991. Why don't all termite species have a sterile worker caste? *Proc. R. Soc. Lond. B.*, 246, 25–29.

Higashi, M., T. Abe, & T. P. Burns. 1992. Carbon-nitrogen balance and termite ecology. *Proc. R. Soc. Lond. B.*, 249, 303–308.

Hölldobler, B., & R. W. Taylor. 1983. A behavioral study of the primitive ant Nothomyrmecia macrops Clark. *Insectes Soc.*, 30, 384–401.

Hölldobler, B., & E. O. Wilson. 1990. *The Ants*. Cambridge, Massachusetts: Harvard University Press.

Ito F., 1993. Social organization in a primitive ponerine ant: Queenless reproduction, dominance hierarchy and functional polygyny in *Amblyopone* sp. (*reclinata* group) (Hymenoptera: Formicidae: Ponerinae). *J. Nat. Hist.*, 27, 1315–1324.

Ito F., & S. Higashi. 1991. A linear dominance hierarchy regulating reproduction and polyethism of the queenless ant *Pachycondyla sublaevis*. *Naturwissenschaften*, 78, 80–82.

Itô, Y. 1989. The evolutionary biology of sterile soldiers in aphids. *TREE*, 4, 69–73.

Jarvis, J. U. M., M. J. O'Riain, N. C. Bennett, & P. W. Sherman. 1994. Mammalian eusociality: a family affair. *TREE*, 9, 47–51.

Kent, D. S., & J. A. Simpson. 1992. Eusociality in the beetle *Austroplatypus incompertus* (Coleoptera: Curculionidae). *Naturwissenschaften*, 79, 86–87.

Kirkendall, L. R., K. F. Raffa, & D. S. Kent. (in press). Interactions among males, females and offspring in bark and ambrosia beetles: the significance of living in tunnels for the evolution of social behavior. In: *Sexual and Social Competition in*

Insects and Arachnids. Volume 2: *Evolution of Social Behavior* (Ed. By J. Choe, & B. J. Crespi). Princeton, New Jersey: Princeton University Press.

Kishino, H., & M. Hasegawa. 1989. Evaluation of the maximum likelihood estimate of the evolutionary tree topologies from DNA sequence data, and the branching order of Hominoidea. *J. Molec. Evol.*, 29, 170–179.

Kukuk, P. K., G. C. Eickwort, M. Raveret-Richter, B. Alexander, R. Gibson, R. A. Morse, & F. Ratnieks. 1989. Importance of the sting in the evolution of sociality in the Hymenoptera. *Ann. Entomol. Soc. Am.*, 82, 1–5.

Kukuk, P., & G. K. Sage. 1994. Reproductivity and relatedness in a communal bee *Lasioglossum* (*Chilalictus*) *hemichalceum*. *Insectes Soc.*, 41, 443–455.

Lande, R., & S. J. Arnold. 1983. The measurement of selection on correlated characters. *Evolution*, 37, 1210–1226.

Lin, N., 1964. Increased parasite pressure as a major factor in the evolution of social behavior in halictine bees. *Insectes Soc.*, 11, 187–192.

Lin, N., & C. D. Michener, 1972. Evolution of sociality in insects. *Q. Rev. Biol.*, 47, 131–159.

Litte, M. 1977. Behavioral ecology of the social wasps, *Mischocyttarus mexicanus*. *Behav. Ecol. Sociobiol.*, 2, 229-246.

Losos, J. B., & D. B. Miles. 1994. Adaptation, constraint, and the comparative method: Phylogenetic issues and methods. In: *Ecological Morphology. Integrative Organismal Biology*. (Ed. by P.C. Wainwright, & S. M. Reilly), pp. 60–98. Chicago: University of Chicago Press.

Maddison, D. R. 1994. Phylogenetic methods for inferring the evolutionary history and processes of change in discretely valued characters. *Annu. Rev. Entomol.*, 39, 267–292.

Maddison, W. P., & D. R. Maddison. 1992. *MacClade: Analysis of Phylogeny and Character Evolution*. Sunderland, Massachusetts: Sinauer and Associates.

Maeta, Y, N. Sugiura, & M. Goubara. 1992. Patterns of offspring production and sex allocation in the small carpenter bee, *Ceratina flavipes* Smith (Hymenoptera, Xylocopinae). *Jpn. J. Entomol.*, 60, 175–190.

Martins, E. P., & T. Garland, Jr. 1991. Phylogenetic analysis of correlated evolution of continuous characters: a simulation study. *Evolution*, 45, 534–557.

Matthews, R. 1968. *Microstigmus comes*: sociality in a sphecid wasp. *Science*, 160, 787–788.

Matthews, R. 1991. Evolution of social behavior in sphecid wasps. In: *The Social Biology of Wasps*. (Ed. by K. G. Ross, & R. W. Matthews), pp. 570–602. Ithaca, New York: Cornell University Press.

Michener, C. D. 1969. Comparative social behavior of bees. *Annu. Rev. Entomol.*, 14, 299–342.

Michener, C. D. 1974. *The Social Behavior of the Bees*. Cambridge, Massachusetts: Belknap Press of Harvard University Press.

Michener, C. D. 1985. From solitary to eusocial: need there be a series of intervening species? In: *Experimental Behavioral Ecology and Sociobiology* (Ed. By B. Hölldobler & M. Lindauer), pp. 293–305. Sunderland, Massachusetts: Sinauer and Associates.

Michener, C. D. 1990a. Reproduction and castes in social halictine bees. In: *Social Insects. An Evolutionary Approach to Castes and Reproduction* (Ed. By W. Engels), pp. 77–121. Berlin: Springer-Verlag.

Michener, C. D. 1990b. Castes in Xylocopine bees. In: *Social Insects: An Evolutionary Approach to Castes and Reproduction* (Ed. by W. Engels) pp. 123–146. Berlin: Springer-Verlag.

Michener, C. D., & F. D. Bennett. 1977. Geographical variation in nesting biology and social organization of *Halictus ligatus*. *Univ. Kans. Sci. Bull.*, 51, 233–260.

Michener, C. D., & D. J. Brothers. 1974. Were workers of eusocial Hymenoptera initially oppressed or altruistic? *Proc. Natl. Acad. Sci. USA*, 71, 671–674.

Møller, A. P., & T. R. Birkhead. 1992. A pairwise comparative method as illustrated by copulation frequency in birds. *Am. Nat.*, 139, 644–656.

Mound, L. A., & B. J. Crespi. 1994. Biosystematics of two new gall-inducing thrips with soldiers (Insecta: Thysanoptera) from *Acacia* trees in Australia. *J. Nat. Hist.*, 29, 147–157.

Mueller, U. G., G. C. Eickwort, & C. F. Aquadro. 1994. DNA fingerprint analysis of parent-offspring conflict in a bee. *Proc. Natl. Acad. Sci. USA*, 91, 5143–5147.

Myles, T. G. 1988. Resource inheritance in social evolution from termites to man. In: *The Ecology of Social Behavior* (Ed. By C. N. Slobodchikoff), pp. 379–423. New York: Academic Press.

Myles, T. G., & W. L. Nutting. 1988. Termite eusocial evolution: A re-examination of Bartz's hypothesis and assumptions. *Q. Rev. Biol.*, 63, 1–23.

Packer, L. 1990. Solitary and eusocial nests in a population of *Augochlorella striata* (Provancher) (Hymenoptera; Halictidae) at the northern edge of its range. *Behav. Ecol. Sociobiol.*, 27, 339–344.

Packer, L. 1991. The evolution of social behavior and nest architecture in sweat bees of the subgenus *Evylaeus* (Hymenoptera: Halictidae). *Behav. Ecol. Sociobiol.*, 29, 153–160.

Pagel, M. 1994. The adaptationist wager. In: *Phylogenetics and Ecology* (Ed. By P. Eggleton, & R. I. Vane-Wright), pp. 29–51. New York: Academic Press.

Peeters, C. 1987. The reproductive division of labor in the queenless ponerine ant *Rhytidoponera* sp. 12. *Insectes Soc.*, 34, 75–86.

Peeters, C. 1993. Monogyny and polygyny in ponerine ants with or without queens. In: *Queen Number and Sociality in Insects* (Ed. by L. Keller), pp. 234–261. Oxford, England: Oxford University Press.

Peeters, C., & R. Crewe. 1984. Insemination controls the reproductive division of labor in a ponerine ant. *Naturwissenschaften*, 71, 50–51.

Queller, D. C. 1989. The evolution of eusociality: Reproductive head starts of workers. *Proc. Natl. Acad. Sci. USA*, 86, 3224–3226.

Queller, D. C. 1994. Extended parental care and the origin of eusociality. *Proc. R. Soc. Lond. B*, 256, 105–111.

Reeve, H. K. 1991. Polistes. In *The Social Biology of Wasps* (Ed. By K. G. Ross, & R. W. Matthews), pp. 99–148. Ithaca, New York: Cornell University Press.

Reeve, H. K. 1993. Haplodiploidy, eusociality and absence of male parental and alloparental care in Hymenoptera: A unifying genetic hypothesis distinct from kin selection theory. *Phil. Trans. R. Soc. Lond. B*, 342, 335–352.

Reeve, H. K., D. F. Westneat, W. A. Noon, P. W. Sherman, & C. F. Aquadro. 1990. DNA "fingerprinting" reveals high levels of inbreeding in colonies of the eusocial naked mole rat. *Proc. Natl. Acad. Sci. USA*, 87, 2496–2500.

Richards, O. W., & M. J. Richards. 1951. Observations on the social wasps of South America (Hymenoptera, Vespidae). *Trans. R. Ent. Soc. Lond.*, 102, 1–170.

Richards, M. H. 1994. Social evolution in the genus *Halictus*: A phylogenetic approach. *Insectes Soc.*, 41, 315–325.

Ridley, M. 1989. Why not to use species in comparative tests. *J. Theor. Biol.*, 136, 361–364.

Roberts, H. 1961. The adult anatomy of *Trachyostus ghanaensis* Schedl (Platypodidae), a W. African beetle, and its relationship to changes in adult behavior. *West Afr. Timber Borer Unit, Report for 1961*, 5, 29–41.

Ross, K. G. & J. M. Carpenter. 1991. Phylogenetic analysis and the evolution of queen number in eusocial Hymenoptera. *J. Evol. Biol.*., 4, 117–130.

Sakagami, S. F. & Y. Maeta. 1984. Multifemale nests and rudimentary castes in the normally solitary bee *Ceratina japonica* (Hymenoptera: Xylocopinae). *J. Kansas. Entomol. Soc.*, 57, 639–656.

Sakagami, S. F. & M. Munakata. 1972. Distribution and bionomics of a transpalaearctic eusocial halictine bee, *Lasioglossum* (*Evylaeus*) *calceatum*, in northern Japan, with special reference to its solitary life cycle at high altitude. *J. Fac. Sci. Hokkaido Uni. Ser. 6 (Zool.)*, 18, 411–439.

Schwarz, M. P., L. X. Silberbauer, & P. S. Hurst. (in press). Intrinsic and extrinsic factors associated with social evolution in allodapine bees. In: *Sexual and Social Competition in Insects and Arachnids*. Volume 2: *Evolution of Social Behavior* (Ed. By J. Choe, & B. J. Crespi). Princeton, New Jersey: Princeton University Press.

Shellman-Reeve, J. (in press). Spectrum of eusociality in termites. In: *Sexual and Social Competition in Insects and Arachnids*. Volume 2: *Evolution of Social Behavior* (Ed. By J. Choe, & B. J. Crespi). Princeton, New Jersey: Princeton University Press.

Sherman, P. W., E.A. Lacey, H. K. Reeve, & L. Keller. 1994. The eusociality continuum. *Behav. Ecol.*., 6, 102–108.

Sinha, A., S. Premnath, K. Chandrashekara, & R. Gadagkar. 1993. *Ropalidia rufoplagiata*: A polistine wasp society probably lacking permanent reproductive division of labour. *Insectes Soc.,* 40, 69–86.

Starr, C. K. 1989. In reply, is the sting the thing? *Ann. Entomol. Soc. Am.*, 82, 6–8.

Stern, D. L. 1994. A phylogenetic analysis of soldier evolution in the aphid family Hormaphididae. *Proc. R. Soc. Lond. B*, 256, 203–209.

Stern, D., & W. Foster. (in press). The evolution of sociality in aphids: a clone's eye view. In: *Sexual and Social Competition in Insects and Arachnids*. Volume 2: *Evolution of Social Behavior* (Ed. By J. Choe, & B. J. Crespi). Princeton, New Jersey: Princeton University Press.

Strassmann, J. E., & D. C. Queller. 1989. Ecological determinants of social evolution. In: *The Genetics of Social Evolution* (Ed. By M. D. Breed, & R. E. Page, Jr.), pp. 81–101. Boulder, Colorado: Westview Press.

Taylor, V. A. 1978. A winged élite in a subcortical beetle as a model for a .prototermite. *Nature*, 276, 73–75.

Templeton, A. R. 1983. Phylogenetic inference from restriction endonuclease cleavage site maps with particular reference to the evolution of apes and humans. *Evolution*, 37, 221–244.

Trivers, R. L., & H. Hare. 1976. Haplodiploidy and the evolution of the social insects. *Science*, 191, 249–263.

Tsuji, K. 1988. Obligate parthenogenesis and reproductive division of labor in the Japanese queenless ant *Pristomyrmex pungens*: a comparison of intranidal and extranidal workers. *Behav. Ecol. Sociobiol.*, 23, 247–255.

Tsuji, K. 1990. Reproductive division of labour related to age in the Japanese queenless ant *Pristomyrmex pungens*. *Anim. Behav.*, 39, 843–849.

Turillazzi, S. 1991. The Stenogastrinae. In: *The Social Biology of Wasps*. (Ed. by K. G. Ross, & R. W. Matthews), pp. 74–98. Ithaca, New York: Cornell University Press.

Velthuis, H. H. W. 1987. The evolution of sociality: Ultimate and proximate factors leading to primitive social behavior in carpenter bees. In: *From Individual to Collective Behavior in Social Insects* (Ed. by J. M. Pasteels, & J.-L. Deneubourg), pp. 405–430. Basel: Birkhauser Verlag.

Wcislo, W. T. (in press a). Behavioral classification as blinders to natural variaton. In: *Sexual and Social Competition in Insects and Arachnids*. Volume 1: *Evolution of Mating Systems* (Ed. By J. Choe, & B. J. Crespi). Princeton, New Jersey: Princeton University Press.

Wcislo, W. C. (in press b). Environments and organizers for sweat bee social behavior (Hymenoptera: Halictidae). In: *Sexual and Social Competition in Insects and Arachnids*. Volume 2: *Evolution of Social Behavior* (Ed. By J. Choe, & B. J. Crespi). Princeton, New Jersey: Princeton University Press.

Wcislo, W. T., A. Wille, & E. Orozco. 1993. Nesting biology of tropical solitary and social sweat bees, *Lasioglossum* (*Dialictus*) *figueresi* Wcislo and *L.* (*D.*) *aeneiventre* (Friese) (Hymenoptera: Halictidae). *Insectes Soc.*, 40, 21–40.

West-Eberhard, M. J. 1978. Polygyny and the evolution of social behavior in wasps. *J. Kansas Ent. Soc.*, 51, 832–856.

Williams, G. C. 1966. *Adaptation and Natural Selection*. Princeton, New Jersey: Princeton University Press.

Wilson, E. O. 1971. *The Insect Societies*. Cambridge, Massachusetts: The Belknap Press of Harvard University Press.

Winston, M. L., & C. D. Michener. 1977. Dual origin of highly social behavior among bees. *Proc. Natl. Acad. Sci. USA*, 74, 1135–1137.

Yanega, D. (in press). Male production in halictine bees (Hymenoptera: Halictidae): demography and sociality. In: *Sexual and Social Competition in Insects and Arachnids*. Volume 2: *Evolution of Social Behavior* (Ed. By J. Choe, & B. J. Crespi). Princeton, New Jersey: Princeton University Press.

CHAPTER 10

Using Comparative Approaches to Integrate Behavior and Population Biology

Daniel E. L. Promislow

In his introduction to a symposium on behavioral mechanisms in evolutionary biology, Les Real argued that "...ecological phenomena and community organization can be viewed, to a large degree, as the immediate consequence of individual actions and behaviors", (Real 1992, p. S1). He suggested that since behavior is influenced by internal molecular and physiological processes, and influences external population- and community-level traits, behavior may provide a framework that unifies all of the biological sciences (Fig. 1).

Others have discussed the need to unify different levels of biological organization (Polis 1981; Hassell & May 1985; Schoener 1986; Lomnicki 1988; Koehl 1989; Garland 1994). In particular, many of these authors try to explain phenomena occurring either at the level of populations or communities in terms of individual behavior patterns. In this chapter, I limit my discussion to population-level traits, by which I mean characters that are measured among a cohort of individuals, rather than within a single organism. For example, density, mortality rate, and population size are undefined at the level of a single individual. In theoretical terms, one can distinguish between individual-state variables (those states, or characteristics, that make up a single individual), and population-state variables (population characteristics that are a function of the distribution of all individuals states) (e.g., Caswell & John 1992).

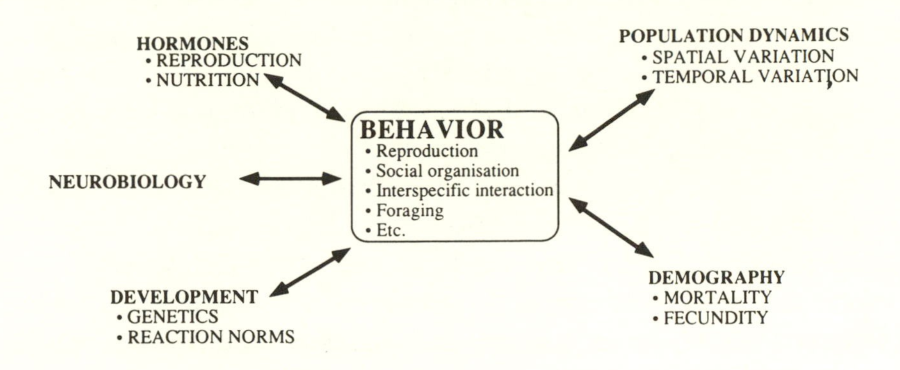

Figure 1. Framework for behavior as the focal point linking biological traits at different levels of organization. Note that in this framework, behavior patterns both affect and are affected by population-level phenomena, which constitute part of the individual's selective environment. Similarly, "lower-level" traits, such as the neurobiology and endocrinology of an organism, may affect and be affected by the organism's behavior. (Reprinted with permission from Real, L. A. 1992. Introduction to the symposium (Behavioral mechanisms in evolutionary biology). *Am. Nat.*, 140, S1–S4. Copyright 1992 University of Chicago Press.)

How might individual behavior patterns influence population level traits, or vice versa? Consider the effects of cannibalistic behavior on population dynamics and mortality rates. Obviously, if some individuals in a population eat others in that same population, mortality rates (at least among the class of individuals being cannibalized) will simply increase. The effects of cannibalism on the dynamics of the population, on the other hand, can be more complex. Some models show that density-dependent cannibalism can stabilize population size (reviewed in Polis 1981), though other models suggest that cannibalism can give rise to fluctuations or even chaos (Hastings & Costantino 1987).

While arguments made by these researchers and others are compelling, explicitly theoretical treatment linking behavior to population biology can be daunting, if not analytically intractable (Caswell & John 1992), and experimental tests of these theories are inherently very difficult. Experimental studies require that we manipulate a trait at one hierarchical level and observe the change at another. Population- or community-level factors operate on large scales and may be difficult to manipulate, and keeping track of factors operating at the level of both individuals and populations poses a particular challenge to the biologist.

In this chapter, I argue that the comparative method provides an excellent way to study how behavior shapes and is shaped by population-level phenomena. I would stress that this chapter is not about the comparison of traits among populations of a single species. Rather, I am interested in using modern comparative methods to compare behavior with population-level traits, measured within species, among different taxa (see also, in this volume, Lauder & Reilly, Chap. 4, for a discussion of trait level of analysis). The comparisons may take place at any taxonomic level, depending on the nature of the question of interest, the taxon or taxa studied, and the particular comparative method used.

In most modern comparative methods, one maps traits onto a phylogeny and then tests for concordant changes among the traits (in this volume, Martins & Hansen, Chap. 2). As such, one can compare any number of biological traits, whether they occur at a molecular, genetic, morphological, behavioral, or population level. And traits from more than one level can be considered simultaneously. Thus, one can use the comparative method to compare easily behavioral and population-level traits.

Within the rubric of comparisons of behavior and population-level phenomena, there lies a vast range of important questions. Figure 2 illustrates a few potential behavior patterns and population-level phenomena that one could consider, as well as the interactions between them. For the purpose of this chapter, I will focus on just a subset of the interactions illustrated in Fig. 2.

One can divide population-level traits into two aspects: demographic phenomena (which includes mortality and fertility schedules) and population dynamic phenomena (temporal and spatial variation in numbers of individuals). Accordingly, the work discussed here falls into these two general themes. First, I discuss the relationship between behavior and demographic parameters, and focus in particular on relationships between behavior and mortality (including factors that directly affect mortality, such as parasitism), and behavior and fecundity. Second, I look at how behavior is related to population

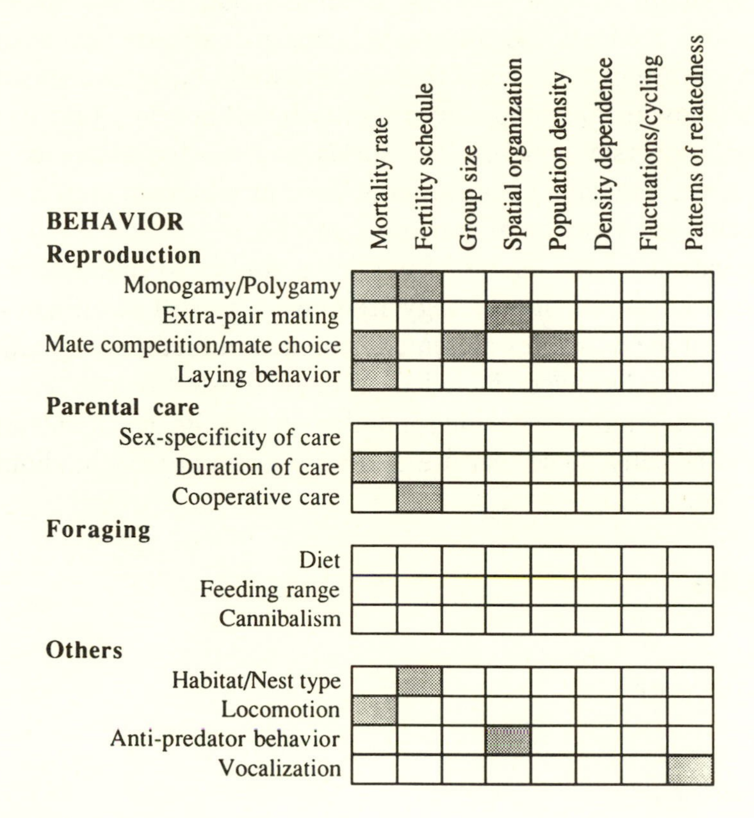

Figure 2. Matrix of selected potential relationships between behavior and population-level phenomena. The shaded boxes indicate those interactions that have been explored using comparative approaches and that are discussed in the text. While many other interactions have been explored either comparatively or experimentally, many boxes in the matrix remain unexamined.

dynamics. I focus here on spatial variation (e.g., density, group size). While there are numerous comparative studies of temporal variation (i.e., density dependence, variation in abundance), these studies have not yet incorporated behavioral differences among taxa.

My intention here is to provide some selected examples to give the reader a feeling for the sorts of problems that can be tackled with comparative methods, to illustrate what we already know, and to encourage work in those areas that are less well understood. In addition, I hope to show that a comparative approach will not only give us a more integrated understanding of the causes and consequences of animal behavior, but may also be of importance in problems of conservation of rare species and invasion of nonnative species.

One caveat — the use of comparative studies to integrate behavior into a general population biology framework is still at an early stage, and many of the examples presented here are not methodologically up to date. These studies were included as good illustrations of the general theme of integration. The interpretation of results for many of these studies, especially those that use a cross-species approach, should be a cautious one.

Examples

A. Demography

1. Mortality rates

The examples in this section focus on comparative studies of the relationship between various behavior patterns and mortality rates among taxa. These examples illustrate three important points. First, comparative approaches sometimes enable us to answer questions for which there is insufficient variation within a species. Second, the comparative analysis of mortality patterns may provide estimates of relative costs among taxa of specific behavior. And third, causal arrows relating behavior and population-level traits may be drawn in both directions.

Optimality models often assume that there is some fitness cost to a behavior, and death is perhaps the clearest currency with which to measure this cost (e.g., Iwasa et al. 1984; Houston et al. 1988). Of course, the causal arrows can go in both directions—while behavioral decisions may affect the risk of mortality over relatively short time scales, over longer time scales the underlying force of mortality may shape the evolution of behavior. (The theme of causal arrows running in both directions will appear throughout this chapter. For a timely and insightful discussion of the philosophical challenges inherent in drawing causal arrows among hierarchical levels in biology, see Inchausti 1994).

Intraspecific studies have considered the effect of behavior on mortality (some recent examples include studies on multiple mating in snakes (Madsen et al. 1992) and beetles (Fox 1993), parental care in mice (Gubernick et al. 1993), and foraging in tadpoles (Werner & Anholt 1993), to name but a few. Comparative studies of mortality patterns in natural populations have only recently become common, with work that focused on the relationship between mortality rates and other life history traits (Sæther 1988; Read & Harvey 1989a; Charnov & Berrigan 1990; Promislow & Harvey 1990; Charnov 1993). Subsequent studies have extended the use of mortality data to behavioral questions, and the number of such studies is sure to grow as more and better data become available.

In some cases, there may be insufficient variation within a species to test a particular hypothesis without resorting to extensive phenotypic or genetic manipulation. Consider the idea, for example, that flying organisms will experience lower mortality than terrestrial ones, all else being equal (Bennett & Ruben 1979). In most species, there is no natural variation with which to compare individuals. To test this hypothesis, Pomeroy (1990) used data from birds and mammals, comparing mortality rates of flying and nonflying species within and between the two classes. He found that volant birds and mammals had consistently lower mortality rates than similar sized terrestrial mammals. It remains to be seen whether these results are supported by patterns in insects, where flying and nonflying species occur within higher taxa, and between sexes within species (Roff 1990; Roff & Fairbairn 1991).

Work such as Pomeroy's attempts to explain variation at the very broadest taxonomic scale, comparing orders or even classes of organisms. But even within much narrower clades, we can find a great deal of variation in mortality rates. Within the primates, life span may vary by over an order of magnitude, from a few years in some lemurs to a century or more in humans (Harvey & Clutton-Brock 1985). In addition to this interspecific variation, comparative approaches may also shed light on variation in mortality rates within species, between sexes, among age classes, or among different castes in insects.

In a series of comparative studies, my coworkers and I have examined among-taxon variation in species-specific sex differences for mortality to assess evidence for costs of sexual selection (Promislow 1992; Promislow et al. 1992; Promislow et al. 1994). Theories of sexual selection typically assume that there is some cost to the production and maintenance of elaborate secondary sexual characteristics. Within species, however, we often find just the opposite. For example, in house finches, more brightly colored males not only have greater than average reproductive success (in accord with expectation) but also have higher than average overwinter survival (contrary to expectation) (Hill 1991). The positive correlation between reproductive success and survival most likely occurs because individuals within a population vary in terms of quality — high-quality individuals may have an advantage in both mating success and survival. To control for variation in quality, previous studies have manipulated individuals experimentally, so that males bear a large or small ornament independent of their quality (Andersson 1982). One can also use the comparative method to control variation in quality, because differences among species in sexually selected traits are unaffected by variation in quality within species. In our studies, using Purvis' (1991) implementation (*Comparative Analysis of Independent Contrasts*, or *CAIC*) of an independent contrast method (Felsenstein 1985; Pagel 1992), we found that taxa with relatively bright and large males had relatively high rates of mortality in males compared to females (Promislow 1992; Promislow et al. 1992; Promislow et al. 1994). In studies on mammals, passerine birds, and waterfowl, we found patterns consistent with a cost either to the production and maintenance of sexually selected traits, or to the behavior associated with them.

As with most comparative studies, we cannot be certain of the direction of causality. We argue that sexually selected traits increase mortality. It is also possible that, over evolutionary time, a sex bias in mortality (and hence a biased operational sex ratio) has increased the strength of sexual selection, and so increased selection for sexual dimorphism. In Pomeroy's work (Pomeroy 1990) as well as my own, one assumes that particular behavior (flying, mate competition) have given rise to mortality patterns. In other studies of mortality patterns, however, the presumed direction of causality has been reversed.

Consider a recent study by Frumhoff and Ward (1992), in which the authors argue that differential survival between queens and colonies could underlie the predicted interspecific relationship between monogyny and worker polymorphism in ants (Oster & Wilson 1978). Oster and Wilson's (1978) hypothesis can be broken down into two component hypotheses (Frumhoff & Ward 1992). First, individual-level selection should favor worker reproduction, and hence monomorphism. This assumes that specialized castes (e.g., soldiers) have a reduced capacity for reproduction. Second, individual-level selection should be relatively stronger in polygynous colonies, where the average relatedness among workers is lower than in monogynous colonies, and the inclusive fitness cost of becoming reproductive (if it suppresses or competes with queen reproduction) is therefore lower. The combination of these two hypotheses thus predicts a correlation between polygyny and monomorphism.

In a study of over 200 species of ants, using comparisons of genera or subfamilies, Frumhoff and Ward (1992) found support for Oster and Wilson's (1978) prediction. However, they noted that within monomorphic species and genera, both polygynous and monogynous colonies occurred. On the basis of this observation, Frumhoff and Ward (1992) reversed the direction of causality in Oster and Wilson's scenario and suggested that polymorphism has driven the evolution of queen number. As one possible explanation for how this might occur, Frumhoff and Ward (1992) suggest that polymorphic taxa would have a greater propensity than monomorphic taxa to accept addition queens into a colony if the ratio of queen to colony survival were relatively low in polymorphic taxa. Adding queens to the colony would ensure continued reproductive output of the colony if the first queen were to die. There are not yet sufficient data to test this hypothesis, which predicts that patterns of survival can drive the evolution of specific behavior patterns (polygyny, in this case).

2. Parasites as a cause of mortality

The first comparative studies of mortality patterns in natural populations focused on life history traits alone (Sæther 1988; Read & Harvey 1989a; Promislow & Harvey 1990). They suggested that in the future, a focus on mortality rates would provide a means to understand the ecological influences on life history evolution (Promislow & Harvey 1991). To do this, one would need to understand the particular factors affecting mortality rates. In this section, I focus on one specific cause of mortality — parasites — and point out how comparative studies have used parasite data to explore two very different sorts of behavioral phenomena. Four important issues arise in this section. First, the result obtained in comparative analyses can depend critically on the method used. Second, in some studies we may not even be able to predict the sign (let alone the direction of causality) of comparative relationships. Third, comparative studies are a valuable tool to test the taxonomic generality of patterns found within a single species. And fourth, comparative approaches can help us to understand interactions between species on an evolutionary time scale (in this case, between parasites and their hosts).

Of late, the locus of comparative studies of parasitism has centered on a single paper published by Hamilton and Zuk (1982). They proposed that female birds use male coloration as a cue in mate choice. Only those males that are resistant to debilitating parasites will be able to produce bright coloration. If this resistance is heritable, as Hamilton and Zuk assumed, then females will choose brightly colored males to ensure that their offspring have genes for resistance to parasites. On the basis of this theory, Hamilton and Zuk (1982) predicted a positive correlation between color and parasite load across species.

Using a cross-species approach, Hamilton and Zuk (1982) found support for their prediction. Later, as more robust comparative approaches became available, Read and Harvey (1989b) used an essentially similar data set and failed to find support for the original hypothesis. The result depended not only the use of a particular method but also on the inclusion of certain taxa (Read & Harvey 1989b). This work signaled the beginning of a more intense effort to test Hamilton & Zuk's hypothesis. In the past 5 years, dozens of theoretical, experimental, and comparative studies have addressed the theory, including several reviews of the evidence both for and against (Anonymous 1990; Read 1990; McLennan & Brooks 1991).

Alas, as Hamilton and Zuk's (1982) theory serves to illustrate, at least in some instances we may not be able to predict unequivocally the sign of the comparative prediction. In their original paper, Hamilton and

Zuk (1982) predicted a positive correlation between parasite burden and coloration. But if females in brightly colored species consistently choose to mate with the most resistant males, then sexual selection could drive down the parasite load within a species over evolutionary time (Read & Harvey 1989c). This point is further developed by Clayton et al. (1992), who note that selection intensity should be greatest at intermediate levels of parasitism. If either all or none of the individuals in a population are parasitized (and assuming the cost of being parasitized does not differ among individuals) then the intensity of selection will equal zero. Despite the power of comparative approaches, in this case the definitive test of Hamilton and Zuk's (1982) hypothesis may come from experimental studies.

Recently, to add further interest to these studies relating the population biology of host-parasite interactions with sexual behavior, several studies have found an additional relationship with patterns of fluctuating asymmetry (FA). In a study on *Drosophila nigrospiracula*, Polak (1993) found that larvae infected with a nematode parasite had increased FA as adults. He also tested a mite parasite that infects adults and found no effect on FA, suggesting that the effect of parasites on FA was due to a disruption of developmental homeostasis. In a similar study, Møller (1992) studied the effect of mite infection in nests of the barn swallow (*Hirundo rustica*) on asymmetry of tail ornaments. Males regrow tail ornaments every year. Møller found that males from parasitized nests had higher FA in tail length the following year compared with males from unmanipulated nests. He found no effect on FA in either shortest tail feather length or wing length, and so suggested that FA in sexually selected traits is a reliable indicator of parasitism.

But can we show that FA plays an important role in the evolution of sexually selected traits for more than these particular species? One of the great strengths of the comparative method is its ability to generalize. Comparative studies have been used to compare FA with sexually selected traits (e.g., Møller & Höglund 1991) and parasitism with sexually selected traits (e.g., Read & Harvey 1989b). If we find a significant relationship between FA and measures of parasitism among taxa, this would suggest that the relationship found within species by Polak (1993) and Møller (1992) holds more generally. Of course, as with Hamilton and Zuk's (1982) original prediction, we are still left with the problem that the predicted sign of the correlation from a comparative study is ambiguous.

One can point to examples where the comparative method is the best way to test parasite-centered evolutionary hypotheses. An elegant comparative study of fig wasps (*Pegoscapus* spp., *Tetrapus* spp.) and their nematode parasites (*Parasitodiplogaster* spp.) by Herre (1989; 1993) nicely illustrates the power of a comparative approach and shows just how behavior and population-level phenomena can be integrated. In doing so, Herre provides us with a better understanding of the behavior and population biology of both host and parasite.

One or more mature female fig wasps enter a fig syconium (which eventually becomes a ripe fig), pollinate the flowers in the syconium, lay eggs, and die. The eggs hatch, develop into adults, and mate before leaving the fig in search of a new syconium. At the same time, nematodes infesting the female consume the wasp, emerge as adults, mate, and lay eggs that will hatch at the same time as the wasps of the next generation emerge. The immature nematodes are carried by the new females to the next syconium, where the cycle is repeated. If only a single wasp enters the syconium, then transmission of the parasite will be vertical, from mother to daughter. If more than one wasp enters a syconium, parasites can be transmitted horizontally. Theory predicts that parasites should be much less virulent in the former case — if the parasites reduce the lone female's fecundity, they also reduce their own chances for transmission to the next generation. In the latter case, a parasite can afford to reduce the host's reproductive output if it increases its probability of horizontal transmission.

Among the 11 species of wasp Herre examined, the average number of single-foundress broods ranged from 24 to 99%. Herre measured virulence in terms of the reproductive success of infected females relative to uninfected females. He predicted that in species with relatively high rates of single-foundress broods (where most parasite transmission is vertical), the reduction in reproductive success should be much lower than in species with few single-foundress broods. This prediction was supported by the data (Fig. 3).

Herre's study provides a dramatic comparative illustration of how population structure can shape the force of natural selection. And at least in this particular system of fig wasps, it would not have been possible to test this theory intraspecifically. It is worth noting that Herre draws his conclusions based on simple cross-species analysis. He does point out, however, that in light of recent molecular phylogenetic studies, it would appear that the patterns of increasing virulence that he observed "have evolved independently several times" (Herre 1993, p. 1445).

But given that Herre's result is correlational, can we be sure of the direction of causality? In his study on virulence, Herre argues that variation in fig wasp behavior (i.e., how many females lay eggs in a single fig) has affected the evolution of nematode virulence. One could argue that because virulent nematodes reduce fecundity in their host, optimum fecundity should increase, thereby increasing the number of emerging females, and the average number of females per host. This scenario seems less likely. In general, however, causal relationships between parasites and host behavior may go in both directions. Behavioral differences among individuals may affect the likelihood that a pathogen infects a host (Murray 1990a) or the pathogen's virulence (Ewald 1983; Ewald 1993). Parasites may, in turn, have an affect on the

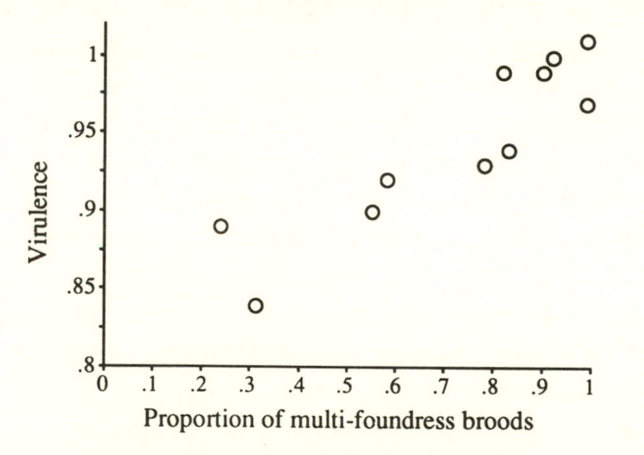

Figure 3. Cross-species relationship of virulence and proportion of figs with multiple foundresses among species of fig wasp. Virulence is measured as the fraction by which fecundity decreases in nematode-infected females compared to uninfected females. As the proportion of figs inhabited by more than one female increases, the virulence of the nematode that parasitizes them also increases. See text for further explanation. (Reprinted with permission from Herre, E. A. 1993. Population structure and the evolution of virulence in nematode parasites of fig wasps. *Science*, 259, 1442–1445. Copyright 1993 American Association for the Advancement of Science.)

behavior of infected individuals, increasing the probability of successful transmission to new hosts (Dobson 1988; Moore & Gotelli 1990).

In their review on the influence of predation on behavior, Lima and Dill (1990) suggested that "research that seriously considers the influence of predation risk will provide significant insight into the nature of animal behavior." I would suggest that to "predation risk" we might add risks due to parasitism, and that comparative studies may be particularly helpful in understanding the relationship between the evolution of behavior and specific mortality risks.

3. Fecundity

Like mortality, we can divide fecundity into various related factors, including clutch size, broods per year, and age at maturity. Data from insects, lizards, birds, and mammals show that these factors are themselves closely correlated with mortality (Dunham et al. 1988; Sæther 1988; Promislow & Harvey 1990; Blackburn 1991; Promislow 1995). Whereas mortality rates can be very difficult to quantify, numbers of offspring are often easily counted, and the literature relating fecundity with behavioral traits is larger and more diverse than for mortality rates. The primary focus of this literature has been clutch size, rather than broods per year or age at maturity. Clutch size is easy to measure in the field, and can vary substantially both within and between species. The following section illustrates that comparative studies of fecundity not only help link population-level traits with individual behavior, but can also provide a compelling way to draw parallels in the evolutionary processes operating among extremely disparate taxa.

While the interest in clutch size dates back at least 2000 years to Pliny (Ridley 1983), the greatest impetus in our age for studies of clutch size came from David Lack's work a half-century ago (Lack 1947; Lack 1948). Lack proposed that the optimum clutch size [often referred to now as “Lack's hypothesis”, or the “Lack clutch” (Lessells 1990; Vanderwerf 1992)] would be that which maximizes the product of per-capita fitness of a single individual in a clutch of a given size, times the size of that clutch. Field studies show that many animals lay clutches that are smaller than the Lack clutch (e.g., Boyce & Perrins 1987; Dhondt et al. 1990; but see Pettifor et al. 1988), which suggests that parameters other than per capita offspring survival influence the evolution of clutch size (Godfray et al. 1991). Recent comparative efforts have used a wide variety of behavior patterns to explain clutch size variation.

As with mortality, the causal arrows that link clutch size and behavioral traits may go in both directions. In some cases, behavioral decisions may drive the evolution of clutch size. In a series of studies on North American passerine and piciform birds, Martin (1993; 1995) examined clutch size variation as a function of the type of nest in which a species raises its offspring. Among the birds he considered, some species nest in open nests, some in hole nests that one or both parents excavate ("excavators"), and some in holes that already exist ("nonexcavators"). Using Purvis' (1991) *CAIC* program, Martin found that nonexcavators have larger clutches than species that excavate new holes (Martin 1993), with open nesters having intermediate clutch sizes (Martin 1995). The most striking difference is that found between nonexcavators and excavators — fecundity differs more than twofold among these two types of behavior (Fig. 4). Martin argues that the striking difference between excavating and nonexcavating hole-nesters cannot be attributed to differences in rates of predation. Rather, he argues, nonexcavating hole-nesting species have been selected to increase reproductive effort in response to limited availability of nesting sites.

Other studies have also linked behavior to clutch size but have drawn the causal arrow in the other direction, with clutch size either driving or constraining the evolution of particular behavior patterns. For example, Ridley (1988; 1993) studied the evolution of mating frequency in insects. He was interesting in testing the idea that in cases where siblings compete, a mother should try to maximize the genetic variance in her offspring to minimize the cost of sib competition. One obvious way to do this is to mate with multiple males. Ridley (1993) assumes that sib competition is likely to be higher in larger clutches. One should note, however, that while increasing the genetic variance will lower the cost of sib competition, there will then be a concomitant increase in competition due to this lower cost. In addition, Godfray and Parker (1992) show analytically that competition does not necessarily increase with clutch size. In accord with his prediction, Ridley found that polyandry occurred in species with larger clutches and longer life span

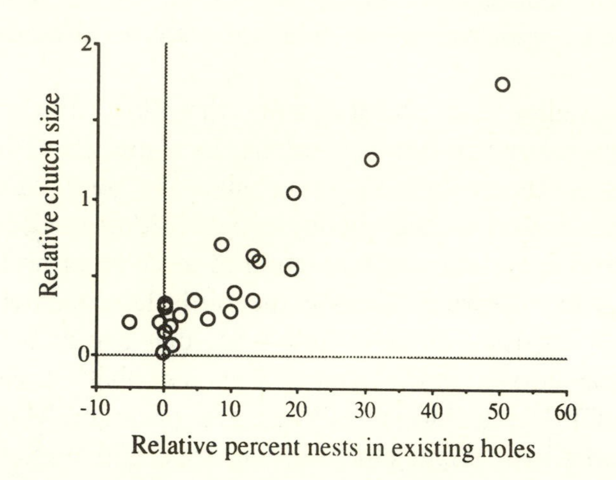

Figure 4. Clutch size versus the percent of existing holes that were settled by the current resident (rather than being excavated anew), after correcting for the effects of phylogeny and body size. Each point represents a phylogenetically independent comparison of the hypothesis (Felsenstein 1985). As the percent of prior existing holes in the population increases, clutch size increases. Martin (1993) suggests that in species that depend on existing sites for nests, the limited availability of nest sites selects for greater reproductive effort in those individuals that find a nest site. (After Martin 1993).

after controlling for the effects of phylogeny. On the basis of this result, Ridley suggested that sib competition in large clutches has driven the evolution of polyandry (Ridley 1993).

A quite different comparative study on mammals also illustrates how offspring numbers may influence behavior. In recent work on mammals, Packer et al. (1992) use two related comparative models (Grafen 1989; Pagel 1992) to test the relationship between litter size and non-offspring nursing. In many mammalian species, females nurse offspring that are not their own. Packer et al. (1992) found that in monotocous species (litter size usually equal to one), non-offspring nursing was associated with parasitic "milk theft", and while it was relatively rare, it was more likely to occur in species with large group size. In polytocous taxa (modal litter size greater than one), by contrast, non-offspring nursing was more common, was not considered milk theft, and decreased with increasing group size.

Surprisingly, one can draw a parallel between Ridley's work on multiple mating in insects, and Packer et al.'s (1992) mammal study. In his work on insects, Ridley (1993) argued that sib competition among unrelated individuals has driven the evolution of polyandry. Similarly, Packer et al. (1992) argue that in polytocous species of mammals living in small groups, relatedness among females in the group is relatively high (given that males typically disperse), so females may be more likely to nurse offspring of close relatives. Competition among relatives may explain why non-offspring nursing is highest in polytocous species found in small groups, where both relatedness and the opportunity for reciprocity are high. It may also explain why it is highest in monotocous taxa found in large groups, where females are unrelated and so more likely to parasitize one another.

B. Population dynamics

The demographic parameters discussed above typically fall within the purview of population biologists, though fecundity and mortality rates are simply composite measures of individual probabilities of reproduction and death. In contrast, phenomena related to population dynamics are not simply reducible to parameters one can measure at the level of individuals. We can divide these population dynamic phenomena into two general kinds of traits. Spatial variation refers to the variety of parameters that define spatial distribution and abundance in a population, including territory and home range size, average group size and the degree of aggregation, and population density. Temporal variation, on the other hand, consists of phenomena related to changes in distribution and abundance over time, including measures of variability, density dependence and population regulation, and rates of colonization and extinction.

In this section, I focus on the role that behavior plays in spatial variation in populations. The comparative method can be of particular value here. Although population dynamic traits may not follow directly from properties of the individuals within a population, we need not view these traits as emergent properties whose causal factors are inexplicable. Furthermore, it may be even more difficult to manipulate population dynamics than demographic parameters. In this case, the comparative method may bridge the gap between theoretical prediction and empirical test. In addition, where models linking individual behavior and population dynamics may be difficult, and in some cases analytically intractable (Caswell & John 1992), results from comparative studies may lead to new theoretical insights.

1. Spatial variation

One of the best studied population-level traits is population density. Density also provides a potentially valuable interface between behavior and population dynamics — at high density, animals may experience elevated encounter rates and increased competition for resources, and in many species there is evidence of density-dependent fluctuations in abundance (Pimm & Redfearn 1988; Hassell et al. 1989; Hanski 1990; Woiwood & Hanski 1992). Similarly, at very low densities, animals may exhibit dysfunctional behavior (Allee et al. 1949). Theoretical and experimental studies have shown density to be an important factor in selection on behavior as disparate as sperm competition (Murray 1990b), aggression (Yokel 1989), dispersal (Cockburn 1988; Li & Margolies 1993), mate searching and mate attraction (Conner 1989; Woolbright et al. 1990; Cade & Cade 1992), extra-pair mating (Møller 1991), mutualistic interactions (Breton & Addicott 1992) and song duration (Eiríksson 1992).

Clearly we know a great deal about the relationship between density and behavior within species. But with the exception of the many comparative studies on the allometry of population density (e.g., Damuth 1981, 1987; Blackburn et al. 1990; Nee et al. 1991), there are few comparative studies that directly address the relationship between density and behavior. Here I discuss studies that successfully illustrate the value of integrating behavior and population dynamics (i.e., spatial distribution) using comparative methods.

Theories of mating systems assume that spatial distribution of a population plays a central role (reviewed in Davies 1991). Dispersion, group size and defensibility in both males and females will have a significant impact on the type of mating systems that can or cannot evolve. Clutton-Brock et al. (1993) were interested in the spatial patterns that may drive the evolution of lek breeding in ungulates. They suggested that, at least in ungulates, lekking can evolve when non-breeding females live in large, unstable groups that are not easily defended by males, whereas estrous females leave the herd to breed. In the absence of a significant, defensible resource, males may benefit by holding clustered territories that attract and keep estrous females. Clutton-Brock et al. (1993) looked at spatial distribution in 12 species of the tribe Reduncinae (Artiodactyla). The three lekking species had consistently higher population densities (in addition to larger female home range and group size) than non-lekking species (see also Balmford et al. 1993). In addition, Bradbury et al. (1986) describe a similar relationship among species of grouse. But as Clutton-Brock et al. (1993) point out, we need to consider not only overall density, but also at the way in which individuals are dispersed.

In general, along with measures of density, we should consider how large average group size is, and the dispersion (i.e., degree of aggregation) of the population. Recent comparative studies in birds (Møller & Birkhead 1992, 1993) and butterflies (Sillén-Tullberg 1988, 1994) illustrate the value of drawing comparisons between solitary and gregarious-living taxa.

Møller and Birkhead (1992) examined mating patterns in colonial and solitary bird species. In an earlier study, Birkhead et al. (1987) argued that in colonially nesting birds, females are more likely to engage in extra-pair copulations. As a result, males will be selected to copulate more frequently with their partners to reduce the probability of being cuckolded. They found that solitary-nesting species do, indeed, mate less frequently than species nesting colonially (Møller & Birkhead 1992). Furthermore, this increase in within-pair mating frequency occurs in concert with an increase in the frequency of extra-pair copulations in colonial species (Møller & Birkhead 1993).

Møller and Birkhead's study shows how spatial distribution might act as a selective force on behavior. By contrast, studies by Sillén-Tullberg (1988) show how a specific individual trait (aposematism) can select for a particular spatial distribution (i.e., gregarious larvae). Sillén-Tullberg (1988) suggested that unpalatability (and aposematic coloration) would have to precede the evolution of gregariousness (which she assumed to be advantageous), since predation on palatable, gregarious larvae would be unbearably high. Sillén-Tullberg compared aposematic coloration and gregariousness in four families of butterflies. She found that whenever unpalatability and gregariousness evolved in the same taxa, and could be separated over evolutionary time, unpalatability appeared before gregariousness. In this case, the spatial distribution of individuals for a given species appears to be caused by particular morphology or behavior. Although subsequent work by Maddison (1990) suggested that Sillén-Tullberg's result was a statistical artifact, a recent study by Sillén-Tullberg (1994) suggests that findings in the original study were correct (see, in this volume, Martins & Hansen, Chap. 2 for details).

Difficulties with the comparative method

Of the many approaches used to study evolutionary questions, each has its advantages and disadvantages. There is much to recommend the comparative approach, yet it is not without its shortcomings. Difficulties in both methodology and interpretation are likely to be encountered in studies that attempt to link phenomena that occur at different levels of biological organization.

A. Correlation vs. Causation

I would place a strong caveat on the types of studies I have discussed here. Of the many examples presented, only Sillén-Tullberg (1994) used an approach that could draw an explicitly directional conclusion from the data. She used Maddison's correlated changes method (Maddison 1990) to show that the evolution of unpalatability in the Lepidoptera precedes gregariousness. Sillén-Tullberg cannot say categorically, however, that there exists a direct causal pathway from unpalatability to gregariousness. Directional studies such as this allow us to infer causal arrows over evolutionary time, but they do not prove the existence of direct causal pathways (Lauder et al. 1993; Leroi et al. 1994).

Nondirectional comparative methods (Harvey & Pagel 1991), which include comparisons across species or higher-level taxa, among pairs of related taxa (e.g., Healy & Guilford 1990; Møller & Birkhead 1992), or within multiple independent nodes of a phylogeny (Felsenstein 1985; Grafen 1989; Pagel 1992), do not allow us to infer causation from correlation. In some instances, it would only be logical to draw the causal arrows in one direction. Healy and Guilford (1990) used a pairwise comparative method as grounds to argue that nocturnal birds have evolved larger olfactory bulbs to compensate for the loss of vision. By contrast, it would seem farfetched to argue that birds with highly developed smell are therefore predisposed to evolve nocturnal habits. However, in many cases there is no reason to favor one hypothesis of causation over another. This was the case, for example, in Herre's study on virulence of nematodes parasitizing fig wasps (Herre 1993). Herre argued that foundress number was driving the evolution of nematode virulence, but as I pointed out, the causal arrows could potentially have gone the other way around. It is worth asking of any comparative study whether the causal arrow could be reversed.

In some cases we may be able to use theoretical models to determine the possible direction of evolutionary trajectories. A model by Godfray (1987) provides such an example. In some parasitoid wasp species, larvae have fighting mandibles with which they can cannibalize their siblings. One offspring hatches per host, having eaten any other siblings. Genes for larger clutches will be selected against, since the extra offspring will simply be eaten. Similarly, genes for nonfighting will not spread, since those individuals carrying the gene will be eaten. So at least in the case of these parasitoids, the evolution from gregarious, nonfighting larvae to solitary fighting larvae is irreversible (see also Bull & Charnov 1985).

In addition, directional comparative studies such as Sillén-Tullberg's (1994) can be used to determine the order of the evolution of traits. And of course, one can also use manipulative experiments to determine the causal relationships of the traits, at least on a microevolutionary scale.

Even if we are able to determine the direction of causality for two traits over evolutionary time, this is no cause for complacency. The actual causal forces linking the two traits may be exceedingly complex. Let us come back to Herre's study on virulence in fig wasps (Herre 1993). The behavior of female fig wasps may drive the evolution of nematode virulence, but the causal pathways involved in this process may include innumerable unexamined processes, including the molecular and cellular biology of virulence, nematode genetics, the biology of the host immune system, and so forth.

B. Method and Madness

1. Discrete and continuous characters

Despite the plethora of comparative methods now available, most are limited either to discrete (categorical) data or continuous data, and do not allow the two to be examined simultaneously (but see, in this volume, Martins & Hansen, Chap. 2, and Ridley & Grafen, Chap. 3). This may be particularly frustrating when trying to link behavior patterns (which are often defined as categorical characters) with population-level phenomena (which are usually measured on a continuous scale). The problem arises when we try to estimate ancestral character states. Categorical comparative methods (e.g., Maddison 1990) use a statistic that minimizes the number of character changes of the trait of interest in the phylogeny. Most continuous comparative methods (e.g., Pagel 1992) define ancestral character values as some weighted mean of the values in the descendant taxa. The two approaches cannot be employed simultaneously in most currently available comparative statistical packages (but see Grafen 1989; and, in this volume, Ridley & Grafen, Chap. 3).

There are several ways around this constraint. We might redefine one of the characters, and change a categorical measure into a continuous one or a continuous variable into categorical data. Unfortunately, the former approach may lead to erroneous attributions of ancestral character values. For example, if we assign lack of parental care, uniparental care, and biparental care values of 0, 1, and 2 respectively, what does it mean to have a value of some intermediate fraction? And how do we decide the ancestral value of a dichotomous trait if either one is equally parsimonious? Similarly, if we code continuous data as categorical, we do so at the cost of lost information and accuracy.

Another way to compare discrete and continuous data simultaneously is through the use of pairwise comparisons of independent taxa (e.g., Healy & Guilford 1990; Møller & Birkhead 1992). Imagine that we are interested in the relationship between a population-level trait (density, for example, measured on a continuous scale) and a behavior with only two states (e.g., open-nesting versus hole-nesting in birds), which we code as 0 and 1 respectively. In a pairwise comparison, we would identify pairs of related species, one of which had behavior 0 and the other behavior 1, and measure the difference in density between the two species. We could repeat this comparison in all taxa for which data were available.

2. Method makes a difference

I hope that these problems, combined with the overwhelming choice of possible comparative methods (in this volume, Martins & Hansen, Chap. 2), do not discourage the reader from using modern comparative approaches in his or her own work. It should be stressed that modern methods, combined with accurate phylogenies, can make a dramatic difference to the conclusions we draw. In the past few years, we have seen numerous cases where the use of a better method or a better phylogeny has altered a previously accepted conclusion. As noted earlier, Read and Harvey (1989b) reanalyzed Hamilton and Zuk's (1982) original work using a better method, and called their original conclusions into question. Many other relationships that were previously thought to be statistically significant have been overturned by modern methods or better phylogenies [e.g., DNA repair rate and life span (Promislow 1994); sexual dimorphism and mating systems (Oakes 1992); or the evolution of sexually-selected traits due to sensory bias (Basolo 1990; Meyer et al. 1994)]. In some instances, comparative biologists have resurrected hypotheses that were previously rejected [e.g., relationships between egg size and clutch size in birds (Blackburn 1990), clutch size and life span in *Drosophila* (Promislow 1995)]. And perhaps most interestingly, by using modern comparative methods, researchers have been able to identify comparative patterns that previously lay hidden. For example, we now know that many evolutionary relationships become stronger, weaker, or even change sign at different taxonomic levels (e.g., Pagel & Harvey 1988; Nee et al. 1991; Promislow 1991a). In others, It remains to be seen to what extent these methodological advances will alter our current understanding of the studies reviewed in this chapter.

Avenues for future research

I began this chapter with a table of some possible comparisons between behavior and population-level traits. I have focused in this review on just a few of the entries in the table, and most of the potential comparisons illustrated in the table have not been examined in any way. In the past, population biologists have called for the integration of behavior and population biology (Polis 1981; Hassell & May 1985; Schoener 1986; Lomnicki 1988; Koehl 1989), and more recently, biologists have begun to generate theory linking individual behavior with population dynamics and life history evolution (e.g., Houston et al. 1988; Mangel & Clark 1988; DeAngelis & Gross 1992).

In spite of this current theoretical impetus, we are still in dire need of concrete predictions, and there are few experimental or comparative tests of existing theory. A more complete theoretical development is likely to take some time, but I think we will soon see major advances in at least four areas as a result of the use of comparative techniques. These include problems concerning kin selection and relatedness, the evolution of senescence, interspecific variation in population fluctuations and density dependence, and issues in conservation biology.

First, the theory of kin selection (Hamilton 1964), which explains how degrees of relatedness can influence the behavior between interacting individuals, revolutionized our approach to the study of behavior. Two of the studies reviewed above (Packer et al. 1992; Ridley 1993) explained the patterns observed on the basis of degrees of relatedness among individuals. While their results are consistent with kin selection theory, neither study measured relatedness directly. With the advent of DNA fingerprinting (Jeffreys et al. 1985), techniques for estimating relatedness have become widely available (Reeve et al. 1992). Students of animal behavior now use DNA fingerprinting as a matter of course to assess relatedness or parentage. This approach is dramatically changing experimental studies of behavior, and is sure to have as forceful an impact on comparative studies.

A recent study provides a good example of just how estimates of relatedness can inform our understanding of both behavior and population-level phenomena. Briskie et al. (1994) wondered why in some species of birds, chicks beg very loudly, while in others the chick can barely be heard. They used a simple kin-selection argument to suggest that in nests where chicks are unrelated, the cost of begging will be reduced. If begging attracts a predator that kills your nestmate but not you, the loss will be more costly if the nestmate is your full sib than if it is your half sib. To test this, they compared begging volume across taxa with the average rate of extra-pair fertilizations (which they assume correlates with the average degree of relatedness). Briskie et al. found that the volume of begging chicks increased linearly and significantly with the average degree of extra-pair paternity, suggesting that the genetic structure of the population may act as a powerful selective force on individual behavior. Such examples are likely to become more common as the information on patterns of relatedness in natural populations increases.

Second, biologists have displayed renewed interest in the evolution of senescence over the past five years (Rose 1991; Partridge & Barton 1993), including several comparative attempts to explain variation among species (Nesse 1988; Finch et al. 1990; Promislow 1991b; Gaillard et al. 1994; Promislow 1995). There is a substantial body of theory relating patterns of aging to life history strategies in general, and both experimental (Rose & Charlesworth 1980; Bell 1984; Luckinbill et al. 1984; Mueller 1987; Service et al. 1988) and comparative (Promislow 1995) evidence to support it. A few experimental studies have examined behavioral correlates of aging. For example, males that are prevented from mating with females live longer than males that are allowed to mate (Partridge & Andrews 1985). In his encyclopedic review of the topic, Finch (1990) suggests at least one behavioral correlate of senescence — many insect species do not eat as adults, and consequential have necessarily limited life spans. However, this constraint on survival is as much morphological as it is behavioral, and no comparative studies have examined behavioral correlates of patterns of aging.

One suggestive comparative example comes from a study of whales by Marsh and Kasuya (1986). Kin-selection theory would predict that in species with extended parental care and cohesive, kinship groups, there may be an advantage to the cessation of reproduction late in life (but see Rogers 1993). Energy might better be devoted to the caring of existing offspring or the offspring of close relatives. In species without extended parental care, selection would not favor a similar decline in reproduction. Marsh and Kasuya (1986) noted that odontocete whales (e.g., killer whales, pilot whales), which have extended parental care, also show signs of reproductive senescence, while mysticete whales (e.g., fin whales) demonstrate neither extended parental care nor reproductive senescence. While their results are preliminary (based on a comparison of only two species), they are in accord with theoretical expectation. It would be of obvious interest to know whether the pattern holds more generally in these and other taxa, and also what the age-specific mortality patterns look like in these two taxa. In contrast to fertility patterns, we might expect age-specific mortality rates to increase less quickly (i.e. slower rate of senescence) in the odontocete whales, where the inclusive fitness benefits of late-age survival are greater than in mysticete whales.

Third, for the past half-century, there has been a concerted effort to understand why so many species of small mammals exhibit periodic fluctuations in abundance. While some of the more popular hypotheses to explain this periodicity provide explicit links between behavior and population dynamics (e.g., Christian & Davis 1964; Chitty 1967; reviewed in Cockburn 1988), there is no compelling evidence for any one hypothesis as a generally applicable explanation (Cockburn 1988). In fact, Ostfeld and Klosterman (1990) argue that the social behavior within species of rodents is so great as to preclude the possibility of comparative studies, aside from the study of the variability itself.

Across species, there is some evidence that cycles increase in period length with body size (Peterson et al. 1984; but see Krukonis & Schaffer 1991). And among populations within species, populations at higher latitudes show greater degree of cyclicity (Hansson & Hentonnen 1985). Despite the plethora of behavioral hypotheses for fluctuations in abundance, no comparative studies have tested these hypotheses.

Biologists have recently begun to amass a wealth of comparative data on temporal variability and density dependence in birds, mammals and insects (Pimm & Redfearn 1988; Hassell et al. 1989; Hanski 1990; Woiwood & Hanski 1992). The tone running through this work is one of description rather than explanation. Surely one can use comparative approaches similar to those discussed above to explain this diversity.

For example, we might expect life history strategies to affect temporal variation. Just on the basis of a simple logistic model (May 1972), we might expect that species with relatively rapid rates of turnover (i.e., high fecundity and mortality) should exhibit greater fluctuations in population abundance. To test this idea, I compiled data on the coefficient of variation in annual population size (Pimm 1984) and natural mortality rates (Dobson 1982; Sæther 1989) in British woodland and farmland birds. Based on contrast scores derived from the CAIC program (Purvis 1991; Pagel 1992), I found a significant negative relationship between the coefficient of variation in population size among years and survival rate. After correcting for the effects of both body size and phylogeny, taxa with relatively high rates of survival tended to exhibit relatively low variation in abundance (Fig. 5).

To my knowledge, this is the first attempt to explain (rather than just describe) variability among taxa in annual population size. However, this correlation is not without problems, and should be interpreted with caution. For example, the negative correlation between population variability and survival rate could be due to a statistical artifact. When survival rates are high, population size may be correlated from one year to the next because many individuals will survive through both years. And this particular result does not examine the possible role of behavior in population variability. In light of this preliminary result, we now need theories to suggest how species-specific behavior may predispose some species to exhibit patterns of density dependent fluctuations (e.g., May et al. 1990), and comparative tests of these theories.

Fourth, while the question of why populations vary over time is of fundamental biological interest, the ability to link behavior and population dynamics is also of obvious value to conservation studies. Conservation biologists are interested both in the likelihood of rare species becoming extinct, and in the propensity of alien species to invade, and potentially extinguish native residents. Theoretical work suggests a link between both demography and population dynamics and the risk of extinction (Simberloff 1986; Goodman 1987; Lande 1988; Pimm et al. 1988). If extinction rates depend on such factors as the temporal variability in abundance, and if this variability is influenced by behavior, then we might expect rare species to be more or less at risk of extinction, depending on their behavioral attributes. Rarity alone does not lead to extinction. Future comparative studies may help us to understand the behavioral attributes which, combined with rarity, put a species at risk of extinction.

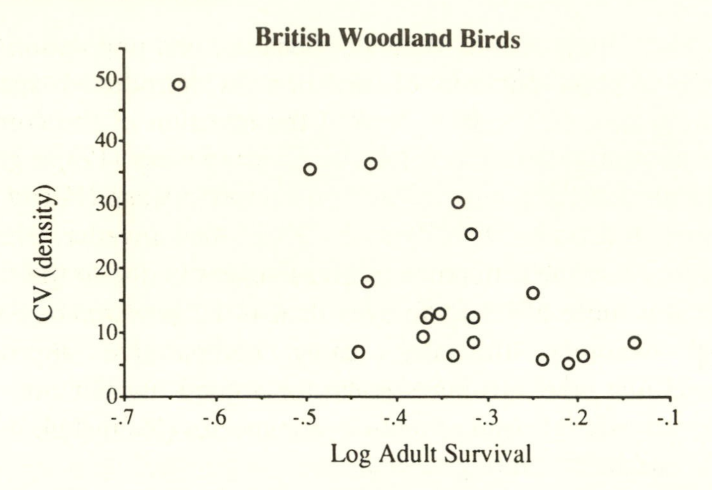

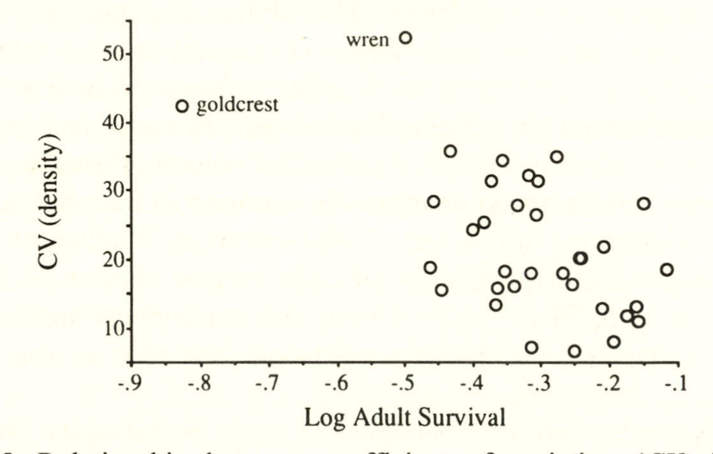

Figure 5. Relationship between coefficient of variation (*CV*) in annual population density and adult survival in natural populations of British birds. Data points in the figures are from single species. Statistical tests were performed using independent contrast scores (Felsenstein 1985) from Purvis' (1991) *CAIC* program, with phylogeny and branch lengths derived from Sibley and Ahlquist's (1990) classification based on DNA-DNA hybridization. Contrast scores from nodes for which the branch length was equal to 2.0 or less (i.e., comparisons between very closely related species) were omitted from the statistical test. Woodland species: two-tailed $t_{23} = 2.97$, $p < 0.01$; Farmland species: two-tailed $t_{13} = 2.03$, $p < 0.06$ (least-squares regression lines fitted through the origin (Garland et al. 1992), and effects of body mass held constant in both cases).

Similarly, comparative studies of behavior and population dynamics may help explain patterns of invasion in introduced species. For example, Harris et al. (1991) studied the invasion of the common and German wasps into New Zealand. The German wasp (*Vespa germanica*) was the first to invade, followed by the common wasp (*Vespa vulgaris*), which then caused local extinction of the first invader. Harris et al. (1991) show that the difference in invasiveness is due to the fact that *V. vulgaris* is a more efficient forager than is *V. germanica*. As a recent botanical example illustrates, using comparative approaches to understand just what attributes make for a good invader may also help us learn to stop the spread of unwelcome species (Rejmanek in press)

Conclusion

There are many cases where biological questions are best addressed using a comparative approach. The above examples illustrate the strengths of comparative methods in examining the relationships between biological phenomena at different hierarchical levels. I have chosen to focus on the relationship between behavior and population-level traits, though as other chapters in this volume illustrate, one could equally relate behavior to interspecific variation in morphology (in this volume, Gittleman et al., Chap. 6; de Queiroz & Wimberger, Chap. 7; Irwin, Chap. 8), biogeography (in this volume, Foster & Cameron, Chap. 5), and so forth. And of course, this hierarchical approach is not limited to studies that include individual behavior as one level of biological organization.

A comprehensive understanding of most evolutionary phenomena will rest on a sound combination of theoretical, experimental, and comparative work. I hope that this chapter has convinced the reader that for certain questions, the comparative method may be the most tractable, fruitful approach. For example, it may be very difficult to gather data on attributes from a variety of hierarchical levels simultaneously. Perhaps more importantly, in some cases there may be little or no genetic or phenotypic variation within species for the trait of interest. Comparative studies may then be the best way to identify causes or consequences of traits that are invariant within species. And finally, where a pattern is found within a single species, the comparative method gives us the power to test the generality of this pattern among higher taxonomic levels, and to discern the extent to which these patterns recur among diverse taxa.

ACKNOWLEDGMENTS

R. D. Montgomerie, P. D. Taylor, J. R. Carey and J. W. Curtsinger provided logistical and financial support during the preparation of this manuscript. Tim Caro, Judy Stamps, Marc Tatar, Emília Martins and three anonymous referees offered many helpful comments on early drafts. This work is dedicated to Sarah Paulina—a peerless, fearless integrator of knowledge.

References

Allee, W. C., A. E. Emerson, O. Park, T. Park, & K. P. Schmidt. 1949. *Principles of Animal Ecology*. Philadelphia: Saunders.

Andersson, M. 1982. Female choice selects for extreme tail length in a widowbird. *Nature*, 299, 818–820.

Anonymous. 1990. Parasites and sexual selection (editorial comment). *Am. Zool.*, 30, 225–352.

Balmford, A., J. C. Deutsch, R. J. C. Nefdt, & T. H. Clutton-Brock. 1993. Testing hotspot models of lek evolution — Data from 3 species of ungulates. *Behav. Ecol. Sociobiol.*, 33, 57–65.

Basolo, A. L. 1990. Female preference predates the evolution of the sword in swordtail fish. *Science*, 250, 808-810.

Bell, G. 1984. Evolutionary and nonevolutionary theories of senescence. *Am. Nat.*, 124, 600–603.

Bennett, A. F., & J. A. Ruben. 1979. Endothermy and activity in vertebrates. *Science*, 206, 649–654.

Birkhead, T. R., L. Atkin, & A. P. Møller. 1987. Copulation behaviour in birds. *Behaviour*, 101, 101–138.

Blackburn, T. 1990. The interspecific relationship between egg size and clutch size in wildfowl. *Auk*, 10, 209–211.

Blackburn, T. M. 1991. A comparative examination of life-span and fecundity in parasitoid Hymenoptera. *J. Anim. Ecol*, 60, 151–164.

Blackburn, T. M., P. H. Harvey, & M. D. Pagel. 1990. Species number, population density and body size relationship in natural communities. *J. Anim. Ecol*, 59, 335–345.

Boyce, M. S., & C. M. Perrins. 1987. Optimizing great tit clutch size in a fluctuating environment. *Ecology*, 68, 142–157.

Bradbury, J., R. Gibson, & I. M. Tsai. 1986. Hotspots and the evolution of leks. *Anim. Behav.*, 34, 1694–1709.

Breton, L., & J. F. Addicott. 1992. Density-dependent mutualism in an aphid-ant interaction. *Ecology*, 73, 2175–2180.

Briskie, J. V., C. Naugler, & S. M. Leech. 1994. Begging intensity of nestling birds varies with sibling relatedness. *Proc. R. Soc. Lond. B*, 258, 73–78.

Bull, J. J., & E. L. Charnov. 1985. On irreversible evolution. *Evolution*, 39, 1149–1155.

Cade, W. H., & E. S. Cade. 1992. Male mating success, calling and searching behaviour at high and low densities in the field cricket, *Gryllus integer*. *Anim. Behav.*, 43, 49–56.

Caswell, H., & A. M. John. 1992. From the individual to the population in demographic models. In: *Individual-Based Models and Approaches in Ecology* (Ed. by D. L. DeAngelis, & L. J. Gross), pp. 36–57. New York: Chapman and Hall.

Charnov, E. L. 1993. *Life History Invariants: Some Explorations of Symmetry in Evolutionary Ecology*. Oxford, England: Oxford University Press.

Charnov, E. L., & D. Berrigan. 1990. Dimensionless numbers and life history evolution: Age of maturity versus the adult lifespan. *Evol. Ecol.*, 4, 273–275.

Chitty, D. 1967. The natural selection of self-regulatory behaviour in animal populations. *Proc. Ecol. Soc. Austr.*, 2, 51–78.

Christian, J. J., & D. E. Davis. 1964. Endocrines, behavior and population. *Science*, 146, 1550–1560.

Clayton, D. H., S. G. Pruett-Jones, & R. Lande. 1992. Reappraisal of the interspecific prediction of parasite-mediated sexual selection: opportunity knocks. *J. Theor.. Biol.*, 157, 95–108.

Clutton-Brock, T. H., J. C. Deutsch, & R. J. C. Nefdt. 1993. The evolution of ungulate leks. *Anim. Behav.*, 46, 1121–1138.

Cockburn, A. 1988. *Social Behaviour in Fluctuating Populations*. London: Croom Helm.

Conner, J. 1989. Density-dependent sexual selection in the fungus beetle, *Bolitotherus cornutus*. *Evolution*, 43, 1378–1386.

Damuth, J. 1981. Population density and body size in small mammals. *Nature*, 290, 699–700.

Damuth, J. 1987. Interspecific allometry of population density in mammals and other animals. *Biol. J. Linn. Soc.*, 31, 193–246.

Davies, N. B. 1991. Mating systems. In: *Behavioral Ecology* (Ed. by J. R. Krebs, & N. B. Davies), pp. 263–294. Oxford, England: Blackwell Scientific Publications.

DeAngelis, D. L., & L. J. Gross. 1992. *Individual-Based Models and Approaches in Ecology*. New York: Chapman and Hall.

Dhondt, A. A., F. Adriansen, E. Matthysen, & B. Kempenaers. 1990. Non-adaptive clutch size in tits. *Nature*, 348,

Dobson, A. P. 1982. *Mortality Rates in British Birds.*. Ph.D. Dissertation. Oxford, England: University of Oxford.

Dobson, A. P. 1988. The population biology of parasite-induced changes in host behavior. *Q. Rev. Biol.*, 63, 139–165.

Dunham, A. E., D. B. Miles, & D. N. Reznick. 1988. Life history patterns in squamate reptiles. In: *Biology of the Reptilia: Defense and Life History* (Ed. by C. Gans, & R. B. Huey), pp. 441–522. New York: Alan R. Liss.

Eiríksson, T. 1992. Density dependent song duration in the grasshopper *Omocestus viridulus*. *Behaviour*, 122, 121–132.

Ewald, P. W. 1983. Host-parasite relations, vectors, and the evolution of disease severity. *Annu. Rev. Ecol. Syst.*, 14, 465–485.

Ewald, P. W. 1993. The evolution of virulence. *Sci. Am.*, 268, 86–93.

Felsenstein, J. 1985. Phylogenies and the comparative method. *Am. Nat.*, 125, 1–15.

Finch, C. E. 1990. *Longevity, Senescence and the Genome*. Chicago: University of Chicago Press.

Finch, C. E., M. C. Pike, & M. Witten. 1990. Slow mortality rate accelerations during aging in some animals approximate that of humans. *Science*, 249, 902–906.

Fox, C. W. 1993. Multiple mating, lifetime fecundity and female mortality of the bruchid beetle, *Callosobruchus maculatus*.(Coleoptera: Bruchidae). *Func. Ecol.*, 7, 203–208.

Frumhoff, P. C., & P. S. Ward. 1992. Individual-level selection, colony-level selection, and the association between polygyny and worker monomorphism in ants. *Am. Nat.*, 139, 559–590.

Gaillard, J.-M., D. Allaine, D. Pontier, N. G. Yoccoz, & D. E. L. Promislow. 1994. Senescence in natural populations of mammals: A comment. *Evolution*, 48, 509–516.

Garland, T. Jr. 1994. Quantitative genetics of locomotor behavior and physiology in a garter snake. In: *Quantitative Genetic Studies of Behavioral Evolution* (Ed. by C. R. B. Boake), pp. 251–277. Chicago: University of Chicago.

Garland, T. Jr., P. H. Harvey, & A. R. Ives. 1992. Procedures for the analysis of comparative data using phylogenetically independent contrasts. *Syst. Biol.*, 41, 18–32.

Godfray, H. C. J. 1987. The evolution of clutch size in invertebrates. *Oxf. Surv. Evol. Biol*, 4, 117–154.

Godfray, H. C. J., & G. A. Parker. 1992. Sibling competition, parent-offspring conflict and clutch size. *Anim. Behav.*, 43, 473–490.

Godfray, H. C. J., L. Partridge, & P. H. Harvey. 1991. Clutch size. *Annu. Rev. Ecol. Syst.*, 22, 409–429.

Goodman, D. 1987. The demography of chance extinction. In: *Viable Populations for Conservation.* (Ed. by M. E. Soulé), pp. 11–34. Cambridge, England: Cambridge University Press.

Grafen, A. 1989. The phylogenetic regression. *Phil. Trans. R. Soc. Lond. B*, 326, 119–157.

Gubernick, D. J., S. L. Wright, & R. E. Brown. 1993. The significance of father's presence for offspring survival in the monogamous California mouse, *Peromyscus californicus*. *Anim. Behav.*, 46, 539–46.

Hamilton, W. D. 1964. The genetical evolution of social behaviour. *J. Theor.. Biol.*, 7, 1–52.

Hamilton, W. D., & M. Zuk. 1982. Heritable true fitness and bright birds: a role for parasites? *Science*, 218, 384–387.

Hanski, I. 1990. Density dependence, regulation and variability in animal populations. *Phil. Trans. R. Soc. Lond. B*, 330, 141–150.

Hansson, L., & H. Hentonnen. 1985. Gradients in density variations of small rodents: the importance of latitude and snow cover. *Oecologia*, 67, 394–402.

Harris, R. J., C. D. Thomas, & H. Moller. 1991. The influence of habitat use and foraging on the replacement of one introduced wasp species by another in New Zealand. *Ecol. Entomol.*, 16, 441–448.

Harvey, P. H., & T. H. Clutton-Brock. 1985. Life history variation in primates. *Evolution*, 39, 559–581.

Harvey, P. H., & M. D. Pagel. 1991. *The Comparative Method in Evolutionary Biology*. Oxford, England: Oxford University Press.

Hassell, M. P., & R. M. May. 1985. From individual behaviour to population dynamics. In: *Behavioural Ecology* (Ed. by R. M. Sibly, & R. H. Smith), pp. 3–32. Oxford, England: Blackwell Scientific Publications.

Hassell, M. P., J. Latto, & R. M. May. 1989. Seeing the wood for the trees - detecting density dependence from existing life-table studies. *J. Anim. Ecol.*, 58, 883–892.

Hastings, A., & R. F. Costantino. 1987. Cannibalistic egg-larva interactions in *Tribolium*: An explanation for the oscillations in population numbers. *Am. Nat.*, 130, 36–52.

Healy, S. D., & T. Guilford. 1990. Olfactory-bulb size and nocturnality in birds. *Evolution*, 44, 339–346.

Herre, E. A. 1989. Coevolution of reproductive characteristics in 12 species of new world figs and their pollinator wasps. *Experientia*, 45, 637–647.

Herre, E. A. 1993. Population structure and the evolution of virulence in nematode parasites of fig wasps. *Science*, 259, 1442–1445.

Hill, G. E. 1991. Plumage coloration is a sexually selected indicator of male quality. *Nature*, 350, 337–339.

Houston, A., C. Clark, J. McNamara, & M. Mangel. 1988. Dynamic models in behavioural ecology and evolutionary ecology. *Nature*, 332, 29–34.

Inchausti, P. 1994. Reductionist approaches in community ecology. *Am. Nat.*, 143, 201–221.

Iwasa, Y., Y. Suzuzi, & H. Matsuda. 1984. Theory of oviposition strategy of parasitoids. I. Effect of mortality and limited egg number. *Theor. Pop. Biol.*, 26, 205–277.

Jeffreys, A. J., V. Wilson & S. L. Thein. 1985. Hypervariable 'minisatellite' regions in human DNA. *Nature*, 314, 67–73.

Koehl, M. A. R. 1989. Discussion: from individuals to populations. In: *Perspectives in Ecological Theory* (Ed. by J. Roughgarden, R. M. May, & S. A. Levin), pp. 39–53. Princeton, New Jersey: Princeton University Press.

Krukonis, G., & W. M. Schaffer. 1991. Population cycles in mammals and birds: Does periodicity scale with body size? *J. Theor. Biol.*, 148, 469–493.

Lack, D. 1947. The significance of clutch-size. *Ibis*, 89, 302–352.

Lack, D. 1948. The significance of clutch size. *Ibis*, 90, 25–45.

Lande, R. 1988. Genetics and demography in biological conservation. *Science*, 241, 1455–1460.

Lauder, G. V., A. M. Leroi, & M. R. Rose. 1993. Adaptations and history. *Trends Ecol. Evol.*, 8, 294-297.

Leroi, A. M., M. R. Rose, & G. V. Lauder. 1994. What does the comparative method reveal about adaptation? *Am. Nat.*, 143, 381–402.

Lessells, C. M. 1990. The evolution of life histories. In: *Behavioural Ecology* (Ed. by J. R. Krebs, & N. B. Davies), pp. 32–68. Oxford, England: Blackwell.

Li, J. B., & D. C. Margolies. 1993. Effects of mite age, mite density, and host quality on aerial dispersal behavior in the 2-spotted spider-mite. *Entomol. Exp. Appl.*, 68, 79–86.

Lima, S. L., & L. M. Dill. 1990. Behavioral decisions made under the risk of predation: A review and prospectus. *Can. J. Zool.*, 88, 619–640.

Lomnicki, A. 1988. *Population Ecology of Individuals*. Princeton, New Jersey: Princeton University Press.

Luckinbill, L. S., R. Arking, M. J. Clare, W. J. Cirocco, & S. A. Buck. 1984. Selection for delayed senescence in *Drosophila melanogaster*. *Evolution*, 38, 996–1003.

Maddison, W. P. 1990. A method for testing the correlated evolution of two binary characters: Are gains or losses concentrated on certain branches of a phylogenetic tree. *Evolution*, 44, 539–557.

Madsen, T., R. Shine, & J. Loman. 1992. Why do female adders copulate so frequently? *Nature*, 355, 440–441.

Mangel, M., & C. W. Clark. 1988. *Dynamic Modeling in Behavioral Ecology*. Princeton, New Jersey: Princeton University Press.

Marsh, H., & T. Kasuya. 1986. Evidence for reproductive senescence in female cetaceans. *Rep. Int. Whal. Comm.*, Special Issue 8, 57–74.

Martin, T. E. 1993. Evolutionary determinants of clutch size in cavity-nesting birds: Nest predation or limited breeding opportunities? *Am. Nat.*, 142, 937–946.

Martin, T. E. 1995. Variation and covariation of life history traits in birds in relation to nest sites, nest predation, and food. *Ecol. Monog.*, 65, 101–127.

May, R. M. 1972. Will a large complex system be stable? *Nature*, 238, 413–414.

May, R. M., S. Nee, & C. Watts. 1990. Could intraspecific brood parasitism cause population cycles? *Acta XX Congr. Int. Ornithol.*, 1012–1022.

McLennan, D. A., & D. R. Brooks. 1991. Parasites and sexual selection: A macroevolutionary perspective. *Q. Rev. Biol.*, 66, 255–286.

Meyer, A., J. M. Morrissey, & M. Schartl. 1994. Recurrent origin of a sexually selected trait in *Xiphophorus* fishes inferred from a molecular phylogeny. *Nature*, 368, 539–542.

Møller, A. P. 1991. Density-dependent extra-pair copulations in the swallow *Hirundo rustica*. *Ethology*, 87, 316–329.

Møller, A. P. 1992. Parasites differentially increase the degree of fluctuating asymmetry in secondary sexual characters. *J. Evol. Biol.*, 5, 691–699.

Møller, A. P., & T. R. Birkhead. 1992. A pairwise comparative method as illustrated by copulation frequency in birds. *Am. Nat.*, 139, 644–656.

Møller, A. P., & T. R. Birkhead. 1993. Cuckoldry and sociality: a comparative study of birds. *Am. Nat.*, 142, 118–1140.

Møller, A. P., & J. Höglund. 1991. Patterns of fluctuating asymmetry in avian feather ornaments: Implications for models of sexual selection. *Proc. R. Soc. Lond. B*, 245, 1–5.

Moore, J., & N. J. Gotelli. 1990. A phylogenetic perspective on the evolution of altered host behaviours: A critical look at the manipulation hypothesis. In: *Parasitism and Host Behaviour* (Ed. by C. J. Barnard, & J. M. Behnke), pp. 193–229. Bristol, Pennsylvania: Taylor & Francis.

Mueller, L. D. 1987. Evolution of accelerated senescence in laboratory populations of *Drosophila. Proc. Natl. Acad. Sci. USA*, 84, 1974–1977.

Murray, M. D. 1990a. Influence of host behaviour on some ectoparasites of birds and mammals. In: *Parasitism and Host Behaviour* (Ed. by C. J. Barnard, & J. M. Behnke), pp. 290-315. Bristol, Pennsylvania: Taylor & Francis.

Murray, M. G. 1990b. Comparative morphology and mate competition of flightless male fig wasps. *Anim. Behav.*, 39, 434–443.

Nee, S., A. F. Read, J. J. D. Greenwood, & P. H. Harvey. 1991. The relationship between abundance and body size in British birds. *Nature*, 351, 312–313.

Nesse, R. M. 1988. Life table tests of evolutionary theories of senescence. *Exp. Gerontol.*, 23, 445–453.

Oakes, E. J. 1992. Lekking and the evolution of sexual dimorphism in birds — Comparative approaches. *Am. Nat.*, 140, 665–684.

Oster, G., & E. O. Wilson. 1978. *Caste and Ecology in the Social Insects*. Princeton, New Jersey: Princeton University Press.

Ostfeld, R. S., & L. L. Klosterman. 1990. Microtine social systems, adaptation, and the comparative method. In: *Social Systems and Population Cycles in Voles* (Ed. by R. H. Tamarin, R. S. Ostfeld, S. R. Pugh, & G. Bujalska), pp. 35–44. Basel: Birkhäusen Verlag.

Packer, C., S. Lewis, & A. Pusey. 1992. A comparative analysis of non-offspring nursing. *Anim. Behav.*, 43, 265–281.

Pagel, M. D. 1992. A method for the analysis of comparative data. *J. Theor. Biol.*, 156, 431–442.

Pagel, M. D., & P. H. Harvey. 1988. The taxon-level problem in the evolution of mammalian brain size: Facts and artifacts. *Am. Nat.*, 132, 344–359.

Partridge, L., & R. Andrews. 1985. The effect of reproductive activity on the longevity of male *Drosophila melanogaster* is not caused by an acceleration of ageing. *J. Insect Physiol.*, 31, 393–395.

Partridge, L., & N. H. Barton. 1993. Optimality, mutation and the evolution of aging. *Nature*, 362, 305–311.

Peterson, R. O., R. E. Page, & K. M. Dodge. 1984. Wolves, moose, and the allometry of population cycles. *Science*, 224, 1350–1352.

Pettifor, R. A., C. M. Perrins, & R. H. McCleery. 1988. Individual optimization of clutch size in great tits. *Nature*, 336, 160–162.

Pimm, S. L. 1984. Food chains and return times. In: *Ecological Communities: Conceptual Issues and Evidence* (Ed. by D. R. Strong Jr., D. Simberloff, L. G. Abele, & A. B. Thistle), pp. 397–412. Princeton, New Jersey: Princeton University Press.

Pimm, S. L., H. L. Jones, & J. Diamond. 1988. On the risk of extinction. *Am. Nat.*, 132, 757–785.

Pimm, S. L., & A. Redfearn. 1988. The variability of population densities. *Nature*, 334, 613–614.

Polak, M. 1993. Parasites increase fluctuating asymmetry of male *Drosophila nigrospiracula*: implications for sexual selection. *Genetica*, 89, 255–265.

Polis, G. 1981. The evolution and dynamics of intraspecific predation. *Annu. Rev. Ecol. Syst.*, 12, 125–151.

Pomeroy, D. 1990. Why fly? The possible benefits for lower mortality. *Biol. J. Linn. Soc.*, 40, 53–65.

Promislow, D. E. L. 1991a. The evolution of mammalian blood parameters: Patterns and their interpretation. *Phys. Zool.*, 64, 393–431.

Promislow, D. E. L. 1991b. Senescence in natural populations of mammals: A comparative study. *Evolution*, 45, 1869–1887.

Promislow, D. E. L. 1992. Costs of sexual selection in natural populations of mammals. *Proc. R. Soc. Lond. B*, 247, 203–210.

Promislow, D. E. L. 1994. DNA repair and the evolution of longevity: A critical analysis. *J. Theor. Biol.*, 170, 291–300.

Promislow, D. E. L. 1995. New perspectives on comparative tests of antagonistic pleiotropy. *Evolution*, 49, 394–397.

Promislow, D. E. L., & P. H. Harvey. 1990. Living fast and dying young: A comparative analysis of life history variation among mammals. *J. Zool.*, 220, 417–437.

Promislow, D. E. L., & P. H. Harvey. 1991. Mortality rates and the evolution of mammal life histories. *Acta Oecologica*, 12, 119–137.

Promislow, D. E. L., R. D. Montgomerie, & T. E. Martin. 1992. Mortality costs of sexual dimorphism in birds. *Proc. Roy. Soc. Lond. B*, 250, 143–150.

Promislow, D. E. L., R. D. Montgomerie, & T. E. Martin. 1994. Sexual selection and survival in North American waterfowl. *Evolution*, 48, 2045–2050

Purvis, A. 1991. *Comparative Analysis by Independent Contrasts*. Version 1.2. Oxford, England: University of Oxford, Department of Zoology.

Read, A. F. 1990. Parasites and the evolution of host sexual behaviour. In: *Parasitism and Host Behaviour* (Ed. by C. J. Barnard, & J. M. Behnke), pp. 117–157. Bristol, Pennsylvania: Taylor & Francis.

Read, A. F., & P. H. Harvey. 1989a. Life history differences among the eutherian radiations. *J. Zool. (Lond.)*, 219, 329–353.

Read, A. F., & P. H. Harvey. 1989b. Reassessment of the comparative evidence for Hamilton and Zuk's theory on the evolution of secondary sexual characters. *Nature*, 339, 618–620.

Read, A. F., & P. H. Harvey. 1989c. Validity of sexual selection in birds. *Nature*, 340, 105.

Real, L. A. 1992. Introduction to the symposium (Behavioral mechanisms in evolutionary biology). *Am. Nat.*, 140, S1–S4.

Reeve, H. K., D. F. Westneat, & D. C. Queller. 1992. Estimating average within-group relatedness from DNA fingerprints. *Mol. Ecol.*, 1, 223–232.

Rejmanek, M. (in press). Global patterns of alien plant species diversity. *Ecol.*

Ridley, M. 1983. *The Explanation of Organic Diversity: The Comparative Method and Adaptations for Mating*. Oxford, England: Oxford University Press.

Ridley, M. 1988. Mating frequency and fecundity in insects. *Biol. Rev.*, 63, 509–549.

Ridley, M. 1993. Clutch size and mating frequency in parasitic Hymenoptera. *Am. Nat.*, 142, 893–910.

Roff, D. A. 1990. The evolution of flightlessness in insects. *Ecol. Monog.*, 60, 389–421.

Roff, D. A., & D. J. Fairbairn. 1991. Wing dimorphisms and the evolution of migratory polymorphisms among the Insecta. *Am. Zool.*, 31, 243–251.

Rogers, A. R. 1993. Why menopause. *Evol. Ecol.*, 7, 406–420.

Rose, M. R. 1991. *Evolutionary Biology of Aging*. Oxford, England: Oxford University Press.

Rose, M. R., & B. Charlesworth. 1980. A test of evolutionary theories of senescence. *Nature*, 287, 141–142.

Sæther, B.-E. 1988. Evolutionary adjustment of reproductive traits to survival rates in European birds. *Nature*, 331, 616–617.

Sæther, B.-E. 1989. Survival rates in relation to body weight in European birds. *Ornis Scand.*, 20, 13–21.

Schoener, T. W. 1986. Mechanistic approaches to community ecology: A new reductionism? *Am. Zool.*, 26, 81–106.

Service, P. M., E. W. Hutchinson, & M. R. Rose. 1988. Multiple genetic mechanisms for the evolution of senescence in *Drosophila melanogaster*. *Evolution*, 42, 708–716.

Sibley, C. G., & J. E. Ahlquist. 1990. *Phylogeny and Classification of Birds: A Study in Molecular Evolution*. New Haven, Connecticut: Yale University Press.

Sillén-Tullberg, B. 1988. Evolution of gregariousness in aposematic butterfly larvae: a phylogenetic analysis. *Evolution*, 42, 293–305.

Sillén-Tullberg, B. 1994. The effect of biased inclusion of taxa on the correlation between discrete characters in phylogenetic trees. *Evolution*, 47, 1182–1191.

Simberloff, D. S. 1986. The proximate causes of extinction. In: *Patterns and Processes in the History of Life* (Ed. by D. M. Raup, & D. Jablonski), pp. 259–276. Berlin: Springer-Verlag.

Vanderwerf, E. 1992. Lack clutch size hypothesis — an examination of the evidence using meta-analysis. *Ecology*, 73, 1699–1705.

Werner, E. E., & B. R. Anholt. 1993. Ecological consequences of the trade-off between growth and mortality rates mediated by foraging activity. *Am. Nat.*, 142, 242–272.

Woiwood, I. P., & I. Hanski. 1992. Patterns of density dependence in moths and aphids. *J. Anim. Ecol.*, 61, 619–621.

Woolbright, L. L., D. J. Greene, & G. C. Rapp. 1990. Density-dependent mate searching strategies of male woodfrogs. *Anim. Behav.*, 40, 135–42.

Yokel, D. A. 1989. Intrasexual aggression and the mating behavior of brown-headed cowbirds: Their relation to population densities and sex ratios. *Condor*, 91, 43–51.

CHAPTER 11

Phylogenetic Interpretations of Primate Socioecology: With Special Reference to Social and Ecological Diversity in *Macaca*

Leslie K. W. Chan

Socioecological explanations have dominated discussion of primate social diversity since the pioneering work of Crook and Gartlan (1966). The authors suggest in this highly influential paper that the social organization of diverse primate species can be predicted on the basis of a number of key ecological variables, most notably predation pressure and food distribution. In the ensuing decades, the notion that ecological variables shape primate societies through natural selection appeared to be vindicated by the discovery of "striking examples of convergence in the social systems of distantly related species with at least superficial similarities in their ecological adaptations" (Wrangham, 1987, p. 282). The search for socioecological convergence between neotropical and Old World primates and its underlying mechanisms has been a major preoccupation of primatological studies ever since (see Terborgh & Janson 1986; Robinson & Janson 1987; Crook 1989; Chan 1992).

With few exceptions, primatologists who employ the socioecological approach have tended to emphasize the adaptive nature of features, such as group size and mating patterns, with little regard to the importance of evolutionary history of the species being compared (e.g., Terborgh 1986; van Schaik & van Hooff 1983; Dunbar 1989).

However, as early as 1969, Thomas Struhsaker raised the possibility that the similarity in the unimale, multifemale group structures and the ecological characteristics found among African *Cercopithecus*, are more likely due to common descent than to adaptation to similar environments. Reaction to Struhsaker's view was far from enthusiastic then, and Rodman (1988, p. 89) still writes:

> Different phylogenetic histories may be called on to explain interspecific differences in social patterns (Struhsaker 1969). This very unsatisfying hypothesis begs the question by asserting in effect that different species are different because they are different species, and it moves the causes of differentiation out of the present and out of reach of observation and functional analysis.
>
> ...phylogenetic hypothesis should be one of the last resort because it leaves us with nothing to do, and because tests of phylogenetic effects should be based on the most thorough understanding of natural behavior possible.

While I agree that "the most thorough understanding of natural behavior" is a prerequisite for phylogenetic analysis, and indeed with any approach, I will argue that far from being the last resort, phylogenetic hypotheses should be viewed as a guide and be integrated into study of primate social evolution and comparative socioecology. The value of a phylogenetic analysis is that it is "an experiment over historical time" (Huey 1987, p. 79) and so it allows us to recover the pattern, distribution and directions of past changes based on comparison of extant species. As O'Hara (1988) points out, we need to establish an "evolutionary chronicle" of the character being studied in order to formulate the sequence of events from which explanations may be sought. Otherwise we risk confounding functional questions about the maintenance of a trait with the evolutionary explanations of its origins (Gould & Vrba 1982).

Primatologists' reliance on functional explanations is in part due to the great faith in the power of natural selection in shaping all aspects of a species' characteristics, including behavior and social organization (Rowell 1979). It may also be due in part to the lack of an alternative methodology for evaluating the origins and diversification of ecological and behavioral characteristics. Although the so-called comparative method is the cornerstone of primate behavioral and ecological studies, relatively few efforts have been directed at the explicit formulation of a set of methods for evaluating "similarities" and "differences" in primate social systems in an evolutionarily meaningful way (but see Clutton-Brock & Harvey 1979; Ridley 1983).

In recent years, however, there has been substantial development in the application of phylogenetic methods to the analysis of behavioral, ecological and social patterns in diverse groups of organisms (e.g., Dobson 1985; Lauder 1986; Prum 1990; Lanyon 1992; Sillén-Tullberg 1988; Peterson & Burt 1992; Langtimm & Dewsbury 1991; McLennan 1991; Edwards & Naeem 1993; Miles & Dunham 1993). One common approach, referred to as character mapping or character optimization method (Brooks & McLennan 1991), consists of establishing genealogical relationships of the species being compared first, and then asking whether similar traits observed among species arose in a common ancestor and have persisted, or whether they arose independently due to convergent adaptation to similar environments (Funk & Brooks 1990). This approach emphasizes the importance of a species' evolutionary history in constraining both the features it currently displays and the subsequent directions and rates of evolution open to it (Seilacher 1990; Riedl 1978; Gould & Lewontin 1979; Alberch 1990; Wanntorp 1983; Stearns 1992; McKitrick 1993).

A key assumption of this approach is that just as the evolution of morphological and life-history traits, behavioral and social evolution take place in the context of phylogeny. It follows that by mapping or superimposing these biological or behavioral features on a well-established phylogenetic hypothesis, we may estimate when, and how many times, an observed characteristic arose in the group being studied. We can also gain some insights as to which features tend to coevolve, which features are highly conservative, and which are most labile. In other words, this approach allows us to compare similar traits across species and to distinguish those similarities that are most likely due to descent from a common ancestor from those that evolve independently.

Unlike the socioecological approach, which is largely concerned with predictive models about current adaptation and pays little attention to the question of evolutionary origin, the character-mapping method stresses that only after examining the hypothesis that the observed similarities are due to common descent should we proceed with adaptive explanations (Coddington 1988; Taylor 1987). Furthermore, the character-mapping method is not fundamentally concerned with the presumed process or causes by which characters became what they are today. However, it is generally agreed that once patterns of character distribution have been established, one can proceed to investigate the processes underlying the observed patterns (McLennan 1991). The main purpose of the character mapping method, then, is to identify the most parsimonious resolution of the observed distribution of traits on a phylogenetic tree (Wenzel 1992). This method is therefore highly dependent on the principle of parsimony, which states that the simplest explanation consistent with the data should be chosen over complex explanations involving more steps or assumptions (Stewart 1993). As such, the principle of parsimony is best seen as a methodological guideline, and the principle by itself says nothing about the underlying processes or mechanisms responsible for the observed pattern of trait distribution. The principle prescribes that complex functional adaptation need not be invoked when phylogenetic effect cannot be ruled out as the primary determinants of observed similarities (Sober 1988). Viewed in this manner, functional and phylogenetic analyses are not dichotomous endeavor, but are complementary components of a complete evolutionary explanation of biological diversity (Wake 1991).

It should also be emphasized that while parsimony analysis is a popular approach for dealing with character data in phylogenetic analysis, it is not necessarily the most reliable way. Other approaches using probability and compatibility methods may be more appropriate, depending on the nature of the data and the hypothesis being tested (see, in this volume, Martins & Hansen, Chap. 2, for review).

Another major prerequisite of the character-optimization approach is that it is dependent on a well-corroborated phylogeny. Such phylogenies, however, are still lacking or poorly understood for most groups of organisms (but see, in this volume, Martins & Hansen, Chap. 2). Primates, in this respect, provide a unique opportunity for the exploration of the role of phylogeny because the evolutionary relationship of the major clades within the Order is relatively well worked out (Fig. 1), although of course there is no shortage of disagreements about the detailed branching patterns within each clade (Fleagle 1988). For this reason, conclusions stemming from phylogenetic character mapping should be viewed as provisional evolutionary hypotheses, subject to further evaluation.

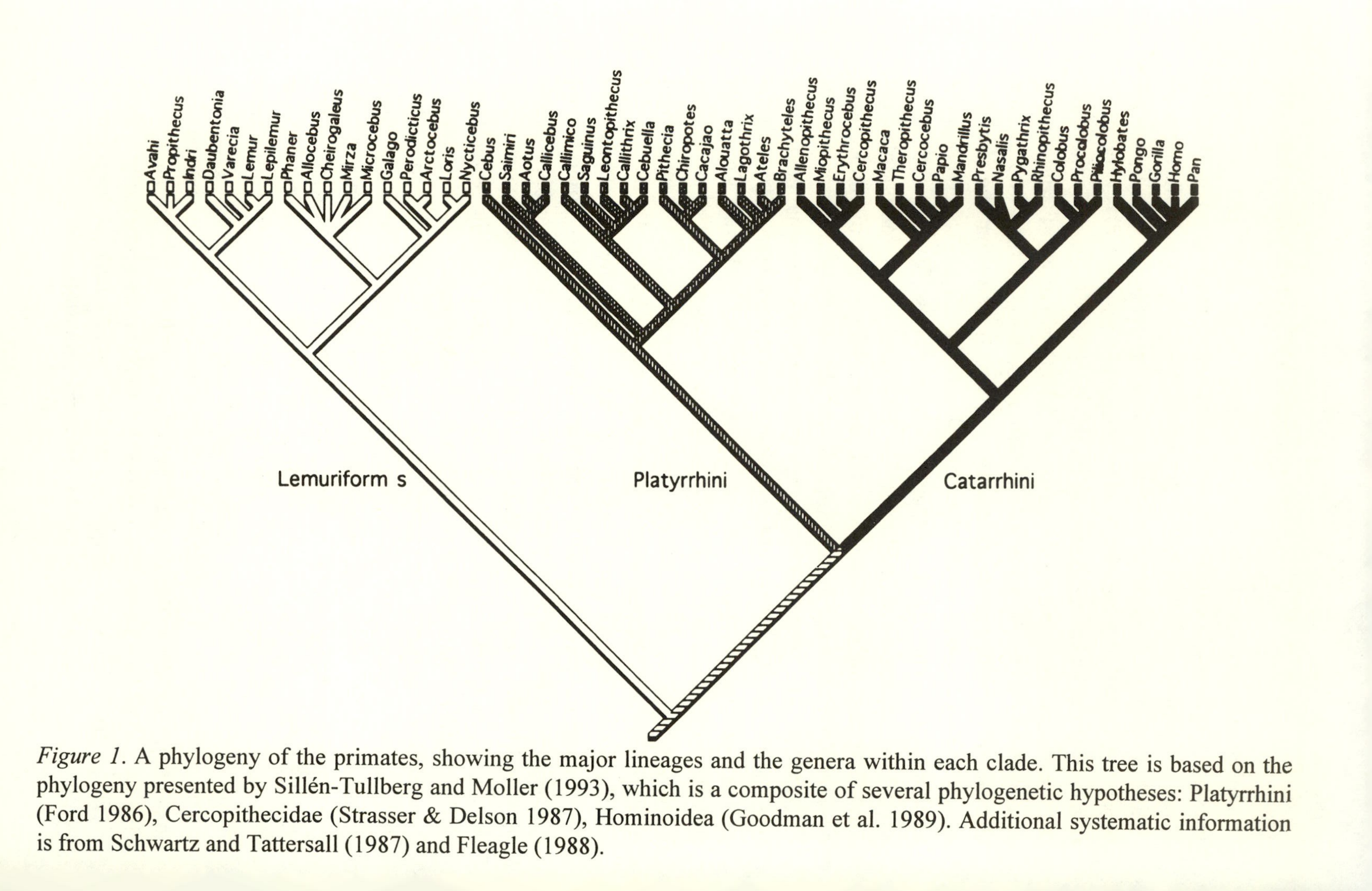

Figure 1. A phylogeny of the primates, showing the major lineages and the genera within each clade. This tree is based on the phylogeny presented by Sillén-Tullberg and Moller (1993), which is a composite of several phylogenetic hypotheses: Platyrrhini (Ford 1986), Cercopithecidae (Strasser & Delson 1987), Hominoidea (Goodman et al. 1989). Additional systematic information is from Schwartz and Tattersall (1987) and Fleagle (1988).

Besides emphasizing the importance of phylogeny or "tree thinking" (O'Hara 1988) in the comparison of primate social diversity, my intent is (a) to show that some of the fundamental assumptions of socioecological explanations are deficient and so phylogenetic alternatives and supplementary viewpoints for understanding the nature of primate social and ecological diversity are needed; (b) to take a close look at the genus *Macaca* as a case study to see how social patterns and ecological adaptation of macaque species can be better understood in a phylogenetic context; and (c) to outline a theoretical framework that treats social organizations as historical entities subjected to the constraint of recent as well as "deep" historical influence. The underlying theme of this discussion is that an understanding of how phylogeny constrains and channels social change is as important in studies of behavioral evolution as it is in developmental biology. Just as knowledge of developmental mechanisms is essential to the understanding of the nature of structural uniformity and diversity (Alberch 1990; Wake 1992; Thomas & Reif 1993), so knowledge of the developmental "rules" of social organization is central to the comprehension of "convergence" and diversification in primate social structure. If the evolution of social organization is constrained by phylogeny, then it is possible that convergence in social organization may represent the result of finite evolutionary pathways (Foley & Lee 1989), or limits in the "design features" (Wake 1991) of social organization .

Assumptions of socioecology and phylogenetic alternatives

While there are a wide variety of models emphasizing the relative influence of ecological phenomena such as predation pressure (e.g., van Schaik 1983; Terborgh 1983) and food distribution (e.g., Wrangham 1980; Rodman 1988) on primate social structures, these models share at least four common, though often implicit, assumptions (for a more in-depth discussion of the assumptions of primate socioecology, see Chan 1992). These are as follows:

1. Social organizations are species-specific and so species constitute independent data points for comparative analysis. As Wilson (1975, p. 255) stated,

> The principal organizing concept in the study of primate societies has been the theory that social parameters are fixed in each species as an adaptation to the particular environment in which the species lived. The parameters include size, demographic structure, home range size and stability, and attention structure.

2. The social systems under study are in stable equilibria with the environment. Explanations of similarities and differences in social systems may, therefore, be sought in the selection forces operating in the current ecological context. Dunbar's (1988) work provides a clear example of this approach.
3. Social organizations are evolutionary "solutions" to "problems" posed by the environment. Socioecological convergence will take place when organisms, regardless of their phylogenetic background, are faced with similar ecological situations. This assumption has been most explicitly stated by Emlen (1980).
4. The mating system is the key component of a social organization since mating patterns are thought to be most directly influenced by ecological factors (See Wrangham 1987 for the large body of literature on this topic).

The validity of assumption 1 will be examined immediately below. Assumptions 2 and 3 will be questioned in the final section of this paper, while assumption 4 will be dealt with in the macaque case study.

In recent years it has been repeatedly pointed out (Clutton-Brock & Harvey 1979; Felsenstein 1985; Ridley 1983) that individual species cannot be treated as independent data points in comparative analyses. This is because species, particularly those who share a recent common ancestor, are likely to share many aspects of their social and ecological adaptation due to descent. As the phylogenetic approach emphasizes that the evolutionary explanation for any particular trait should be sought at the appropriate level in the phylogenetic tree, it follows that the level at which an explanation is sought will vary depending on the trait under consideration (Wenzel 1992; in this volume, see Lauder & Reilly, Chap. 4, and Foster & Cameron, Chap. 5).

The assumption that social characteristics are species-specific and the problem of non-independence can be easily discerned when we consider the occurrence of monogamy in primates. It is commonly accepted that a monogamous, pair-bonded, nuclear-family social group is rare in mammals and, for example, found in only "3% of mammals and 14% of primates" (Rutberg 1983, cited in Leighton 1987, p. 142). Using data published in Smuts et al. (1987, p. 501-505), Wrangham (1987, p. 287) estimates that between 13 and 22% of primates are monogamous. It is clear from the tables presented by Wrangham that he arrived at these percentage figures by counting each primate species as an independent data point. Another look at the information presented in Smuts et al. suggests that the distribution of monogamy in primates is not independent of phylogeny but is concentrated instead in certain taxonomic groups. Among New World monkeys, the eight cases of monogamous species in the family Cebidae all belong to the subfamily Pitheciinae, within which all four species of the genus *Pithecia* (commonly known as saki) are monogamous. The other four monogamous species within this family are found in two closely related genera, *Callicebus* (three species) and *Aotus* (one species). Whether a monogamous social group is a shared derived feature of the subfamily Pitheciinae is an interesting hypothesis that remains to be tested. Similarly, the only two cases of monogamy in Old World Monkey are found in two closely related genera of colobines: *Presbytis potenziani* and *Nasalis concolor* (Fig. 1). While monogamous mating pattern is also rare among the hominoids, all nine gibbon species are monogamous, strongly suggesting that this is a shared feature for the entire *Hylobates* genus (Leighton 1987).

The brief example above illustrates that knowing the percentage of species having a particular trait such as monogamy is not particularly informative. This is because estimates such as those made by Rutberg and Wrangham are based on a model of explanation that assumes monogamy as randomly distributed with respect to phylogeny. Since species are connected to each other through descent, we need to make reference to phylogeny in order to determine how often monogamous social group evolves, and whether the evolution of the trait constitutes one or more evolutionary events (O'Hara 1988). Making reference to phylogeny also allows us to question whether monogamy, once evolved, is likely to persist in descendant species. We can then proceed with questions such as whether the origin of this trait repeatedly coincides with any particular ecological or behavioral shifts (Brooks & McLennan 1991). It is interesting to note however, that two recent theoretical treatments of the evolution of monogamy in primates (Snowdon 1990; van Schaik & Dunbar 1990) continue to pay primary attention to possible ecological correlates of monogamy and little or no attention to phylogenetic effects and the point at which monogamy originated in the genealogical hierarchy. This is in keeping with the primatological tradition of emphasizing the diversity of primate behavior and social systems and to explain this diversity in functional terms, while paying little corresponding attention to the underlying uniformity across taxa, or to the limitations of what can possibly evolve, even though such uniformity and constraint have long been recognized.

It is not surprising that, among primates, some biological and social traits are broadly distributed and highly conservative, while others are unique to a small number of taxa. For example, female dominance over males occurs in all the lemuriforms so far studied (Richard 1987, p.32), but such a relationship between the sexes is seldom or never found in other primates. In colobines, aggression among females is usually infrequent, and female dominance hierarchies are either weakly developed or not apparent (Struhsaker & Leland 1987). In contrast, female dominance hierarchy and stable social relationship among females are prominent features of cercopithecine societies. Grooming patterns of males and females are remarkably similar in all colobines and in most cercopthecines: females engage in most of the grooming while males tend to receive most of the grooming relative to their representation in the group. When adult males do groom, the activity is generally directed to adult females rather than toward adult males. According to Struhsaker & Leland (1987, p.91), "this pattern prevails, regardless of ecology, group size, and composition."

All the species in the subfamily Callitrichinae (tamarins and marmosets) show strong paternal care, although this behavior is rare in the other primates. Allomothering behavior is common among the Colobines, and while such behavior is often observed in Old World primates, allomothering behavior in Cercopithecines is not as widespread as among colobines (McKenna 1979). This has prompted Moore (1992, p.372) to ask: "Why is alloparenting so prominently a *colobine* trait?" There is as yet no satisfactory ecological explanation for this question.

Among cercopithecines, females never or seldom disperse from their natal group, while female dispersal is common in the hominoids and in some species of New World monkeys (Pusey & Packer 1987). Although dispersal by female is also "rare" in New World primates, it is found in four species of *Ateles* and one species of *Brachyteles*, two genera that are also closely related. In howler, another closely related group, both male and female transfer take place regularly (Moore 1992). Why then are there so many forms of dispersal in New World primates, while the pattern of dispersal appears more limited in Old World monkeys? Despite the tremendous morphological, genetic, and ecological diversity of Old World monkey species, male-bias dispersal and strong female bonds remain the central core of Cercopithecine social structure. Why these features remain so persistent and invariable in Cercopthecines remains a puzzling problem (Moore 1992), and the possible constraints these features may have on the evolution of other social characteristics also remain poorly understood.

In the case of life history traits, no definite case of strict birth seasonality, for example, has been described for any Colobinae. In contrast, Madagascan primates for which data are available are all seasonal breeders. The reproductive seasons of African lorises, however, is highly variable with different species exhibiting clear-cut seasons, annual peaks, and births evenly distributed throughout the year (Lindburg 1987). Within the Cercopithecines, while some life history traits such as age at maturity, birth seasonality, and interbirth interval are highly variable and are most likely related to food availability and social factors, other characteristics such as gestation length appear to show little variation, despite considerable differences in body size and habitats between many species (Harvey et al. 1987).

The average length of gestation in the 15 Cercopthecine species (including baboons, macaques, vervets, mangabeys, and talapoins) surveyed by Melnick and Pearl (1987) varies little, ranging from 160 to 180 days. "This is surprising," according to Melnick and Pearl (1987, p.125), "given the positive correlation between body size and gestation length in primates and the enormous difference in body size between an adult female talapoin (body weight 1.1 kg, gestation period 165 days; Gautier-Hion 1973) and an adult female yellow baboon (approximately 12 kg, gestations period 175 days; Altmann et al. 1977)."

This example may not be as puzzling, if we consider Stearns' (1992, p.94) statement that "the basic design common to a higher taxon shapes the rough boundaries within which the life cycles of all its constituent species evolve." The question then becomes why is gestation length in Cercopithecine so resistance to change and what effects this have on other life history traits? From the above examples, it appears that there are limits to the range of variation or expression in certain social behavior and biological traits, and this may be due to the fact that species are subject to historical constraint and are not free to evolve any possible biological structure or social arrangement. To use another example, there are no primate social systems in which males defend food resources while females visit the territories of a number of different males, although this system is widespread in other species, such as ungulates (Owen-Smith 1977, cited in Wrangham 1987). The theme that some aspects of the current social system of a species is highly constrained by phylogeny and internal mechanisms will be further developed below.

Case study: Macaque phylogeny and socioecology

To illustrate further how phylogenetic analysis can illuminate functional interpretations and help reveal patterns that are otherwise unrecognized, I will look at the possible relationship between ecological adaptation and male copulatory behavior in macaques. Copulatory behavior is chosen as the character for analysis because if mating behavior is strongly influenced by ecological factors, as many classic socioecological models predict (Emlen 1980; Wrangham 1980), we should expect a poor correspondence between mating behavior and phylogeny but a high correlation between mating patterns and ecological conditions.

A. Some brief background on macaques

Macaques are particularly well suited for the evaluation of phylogenetic influence and ecological adaptation. This is because the genus, which consists of a large number of species (16 to 20 depending on the number of Sulawesi macaques one recognizes, Groves 1989), is the most geographically widespread and ecologically diverse nonhuman primate genus (Lindburg 1991). In Asia, macaques range from Japan in the East to Afghanistan in the West, and as far north as 42^0 N in Japan and into 9^0 S in South East Asia. A relict population of *M. sylvanus* also survives in fragmented habitat in Morocco and Algeria (Fooden 1980). Macaque species are found in broadleaf evergreen forests as well as in coniferous, deciduous, riverine, mangrove, and secondary forest. Some species, such as *M. mulatta* and *M. fascicularis*, also thrive in highly disturbed nonforested, and even urban areas (Fooden 1982).

Macaques are also well suited for the study of the relative role of historical factors and ecological adaptation because the phylogenetic relationship of macaques species have been intensely studied over the last two decades (see review by Fa 1989), and there is a high degree of concordance between estimates generated by morphological characters (Fooden 1980; Delson 1980; Groves 1989) and molecular markers (Fooden & Lanyon 1989; Melnick & Kidd 1985, Hoelzer & Melnick 1992).

Further, macaques are good subjects for comparative analysis because the behavior, reproductive physiology and social organization of many macaque species have been well documented. Species such as rhesus (*M. mulatta*), Japanese macaques (*M. fuscuta*), and Barbary macaques (*M. sylvanus*) are particularly well studied, both in captivity and in the field, and long-term data on ecology and social organization are available for these species (for references, see Richard 1985). Notable exceptions are the various Sulawesi macaques, which are poorly known in terms of their behavior and ecology. Species such as stumptails macaques (*M. arctoides*) are fairly well studied in the laboratory, but the species remain poorly studied in the field (Rhine in press). The quality of information across macaque species is therefore far from even. Nonetheless, there is a wealth of published information on various macaque species for testing propositions pertaining to the evolutionary history of the genus. In particular, I will examine Caldecott's (1986) hypothesis that specifically relates the different patterns of copulatory behavior in male macaques with varying ecological conditions.

Within the genus *Macaca*, as in other male primates, two distinct patterns of copulation have been identified. Macaque species can be characterized according to whether the males typically ejaculate at the end of one isolated intromissive mount or only during the last of a series of such mounts. Members of the first group are termed single-mount ejaculators (SME) and those of the second group multiple-mount ejaculators (MME) (Fooden 1980; Shively et al. 1982). Despite the well-founded distinction of these two patterns of copulation, little is known about their origins and functional significance (Dewsbury & Pierce 1989).

In addition to this well-known difference in copulatory patterns, macaques species also differ in other reproductive characteristics such as seasonality of breeding or birth (Lindburg 1987), female sexual skins swelling (Hrdy & Whitten 1987), selectivity in female mate choice and length of consortship (Smuts 1987). Macaques also exhibit species differences in intermale tolerance, and in what de Waal and Luttrell (1989) refer to as dominance "styles." For example, it is generally agreed among macaque researchers that intermale relationship in species such as Barbary macaques, bonnets (*M. radiata*) and stumptails are more "relaxed" (de Waal & Luttrell 1989, p.100) and "easygoing" (Shively et al. 1982) than those of other macaques, such as rhesus, Japanese, and pig-tail macaques. The level of aggression is lower among the first three species, and males in these species are more likely to reconcile after a conflict (de Waal & Luttrell 1989).

In addition to these well-known differences between species, there are also strong indications that aspects of macaque reproductive physiology and social systems are highly uniform or conservative across species. For example, all macaques live in groups consisting of adults of both sexes, though the adult sex ratios tend to be female-biased (Andelman 1986; Caldecott 1986). Most notably, all macaque societies are centered on what Wrangham (1980) called the female-bonded group, in which females form the stable core of the group and retain close and often lifelong alliances with female kin. On the other hand, adult males in a group tend to be unrelated, since at maturity males typically migrate from their natal group and into a new social group. These patterns are remarkably uniform across macaque species, despite the fact that they inhabit a wide variety of environments.

Why do some characteristics vary among macaque species? Do these characteristics tend to covary? And why are some characteristics of macaque so persistent through space and time? Linking together a variety of physiological and behavioral traits, Caldecott (1986) proposed an ecological explanation for why macaque species might be different. The main assumption of Caldecott's proposal is that resource distribution, in constraining female reproductive strategies, ultimately affects the number of adult males within macaque groups. He suggests that species living in relatively poor macaque habitats such as dipterocarp forests will develop breeding system in which females are more selective, thereby leading to more intense intermale competition. Under this condition, multiple-mount copulation and prolonged consortship would be likely, since these behavior patterns would improve the male's reproductive success. By the same token, those species with single-mount copulation, promiscuity, and short-duration consortship should be found in higher-quality environments. This would account for male stumptail, bonnet, and Barbary macaques being less aggressive, since historically they lived in richer habitats than other species, such as rhesus and Japanese macaques.

In short, there are ecological and hence functional reasons why male macaques of different species engage in different patterns of copulation. What is more important, the habitat to which each species is adapted may be more reliable in predicting copulatory patterns than phylogenetic affinity since copulatory patterns appear to cut across phyletic lines (Fa 1989). Caldecott further suggests that copulatory patterns may vary on a subspecies level; so in the wide-ranging *M. fascicularis*, the use of both multiple and single mounting may be a function of the diversity of the habitats occupied by different populations of the same species.

B. Methodological considerations

To evaluate Caldecott's thesis phylogenetically, I first constructed a phylogeny of the macaques (Fig. 2). The phylogeny is derived from four systematic studies of the *Macaca*: Fooden (1980) and Delson (1980) using morphological data, Melnick and Kidd (1985) and Fooden and Lanyon (1989) using molecular genetic data. These data sets were entered into the program *MacClade 3.0.4* (Maddison & Maddison 1992) and the phylogenetic trees were reconstructed by manually manipulating the tree branches. The saved trees were then imported into the program *PAUP 3.1.1* (Swofford 1991) and a consensus tree of the macaques was built using the 50% majority rule. The resultant tree broadly agrees with Fooden's (1980) phylogenetic hypothesis, and the four monophyletic groups within the genus correspond to Fooden's four species groups (i.e. *sylvanus-silenus*, *fascicularis*, *sinica*, and *artoides*). I then mapped the distribution of MME and SME along with a number of other social, physiological characteristics as defined by Caldecott (see Table 11.1A and B) onto the consensus tree using *MacClade 3.0.4* (Maddison & Maddison 1992) and the parsimony algorithm.

MacClade allows the user to seek trees that require the fewest character changes, thereby minimizing the need to invoke convergence or homoplasy in explaining the data. The total number of reconstructions is reported by the program, along with the Consistency Index (*CI*) and Retention index (*RI*) of the characters being analyzed. The *CI* is essentially a goodness-of-fit measure designed to indicate the degree of support of a particular data set for a particular tree. The *CI* indicates the amount of homoplasy or convergence. The lower the index, which ranges from zero to one, the greater the likelihood of convergence, parallelism and possible reversal. In other words, when the characters of interest are mapped onto a highly resolved cladogram, the lower the consistency index, the greater the possibility that the characters are evolutionarily labile with a weak phylogenetic link. Likewise, a high *CI* is indicative of a strong phylogenetic influence and the likelihood of similarities due to common descent (for a more detail discussion of *CI* and its interpretation, see de Queiroz & Wimberger 1993 and Chap. 7 in this volume). The *RI* is a ratio from zero to one relating the range of possible character state changes a trait could require on any conceivable tree, to the actual number of changes observed on the given tree. A value close to zero means a very poor fit to the tree, while a value close to one indicates a strong likelihood that the character state would be retained once it had evolved (Farris 1988).

The results (see Table 11.2) suggest that, contrary to Caldecott's suggestion, copulatory patterns (*CI* = 0.75; *RI* = 0.5) do appear to be strongly influenced by phylogeny (Fig. 3), and thus, the possibility that male macaque mating behavior is due to historical influence should not be excluded. This conclusion is further strengthened by the examination of the distribution of male copulatory patterns in primates. When the two patterns of copulation were mapped onto the primate phylogeny (Fig. 4), the resulting indexes (*CI* = 0.78; *RI* = 0.67) also suggest the possibility of strong phylogenetic influence. Thus the ecological hypothesis proposed by Caldecott should be more closely scrutinized.

The results also indicate that, while some characteristics, such as breeding seasonality and intermale tolerance, are indeed highly variable

Table 11.1A. List of macaque reproductive, social and ecological characters and their alternative states[a]

Character	Character States		
	0	1	2
1. Multiple mount ejaculation	No (SME)	Yes (MME)	
2. Multiple ejaculations	No	Yes	
3. Sexual skin swelling	Pronounced	Mildly visible	No
4. Seasonal breeding	No	Yes	
5. Female selectivity[b]	Low	High	
6. Male dispersal[b]	Common	Infrequent	
7. Intramale tolerance[b]	Low	High	
8. Consortship[b]	Brief	Prolonged	
9. Forest type	Broadleaf evergreen	Non B-E	
10. Ecological strategy	Nonweed	Weed	

[a]*Sources*: (1, 2) Chevalier-Skolnikoff (1975), Shively et al. (1982), Dewsbury and Pierce (1989); (3) Hrdy and Whitten (1987); (4) Lindburg (1987); (5–8) Caldecott (1986); (9) Fooden (1982); (10) Richard et al. (1989). [b]Note the character state coding of these four characters followed that assigned by Caldecott (1986, p. 215).

across species, others — such as external sign of ovulation, female selectivity of mates, and consortship formation — appeared to be influenced by phylogeny. These results should be interpreted with caution, since features such as female selectivity and length of consortship are thought to be highly variable within species (Hrdy & Whitten 1987), and the coding of these complex behavior patterns into two dichotomous states may severely undermine the complexity of these behavior. It is therefore likely that different coding method may yield different results. This is a problem that will be addressed in future analysis.

While the fact that many aspects of macaque social organization are conservative may not be surprising, it is interesting that the ecological "strategies" of macaques, which are commonly assumed to be highly flexible, may also be under strong phylogenetic influence. I reached this opinion after overlaying the ecological strategies of macaques as defined by Richard et al. (1989) onto the macaque phylogeny (Fig. 5). The authors point out that while macaque species are uniform in many aspects of their ecological requirements, they differ markedly in the degree to which each species tolerates and even prospers in close association with humans. In particular, macaque species differ in their ability to establish themselves in disturbed habitats and in their reliance upon humans as a voluntary or involuntary source of food. Thus, according to Richard et al., macaque species can be divided into two main groups, "weed" and "non-weed" macaques, on the basis of their ability to live and even thrive in close relationship with human settlements.

Table 11.1B. Character-state matrix for the genus *Macaca*[a]

	Character									
Species	1	2	3	4	5	6	7	8	9	10
M. sylvanus	0&1	1	0	1	0	0	1	0	1	0&1
M. silenus	1	1	1	0	1	0	0	1	0	0
M. nemestrina	1	1	0	0	1	0	0	1	0	0
M. nigra	1	1	0	0	?	0	0	?	0	0
M. mulatta	1	1	1	1	1	0	0	1	1	1
M. cyclopis	1	1	1	1	?	0	?	?	?	0&1
M. fuscata	1	1	1	1	?	0	?	?	?	0&1
M. fasicularis	0&1	1	2	0	1	0	0	1	1	1
M. arctoides	0	1	1	0	?	0	1	0	0	0
M. radiata	0	1	2	1	0	0	1	1	1	1
M. sinica	0	1	2	1	?	0	0	1	0	1
M. assamensis	?	1	?	?	?	0	?	?	0	0
M. thibetana	1	1	1	?	?	0	?	?	0	0

[a]"0&1" denotes polymorphic state, and "?" refers to missing data. Only one species of Sulawesi macaque, *M. nigra*, is used as the representative for the entire group since data for other species are not available.

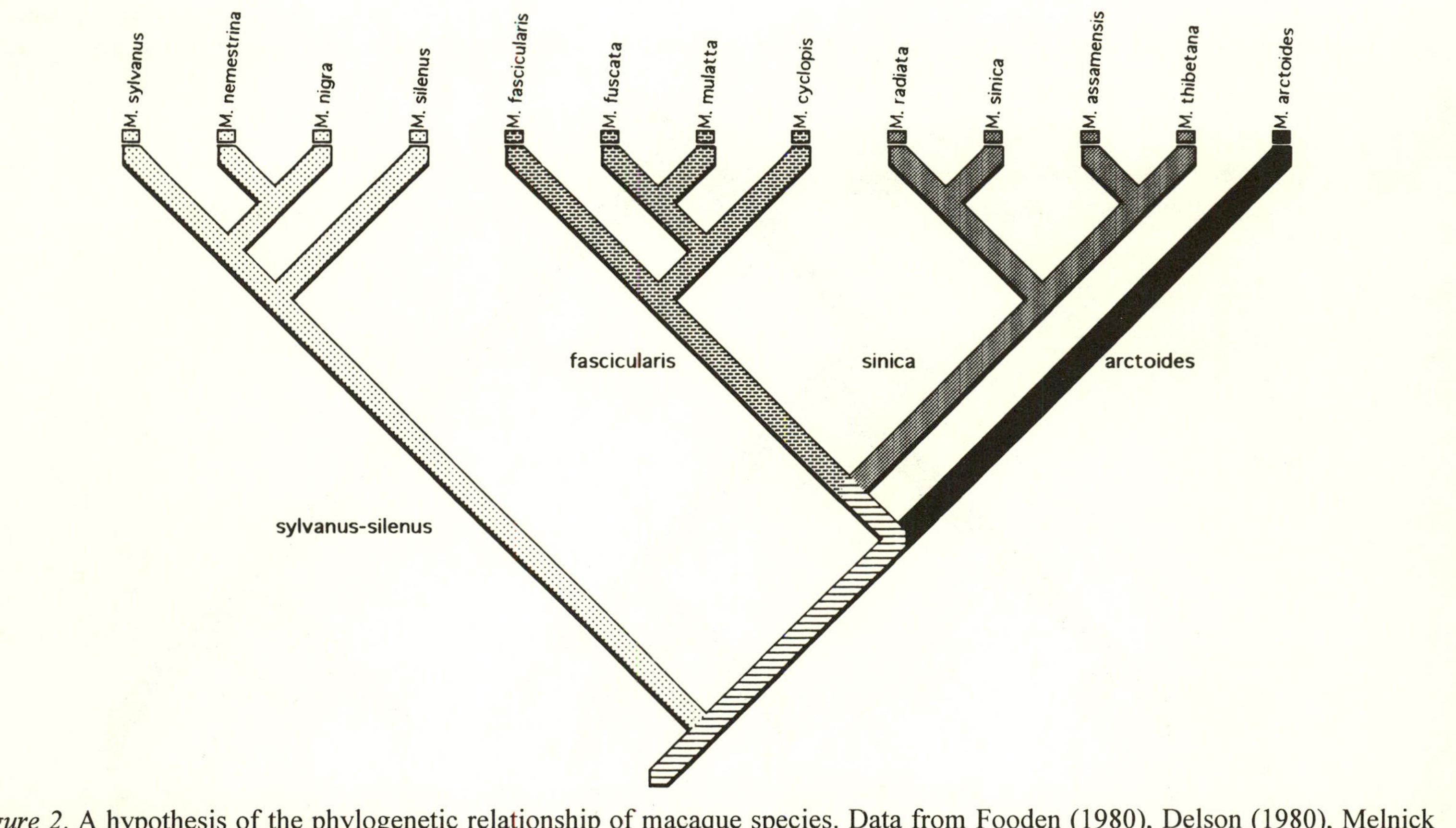

Figure 2. A hypothesis of the phylogenetic relationship of macaque species. Data from Fooden (1980), Delson (1980), Melnick and Kidd (1985) and Fooden and Lanyon (1989). See text for details.

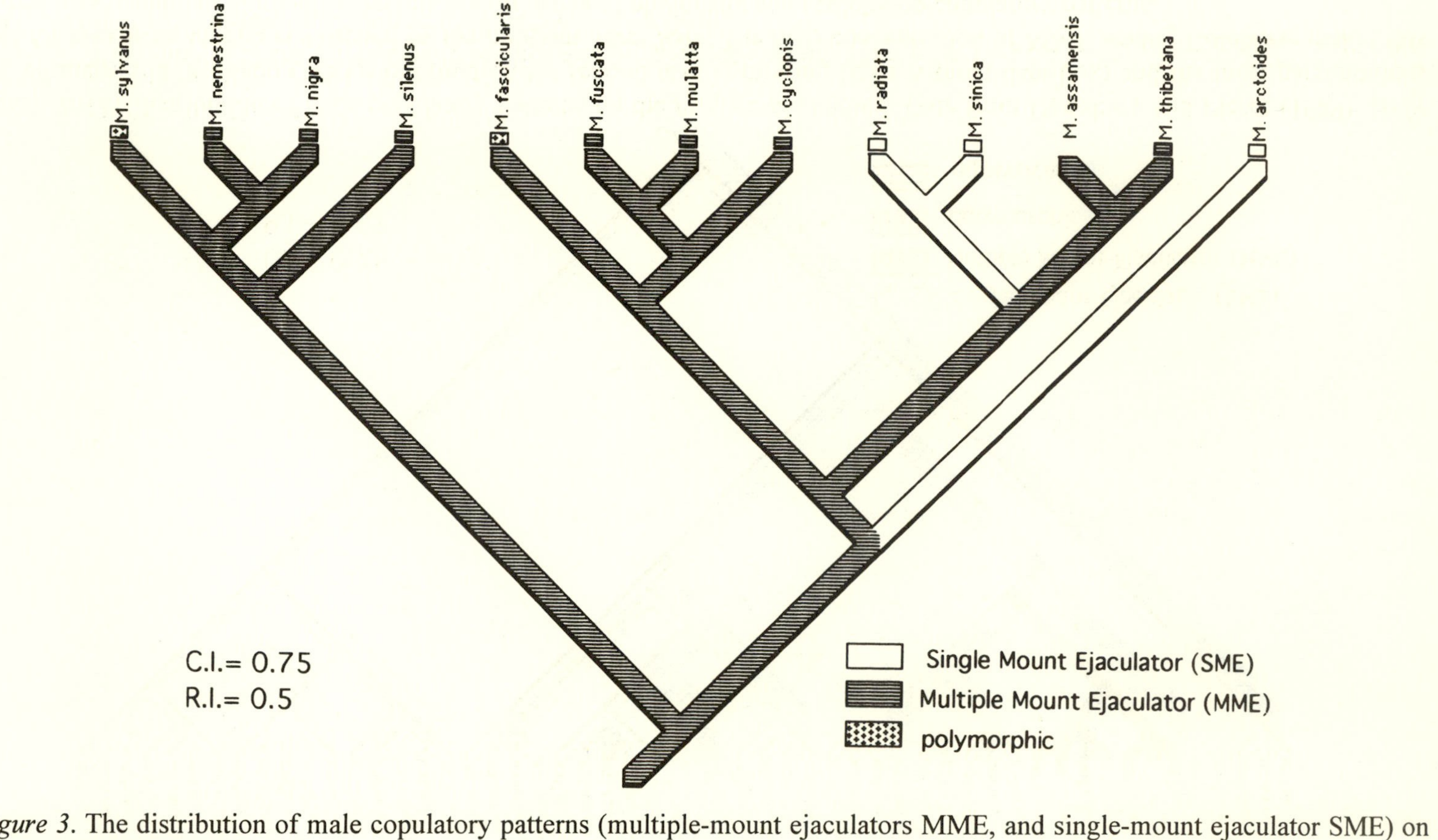

Figure 3. The distribution of male copulatory patterns (multiple-mount ejaculators MME, and single-mount ejaculator SME) on the macaque phylogenetic tree.

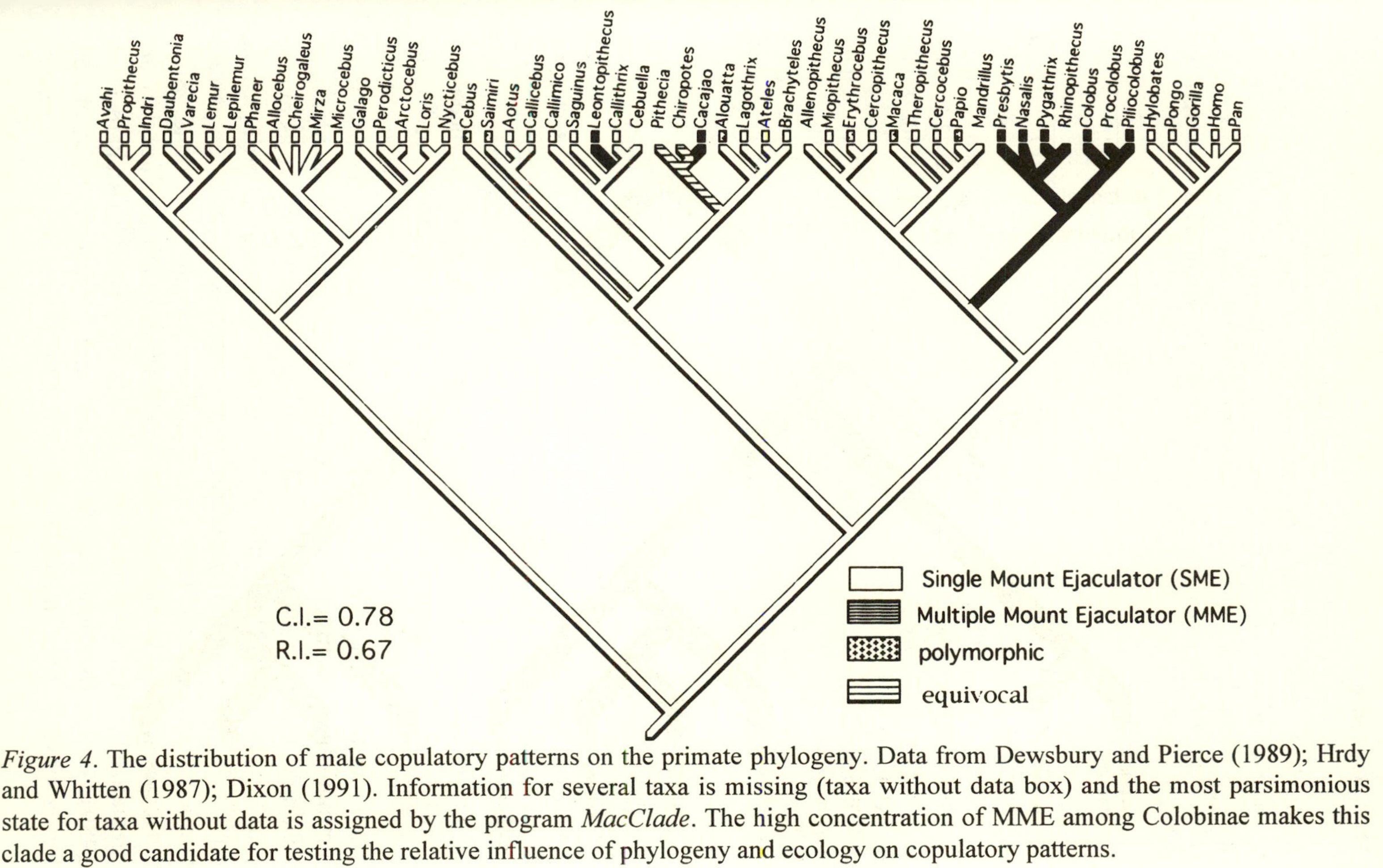

Figure 4. The distribution of male copulatory patterns on the primate phylogeny. Data from Dewsbury and Pierce (1989); Hrdy and Whitten (1987); Dixon (1991). Information for several taxa is missing (taxa without data box) and the most parsimonious state for taxa without data is assigned by the program *MacClade*. The high concentration of MME among Colobinae makes this clade a good candidate for testing the relative influence of phylogeny and ecology on copulatory patterns.

Table 11.2. Table showing the assumptions of the characters and the consistency index (*CI*) and retention index (*RI*) derived from the parsimony analysis using *MacClade 3.0.4*. Characters 2 and 6 were excluded from the calculation of these indexes because these characters show no variation between species.

Character	Type	Weight	States	Steps	*CI*	*RI*
1. Multiple mount ejaculation	Unordered	1	2	4	0.75	0.50
2. Multiple ejaculations	(Excluded)	-	-	-	-	-
3. Sexual skin swelling	Unordered	1	3	3	0.67	0.75
4. Seasonal breeding	Unordered	1	2	3	0.33	0.60
5. Female selectivity	Unordered	1	2	2	0.50	0.00
6. Male dispersal	(Excluded)	-	-	-	-	-
7. Intramale tolerance	Unordered	1	2	3	0.33	0.00
8. Consortship	Unordered	1	2	2	0.50	0.00
9. Forest type	Unordered	1	2	3	0.33	0.50
10. Ecological strategy	Unordered	1	2	5	0.80	0.67

The ecological division of macaques proposed by Richard et al. differs in a significant way from Caldecott's classification of macaque societies based on ecological adaptation. While the latter emphasizes the kinds, quality and distribution of food sources, the weed/non-weed distinction emphasizes the means by which foods are obtained, as well as the behavioral characteristics associated with being a successful "weed" species. Richard et al. point out that macaques need other behavioral attributes to live as "weeds" in close association with people. They suggest that even if weed macaques have life history traits distinguishing them from other macaques (though such evidence is lacking), "weed" species need additional "qualities" to survive as human camp followers. "These include such hard-to-measure attributes as curiosity, behavioral adaptability, an aggressive and gregarious temperament and, not least, speed and agility on the ground. Such attributes are hard to measure, and it is difficult to tell whether they are specialization for, or consequences of life as a weed." (Richard et al. 1989) As the distribution of the "weed" and "nonweed" traits closely follow phyletic lines, it would appear that the behavioral characteristics of "weed" species outlined above were, indeed, specializations for life as a "weed." If these characteristics are consequences of life as a "weed," then the traits must have evolved independently several times, a conjecture that is not as parsimonious as the hypothesis of persistent ancestral influence.

The preceding analysis raises an interesting question about the causal relationship between social behavior and biological diversity. Traditional neo-Darwinian models are infused with a strong element of ecological determinism which sees ecological or environmental conditions as "driving" evolutionary change (Emlen 1980; Cracraft 1991). This has led to the tacit assumption that the characteristics of a species, including biological as well as behavioral traits, are "solutions" evolved to solve the "problems" posed by the environment (Gould & Lewontin 1979). This emphasis on the determining role of the environment further reinforces the notion that behavior and social structure are products rather than the causes of evolutionary change (Plotkin 1988). However, the macaque analysis indicates that speciation events, as measured in terms of morphological and genetic differentiation, are not necessarily accompanied by ecological and behavioral innovations. Instead, it appears that it is the persistent behavioral and social traits that underlie subsequent biological diversity. If "weed" behavior evolved in the stem species of the *fascicularis* clade (Fig. 5), and if this complex character is indeed conservative, then the subsequent diversification of *M. cyclopsis, M. fuscata, M. mulatta*, and *M. fascicularis* may have been the consequent of this behavioral complex. This may serve as an example of how behavioral and social traits can act as initiators or precursors of subsequent biological change rather than simply as products of ecological adaptation (Gray 1988; Oyama 1985 1989; Kitchell 1990; Bateson 1988; Hailman 1982). The behavioral repertoire and social structure of a species is made up of mosaic features, some may be recent innovations due to adaptation, while others may be highly conservative traits that set the bounds for future evolutionary change. This theme is further explored below.

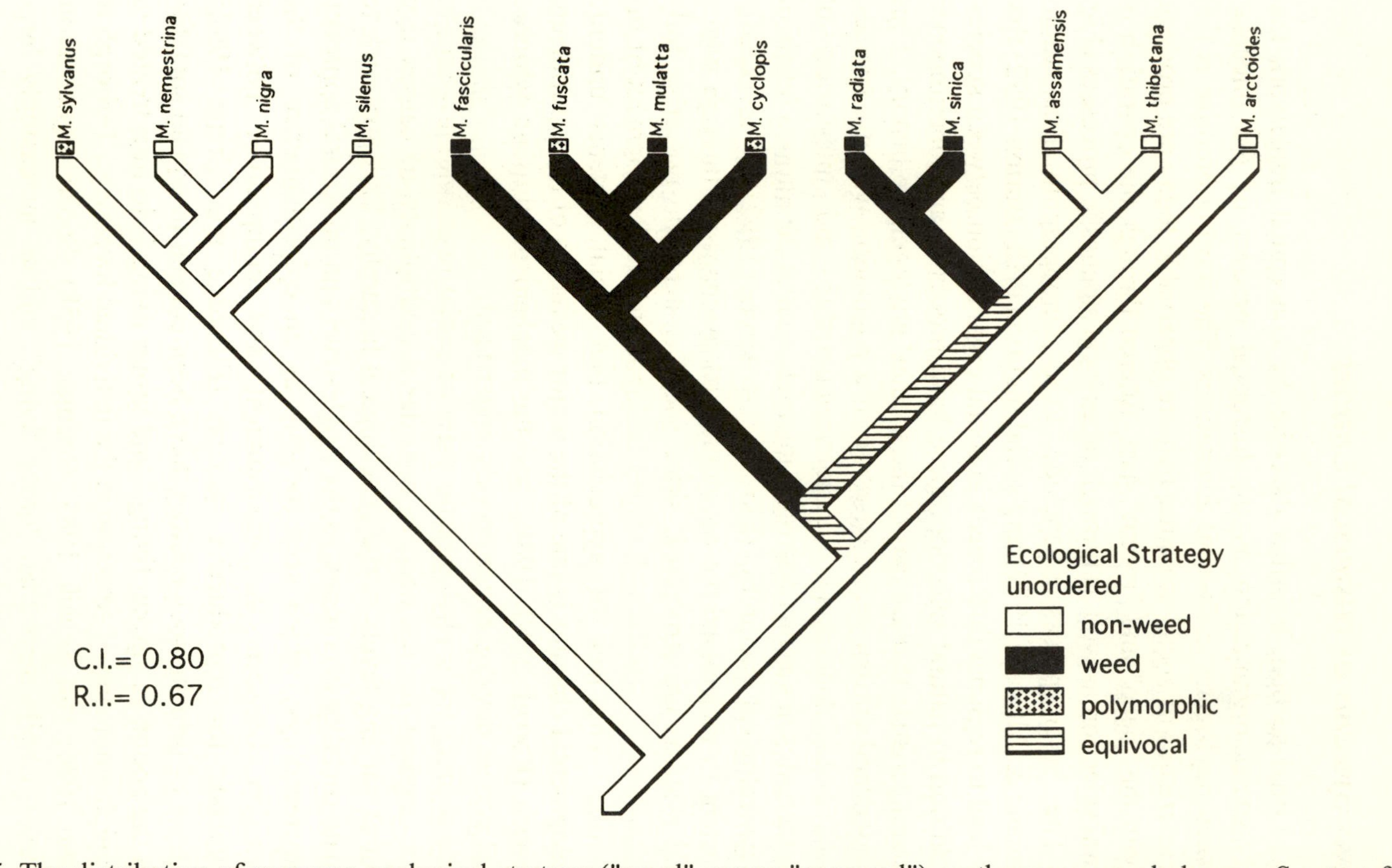

Figure 5. The distribution of macaque ecological strategy ("weed" versus "nonweed") on the macaque phylogeny. See text for detail discussion of the significance of the ecological pattern revealed by phylogenetic analysis.

Social Systems as Historical Entities

In this final section, I further describe the conceptual framework that treats social organization as the historical product of the interplay between phylogeny and recent adaptations. The starting premise is that characteristics of social structure that are observable today are products of both the "deep" and "recent" past, filtered through the operation of natural selection as well as through other local stochastic processes (Fig. 6). The latter processes include local demographic fluctuations (Altmann & Altmann 1979) and social tradition drift (Burton 1992) that may lead to regional differences in social organization between species. In addition to natural selection, these local processes are superimposed on characteristics that are part of the species' phylogenetic history. Thus to understand current social structure of a population within a species, we have to understand the interrelationship and interaction of characteristics at various historical intervals, and at the various levels of the genealogical hierarchy (Eldredge & Grene 1992). It is equally important to understand the developmental mechanisms through which social systems are structured, since social systems do not arise fully formed, but must go through stages analogous to the development of the individual. The lack of appreciation that social systems undergo developmental changes is one of the major weakness of the equilibrium approach (Rowell 1983, 1988), and the mechanisms through which a social system develop remain poorly understood.

In arguing for a phylogenetic view of social organization I am, in effect, arguing for the need to consider a multiplicity of causes that occur at varying levels in the genealogical hierarchy. This approach to primate sociality is contrary to the single-level and uni-causal approach of socioecology, which sees sociality as an epiphenomenon of the interaction between the environment and competition between individuals for reproductive success (Eldredge and Grene 1992). Students of primate socioecology have been content to equate sociality with the notion of "group living," and group living is in turn treated as the epiphenomenon of selection at the individual level (e.g., Terborgh & Janson 1986; van Schaik 1983; Rodman 1988). Social organization, however, entails more than "group living", and an increasingly large body of literature indicates that social systems in non-human primates emerge as a consequence of the interaction between individuals for a multiplicity of reasons unrelated to reproduction (Rowell 1991, 1993). Both recent history, in the form of social traditions (Burton 1992), and phylogeny play a role in the current expression of a social system.

A phylogenetic perspective also encourages a definition of sociality and social organization in terms of the temporal dimension, both in the recent and the evolutionary past, of social relationship. This historical perspective emphasizes the enduring nature of social relationships, while the socioecological approach emphasizes functional interaction in the here and now. In addition, the phylogenetic view is closely allied to the anthropological definition of sociality that emphasizes the structural components of social relationship and the rules that regulate such relationships (e.g., Hinde & Steveson-Hinde 1976; Hinde 1983). In socioecological descriptions of what groups do, the social relationships of individual group members do not seem to be particularly relevant. Instead, the more interesting questions seem to be about how groups interact with their environment, how much and what kinds of food are eaten and the amount of time spent feeding, how much effort is spent on competition for food with other groups. Rarely are there description and explanation of the internal mechanisms through which groups themselves are maintained.

The increasing recognition of the considerable intraspecific variation in primate social organization and behavior is not antithetical to phylogenetic analysis, but rather complements the idea that intraspecific variations are the manifestation of the recombination of a set of common underlying themes expressed in varying or novel contexts (Burghardt & Gittleman 1990). This is because social organization, like an individual organism, is composed of differing structures, some of which are ancient, while others are newly derived; some features of social system may be highly constrained by phylogeny, while others may be loosely integrated and can be easily modified. Like morphological evolution where old structures are often co-opted for new purposes in a novel context (Gould & Vrba 1982), so social organization may embody evolutionarily primitive features that assume new functions in varying contexts. The identification of these features and how they are integrated as a functioning system in the current environment should be a central task for the understanding of behavioral evolution.

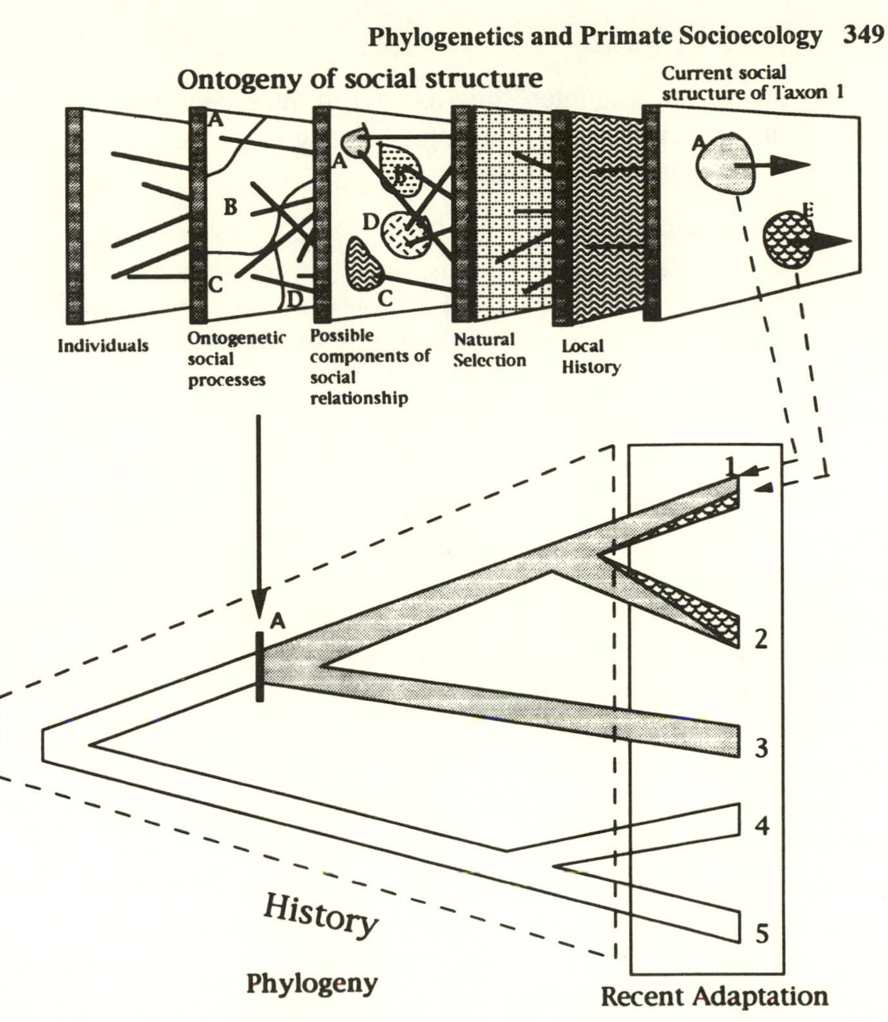

Figure 6. The relationship between phylogeny and ontogeny of a species' social system. Social systems that are observable today are products of the interaction between phylogeny and recent adaptation. For simplicity, the complexity of social system is represented by only two characters (top half). The "material" presented to natural selection is subjected to prior developmental and phylogenetic constraints. The vertical arrow linking the top half of the diagram to the phylogeny indicates that there are tightly integrated components of social system (represented by character A) that are difficult to disassemble during evolution and so the character presents itself as a constraint on, or an opportunity for, future change. Character A remained unchanged in taxon 3. Character E appeared in taxon 1, in conjunction with character A, as a result of recent adaptation, and totally replaced character A in taxon 2 due to recent adaptation. This diagram is based on Alberch (1982, p. 1990) and Wanntorp et al. (1990, p. 129).

In this regard, it is interesting to refer to the recent paper by Dunbar (1993) on "Co-evolution of neocortex size, group size and language in humans." Dunbar suggests that

> although the size of the group in which animals live in a given habitat is a function of habitat-specific ecologically-determined costs and benefits, there is a species-specific upper limit to group size which is set by purely cognitive constraints: animals cannot maintain the cohesion and integrity of groups larger than a size set by the information processing capacity of their neocortex.

As group sizes are closely related to social complexity, there are, probably, also limits on how complex a social organization can evolve. Dunbar's emphasis on cognitive constraints is a good example of the need to identify the boundary conditions of evolutionary change. This emphasis on constraint of and limits to social change is not a common theme in primatology, because it is generally accepted that phylogenetic constraint "is unlikely to be common because major aspects of social organization are known to be labile throughout the primate order" (Wrangham 1987, p.282).

While primates are highly intelligent, and can often adjust quickly to novel situations, it is well known that there are limitations to what each different primate species can learn (Burton 1984; in this volume, see Parker, Chap. 12). Monkeys, to use a simple example, evidently do not recognize themselves in the mirror, whereas great apes often learn to do so (Povinelli 1993). Cognitive differences have enormous implications for social relationships. Knowing the limits of what a species can learn is therefore just as telling from an evolutionary perspective (Wenzel 1992). If constraint is indeed a general phenomenon in evolution, then there is a need to better define and measure constraint and to understand its various manifestation.

Primatologists and evolutionary biologists are justifiably cautious about invoking the notion of constraint because the concept is often poorly defined and it seems to embody a great variety of meanings. In a recent review, McKitrick's (1993, p.309) defined phylogenetic constraint as "any result or component of the phylogenetic history of a lineage that prevents an anticipated course of evolution in that lineage." In other words, the evolutionary pathways that are open to a species or group of related species are contingent upon prior evolutionary history.

As Ligon (cited in McKitrick 1993, p.309) puts it, "yesterday's adaptation may be today's constraint." The ancestral states of the social systems will strongly affect the number of possible adaptive "solutions" open to species facing different environmental "problems." This raises the possibility that current social organization may not represent the optimal solution to present ecological problem, but rather represents retention of ancestral features that are tolerated by the current environment (Hailman 1982). It should also be emphasized that history does not only serve to rule out possible changes but it also serves the constructive role of making some changes more likely. Thus the presence of a particular ancestral feature may be a necessary prerequisite for a particular new adaptive change, adding to the complexity of social structure in the process (Gould 1991).

Conclusions

The purpose of this paper has not been to provide answers to why primate societies are so diverse but to point to alternative directions in research, and the need to consider (a) phylogenetic hypotheses as an integral part of any study on the mechanisms and evolution of social structure and (b) the possible role of phylogenetic constraint in social evolution.

Despite the enormous popularity of the socioecological approach, many empirical and theoretical problems regarding primate sociality remain to be solved (Wrangham 1987; Rowell 1993). At the most fundamental level, socioecology has had difficulty explaining why all species living in a given habitat did not converge on the same solutions. This is because organisms are not evolutionary blank slates, and their "solutions" to external "problems" depend both on the species' biology and on the evolutionary and local history of the social group in which the organisms are found.

Another major stumbling block for socioecology has been that many predictions of socioecological models are simply impossible to test, while many of those that are have never been tested. Plausible scenarios or narratives about reproductive advantages and functional fits of social groups to the environment are often substituted for proof. This situation has led Rowell (1991, p.264) to lament that:

> Evolutionary discussions of groups are entertaining but ultimately unsatisfactory in that they cannot be tested. There is no way that the order of adaptations that have led to modern social systems can be discovered, and it may be impossible to unravel the interlocking effects of different aspects of the environment (past or present) on a particular group, since anything which changes in that environment, by experiment or by chance, will affect several aspects of it.

The situation may not be as gloomy as Rowell depicts, for the purpose of this volume is to illustrate that reference to phylogeny in comparative studies will help reveal the pattern and distribution of past and current adaptations. The identification of such patterns, however, is only the first step; understanding the mechanisms responsible for the persistence of certain traits and the modifiability of others remains a considerable challenge.

Phylogenetic approaches are now common in studies of ecology and behavior (see recent reviews by Wenzel 1992; Miles & Dunham 1993, and, in this volume, Martins & Hansen, Chap. 2), but their impact on field primatology has been minimal. Nevertheless, there are signs that primatologists are rediscovering the importance of phylogeny in understanding primate ecology and behavior (Foley 1989; Richard & Dewar 1991; Sillén-Tullberg & Moller 1993; Chan 1993; DiFiore & Rendall 1994; Garber 1994). Rosenberger and Strier (1989), for example, found that phylogenetic affinity is a good predictor of social behavior for various species of Atelinae (spider and woolly monkeys). Likewise, Fedigan (1993, p.873) found many similar patterns of male and female behavior and patterns of interactions between the sexes in all capuchin species. And Fedigan (1993, p.874) concluded that, as among the macaques, "ecological similarities among the four species of *Cebus* can be interpreted as an integral part of their common phylogenetic heritage" rather than due to convergent adaptation. Fedigan (1993, p.875) further points out that "the pattern of female philopatry/male dispersal has consistent and important effects on sex differences and relations between the sexes, even across major phyletic categories." So similarities between old world forest monkeys such as *Cercopithecus* and new world *Cebus* may be due to persistence of the male and female pattern of relationship rather than to socioecological convergence.

Of further interest is Fedigan's suggestion that a more instructive form of ecological comparison should be based on the behavioral strategy(ies) of the species or genus for the exploitation of resources, in stead of on the quantity of the resources taken or the time spent on food gathering. This is because the quantity and proportion of different foods gathered are contingent upon local conditions and so can vary between species within a genus, and even between groups within a species.

> Unfortunately, foraging strategies do not lend themselves readily to quantitative comparisons, even though "opportunistic, extractive foraging" as a description, may capture more of the characteristic ecological pattern of capuchins than the relative proportion of fruit/insects/leaves in their diet does. [Fedigan 1993, p.873]

Richard et al.'s (1989) distinction of weed/nonweed macaques is based on similar reasoning. However, the bias in socioecological studies in favor of more quantitative and functional approaches often obscures the qualitative aspects of behavioral strategy. Quantitative comparison alone may not reveal much about the origin and diversification of behavior, but identifying the distribution and temporal sequence of different traits on a phylogeny will help generate testable hypotheses about origins and directions of evolutionary change. Recently, Moore (1992, p.366) suggests that "a major direction of primate socioecology in the next decade" will be "the investigation of ecological bases for what have been called 'temperament' or 'style' differences among species." Perhaps phylogenetic analysis of these qualities and other ecological strategies as identified by Fedigan, Richard and colleagues will be an integral part of this new direction.

ACKNOWLEDGMENTS

I wish to thank Emília Martins for organizing and for inviting me to participate in the symposium *Phylogenies and the Comparative Method in Animal Behavior*. I would like to thank Irving Bernstein, Linda Wolfe, Linda Fedigan, Emília Martins and two anonymous reviewers for their very helpful comments on an earlier version of this manuscript. Deirdre Breton provided much-needed editorial assistance. I am responsible for any remaining errors or misinterpretation. My research is supported in part by a University of Toronto Open Fellowship and the Ontario Graduate Scholarship.

References

Alberch, P. 1982. Developmental constraints in evolutionary processes. In: *Development and Evolution* (Ed. by J. T. Bonner), pp. 4–26. Berlin: Springer-Verlag.

Alberch, P. 1990. Natural selection and developmental constraints: External versus internal determinants of order in nature. In: *Primate Life History and Evolution* (Ed. by J. de Rosseau), pp. 15–35. New York: Wiley-Liss.

Altmann, S. A., & J. Altmann. 1979. Demographic constraints on behavior and social organization. In: *Primate Ecology and Human Origins* (Ed. by I. S. Bernstein, & E. O. Smith), pp. 47–62. New York: Garland.

Andelman, S. J. 1986. Ecological and social determinants of Cercopithecine mating patterns. In: *Ecological Aspects of Social Evolution* (Ed. by P. U. Press), pp. 201-216.

Bateson, P. 1988. The active role of behavior in evolution. In: *Evolutionary Processes and Metaphors* (Ed. by M.-W. Ho, & S. W. Fox), pp. 191-207. New York: John Wiley & Sons.

Brooks, D. R., & D. A. McLennan. 1991. *Phylogeny, Ecology, and Behavior: A Research Program in Comparative Biology*. Chicago: The University of Chicago Press.

Burghardt, G. M., & J. L. Gittleman. 1990. Comparative behavior and phylogenetic analyses: New wine, old bottles. In: *Interpretation and Explanation in the Study of Animal Behavior* (Ed. by M. Bekoff, & D. Jamieson), pp. 192–225. Boulder, Colorado: Westview Press.

Burton, F. D. 1984. Inferences of cognitive abilities in Old World Monkeys. *Semiotica*, 50, 69–81.

Burton, F. D. 1992. The social group as information unit: Cognitive behavior, cultural processes. In: *Social Processes and Mental Abilities in Non-Human Primates: Evidences from Longitudinal Field Studies* (Ed. by F. D. Burton), pp. 31–60. Lewiston, New York: The Edwin Mellon Press.

Caldecott, J. O. 1986. Mating patterns, societies and the ecogeography of macaques. *Anim. Behav.*, 34, 208–220.

Chan, L. K. W. 1992. Problems with socioecological explanations of primate social diversity. In: *Social Processes and Mental Abilities in Non-Human Primates: Evidences from Longitudinal Field Studies* (Ed. by F. D. Burton), pp. 1–30. Lewiston, New York: The Edwin Mellon Press.

Chan, L. K. W. 1993. A phylogenetic interpretation of reproductive parameters and mating patterns in *Macaca*. *Am. J. Primatol.*, 30, 303–304.

Chevalier-Skolnikoff, S. 1975. Heterosexual copulatory patterns in stumptail macaques (*M. arctoides*) and in other macaque species. *Arch. Sex. Behav.*, 4, 199–220.

Clutton-Brock, T. H., & P. H. Harvey. 1979. Comparison and adaptation. *Proc. R. Soc. Lond. B*, 205, 547–565.

Coddington, J. A. 1988. Cladistic tests of adaptational hypotheses. *Cladistics*, 4, 3–22.

Cracraft, J. 1991. Review of *Phylogenetic Systematics as the Basis of Comparative Biology*, by V. A. Funk & D. R. Brooks. *Auk*, 108, 742–743.

Crook, J. 1989. Socioecological paradigms, evolution and history: perspectives for the 1990s. In: *Comparative Socioecology: the Behavioral Ecology of Humans and Other Mammals* (Ed. by V. Standen & R. Foley), pp. 1–36. Oxford, England: Blackwell Scientific Publications.

Crook, J. H., & J. S. Gartlan. 1966. Evolution of primate societies. *Nature*, 210, 1200–1203.

de Queiroz, A., & P. H. Wimberger. 1993. The usefulness of behavior for phylogeny estimation: Levels of homoplasy in behavioral and morphological characters. *Evolution,* 47, 46–60.

de Waal, F. B. M., & L. M. Luttrell. 1989. Toward a comparative socioecology of the genus Macaca: Different dominance styles in rhesus and stumptail monkeys. *Am. J. Primatol.*, 19, 83–109.

Delson, E. 1980. Fossil macaques, phyletic relationships and a scenario of deployment. In: *The Macaques: Studies in Ecology, Behavior and Evolution* (Ed. by D. G. Lindburg), pp. 10–29. New York: Van Nostrand Reinhold.

Dewsbury, D. A., & J. D. Pierce. 1989. Copulatory patterns of primates as viewed in broad mammalian perspective. *Am. J. Primatol.*, 17, 51–72.

Di Fiore, A. & D. Rendall. 1994. The evolution of social organization: A reappraisal for primates by using phylogenetic methods. *Proc. Natl. Acad. Sci. USA*, 91, 9941–9945.

Dixon, A. F. 1991. Sexual selection, natural selection, and copulatory patterns in male primates. *Folia Primatol.*, 57, 96–101.

Dobson, F. S. 1985. The use of phylogeny in behavior and ecology. *Evolution*, 39, 1384–1388.

Dunbar, R. I. M. 1988. *Primate Social Systems*. Ithaca, New York: Cornell University Press.

Dunbar, R. I. M. 1989. Ecological modeling in evolutionary context. *Folia Primatol.*, 53, 235–246.

Dunbar, R. I. M. 1993. Co-evolution of neocortex size, group size and language in humans. *Behav. Brain. Sci.*

Edwards, S. V., & S. Naeem. 1993. The phylogenetic component of cooperative breeding in perching birds. *Am. Nat.*, 141, 754–789.

Eldredge, N., & M. Grene. 1992. *Interactions: The Biological Context of Social Systems*. New York: Columbia University Press.

Emlen, S. T. 1980. Ecological determinism and sociobiology. In: *Sociobiology: Beyond Nature/Nurture* (Ed. by G. B., & J. Silverberg), pp. 125–140. AAAS selected symposium.

Endler, J. A. 1982. Problems in distinguishing historical from ecological factors in biogeography. *Am. Zool.*, 22, 441–452.

Fa, J. E. 1989. The genus *Macaca*: A review of taxonomy and evolution. *Mammal. Rev.*, 19, 45–81.

Farris, J. S. 1988. The retention index and homoplasy excess. *Syst. Zool.*, 38, 406–407.

Fedigan, L. 1993. Sex differences and intersexual relations in adult white-faced capuchins (*Cebus capucinus*). *Int. J. Primatol.,* 14, 853–877.

Felsenstein, J. 1985. Phylogenies and the comparative method. *Am. Nat.*, 125, 1–15.

Fleagle, J. G. 1988. *Primate Adaptation and Evolution.* New York: Academic Press.

Foley, R. 1989. The evolution of hominid social behavior. In: *Comparative Socioecology: The Behavioral Ecology of Humans and Other Mammals* (Ed. by V. Standen, & R. Foley), pp. 473–494. Oxford, England: Blackwell Scientific Publications.

Foley, R., & P. C. Lee. 1989. Finite social space, evolutionary pathways, and reconstructing hominid behavior. *Science*, 243, 901–906.

Fooden, J. 1980. Classification and distribution of living macaques (*Macaca lacepede)*, 1799. In: *The Macaques: Studies in Ecology, Behavior and Evolution* (Ed. by D. G. Lindburg), pp. 1–9. New York: Van Nostrand Reinhold.

Fooden, J. 1982. Ecogeographic segregation on macaque species. *Primates*, 6, 574–579.

Fooden, J. & S. Lanyon. 1989. Blood-protein allele frequencies and phylogenetic relationships in Macaca: A review. *Am. J. Primatol.*, 17, 209–241.

Ford, S. 1986. Systematics of the New World monkey. In: *Comparative Primate Biology*. Vol. 1. *Systematics, evolution, and anatomy* (Ed. by D. R. Swindler, & J. Erwin), pp. 73–135. New York: Liss.

Funk, V. A., & D. R. Brooks. 1990. *Phylogenetic Systematics as the Basis of Comparative Biology*. Washington, D. C.: Smithsonian Institution Press.

Garber, P. A. 1994. Phylogenetic approach to the study of Tamarin and Marmoset social systems. *Am. J. Primatol.*, 34, 199–219.

Goodman, M., B. F. Koop, J. Czelusniak, A. Fitch, D. A. Tagle, & J. L. Slingtom. 1989. Molecular phylogeny of the family of apes and humans. *Genome*, 31, 316–335.

Gould, S. J. 1989. *Wonderful Life*. New York: W. W. Norton.

Gould, S. J. 1991. Exaptation: A crucial tool for an evolutionary psychology. *J. Social Issues*, 11, 43–65.

Gould, S. J. & R. C. Lewontin. 1979. The spandrels of San Macro and the Panglossian paradigm: A critique of the adaptationist programme. *Proc. R. Soc. Lond. B.*, 205, 581–98.

Gould, S. J. & E. S. Vrba. 1982. Exaptation — a missing term in the science of form. *Paleobiology*, 8, 4–15.

Gray, R. 1988. Metaphors and methods: behavioral ecology, panbiogeography and the evolving synthesis. In: *Evolutionary Processes and Metaphors* (Ed. by M.-W. Ho, & S. W. Fox), pp. 209–242. New York: John Wiley & Sons.

Groves, C. P. 1989. *A Theory of Human and Primate Evolution.* Oxford, England: Clarendon Press.

Hailman, J. 1982. Evolution and behavior: An iconoclastic view. In: *Learning, Development, and Culture* (Ed. by H. C. Plotkin), pp. 205–54. Chichester, England: Wiley.

Harvey, P. H., R. D. Martin, & T. H. Clutton-Brock. 1987. Life histories in comparative perspective. In: *Primate Societies* (Ed. by B. B. Smuts, D. L. Cheney, R. M. Seyfarth, R. W. Wrangham, & T. T. Struhsaker), pp. 181–196. Chicago: University of Chicago Press.

Hinde, R. A. 1983. *Primate Social Relationships*. Oxford, England: Blackwell.

Hinde, R. A., & J. Steveson-Hinde. 1976. Towards understanding relationships: dynamic stability. In: *Growing Points in Ethology* (Ed. by P. P. G. B. a. R. a. Hinde), pp. 451-479. Cambridge, England: Cambridge University Press.

Hoelzer, G., & D. Melnick. 1992. Genetic and evolutionary relationships of the macaques. In: *Evolutionary Ecology and Behavior of the Macaques* (Ed. by J. Fa, & D. G. Lindburg), pp. Cambridge, England: Cambridge University Press.

Hrdy, S. B., & P. L Whitten. 1987. Patterning of sexual activity. In: *Primate Societies* (Ed. by B. B. Smuts, D. L. Cheney, R. M. Seyfarth, R. W. Wrangham, & T. T. Struhsaker), pp. 370–384. Chicago: University of Chicago Press.

Huey, R. B. 1987. Phylogeny, history, and the comparative method. In: *New Directions in Ecological Physiology* (Ed. by M. E. Feder, A. F. Bennett, W. W. Burggren, & R. B. Huey), pp. 76–98. Cambridge, England: Cambridge University Press.

Kitchell, J. A. 1990. The reciprocal interaction of organism and effective environment: Learning more about "and." In: *Causes of Evolution: A Paleontological Perspective* (Ed. by W. D. Allmon, & R. M. Ross), pp. 151–172. Chicago: University of Chicago Press.

Langtimm, C. A., & D. A. Dewsbury. 1991. Phylogeny and evolution of rodent copulatory behavior. *Anim. Behav.*, 41, 217–225.

Lanyon, S. M. 1992. Interspecific brood parasitism in blackbirds (Icterinae): A phylogenetic perspective. *Nature*, 255, 77–79.

Lauder, G. 1986. Homology, analogy, and the evolution of behavior. In: *Evolution of Animal Behavior: Paleontological and Field Approaches* (Ed. by M. H. Nitecki, & J. A. Kitchell), pp. 9–40. New York: Oxford University Press.

Leighton, D. R. 1987. Gibbons: territoriality and monogamy. In: *Primate Societies* (Ed. by B. B. Smuts, D. L. Cheney, R. M. Seyfarth, R. W. Wrangham, & T. T. Struhsaker), pp. 135–145. Chicago: University of Chicago Press.

Lindburg, D. G. 1991. Ecological requirements of macaques. *Lab. Anim. Sci.*, 41, 315–322.

Lindburg, D. L. 1987. Seasonality of reproduction in primates. *Comp. Primate Biol.*, 2, 167–218.

Maddison, W. P., & D. R. Maddison. 1992. *MacClade: Analysis of phylogeny and character evolution. Version 3* Sunderland, Massachusetts: Sinauer and Associates.

McKenna, J. J. 1979. The evolution of allomothering behavior among colobine monkeys: Function and opportunism in evolution. *Am. Anthropol.*, 81, 818–840.

McKitrick, M. C. 1993. Phylogenetic constraint in evolutionary theory: Has it any explanatory power? *Annu. Rev. Ecol. Syst.*, 24, 307–330.

McLennan, D. A. 1991. Integrating phylogeny and experimental ethology: from pattern to process. *Evolution*, 45, 1773–1789.

Melnick, D. J., & K. K. Kidd. 1985. Genetic and evolutionary relationships among Asian macaques. *Int. J. Primatol.*, 6, 123–160.

Melnick, D. J., & M. C. Pearl. 1987. Cercopithecines in multimale groups: Genetic diversity and population structure. In: *Primate Societies* (Ed. by B. B. Smuts, D. L. Cheney, R. M. Seyfarth, R. W. Wrangham, & T. T. Struhsaker), pp. 282–296. Chicago: University of Chicago Press.

Miles, D. B., & A. E. Dunham. 1993. Historical perspectives in ecology and evolutionary biology: The use of phylogenetic comparative analyses. *Annu. Rev. Ecol. Syst.*, 24, 587–619.

Moore, J. 1992. Dispersal, nepotism, and primate social behavior. *Int. J. Primatol.*, 13, 361–377.

O'Hara, R. J. 1. 1988. Homage to Clio, or, toward an historical philosophy for evolutionary biology. *Syst. Zool.*, 37, 142–155.

Owen-Smith, N. 1977. On territoriality in ungulates and an evolutionary model. *Q. Rev. Biol.*, 52, 1–38.

Oyama, S. 1985. *The Ontogeny of Information*. Cambridge, England: Cambridge University Press.

Oyama, S. 1989. Ontogeny and the central dogma: Do we need the concept of genetic programming in order to have an evolutionary perspective. In: *Systems and Development* (Ed. by M. R. Gunnar, & E. Thelen), pp. 1–34. Hillsdale, New Jersey: Erlbaum.

Peterson, A. T., & D. B. Burt. 1992. Phylogenetic history of social evolution and habitat use in the Aphelocoma jays. *Anim. Behav.*, 44, 859–866.

Plotkin, H. C. 1988. Behavior and evolution. In: *The Role of Behavior in Evolution* (Ed. by H. C. Plotkin). Cambridge, Massachusetts: MIT Press.

Povinelli, D. J. 1993. Reconstructing the evolution of mind. *Am. Psychol.*, 48, 493–509.

Prum, R. O. 1990. Phylogenetic analysis of the evolution of display behavior in the neotropical manakins (Aves: Pipridae). *Ethology*, 84, 202–231.

Pusey, A. E., & C. Packer. 1987. Dispersal and Philopatry. In: *Primate Societies* (Ed. by B. B. Smuts, D. L. Cheney, R. M. Seyfarth, R. W. Wrangham, & T. T. Struhsaker), pp. 250–266. Chicago: University of Chicago Press.

Rhine, G. 1995. Review of behavior of stumptail macaques. In: *Evolutionary Ecology and Behavior of the Macaques* (Ed. by J. Fa, & D. G. Lindburg). Cambridge, England: Cambridge University Press.

Richard, A. 1987. Malagasy prosimians: Female dominance. In: *Primate Societies* (Ed. by B. B. Smuts, D. L. Cheney, R. M. Seyfarth, R. W. Wrangham, & T. T. Struhsaker), pp. 25–33. Chicago: University of Chicago Press.

Richard, A., & R. Dewar. 1991. Lemur ecology. *Annu. Rev. Ecol. Syst.*, 22, 145–175.

Richard, A. F., S. J. Goldstein, & R. E. Dewar. 1989. Weed macaques: The evolutionary implications of macaque feeding ecology. *Int. J. Primatol.*, 10, 569–597.

Richard, A. J. 1985. *Primates in Nature*. New York: Freeman.

Ridley, M. 1983. *The Explanation of Organic Diversity: The Comparative Method and Adaptations for Mating*. Oxford, England: Clarendon Press.

Riedl, R. 1978. *Order in Living Organisms*. New York:.

Robinson, J. G., & C. H. Janson. 1987. Capuchins, squirrel monkeys, and atelines: Socioecological convergence with Old World primates. In: *Primate Societies* (Ed. by B. B. Smuts, D. L. Cheney, R. M. Seyfarth, R. W. Wrangham, & T. T. Struhsaker), pp. 69–82. Chicago: University of Chicago Press.

Rodman, P. 1988. Resources and group sizes in primates. In: *The Ecology of Social Behavior* (Ed. by C. N. Slobodchikoff), pp. 83–107. New York: Academic Press.

Rose, S., L. Kamin, & R. C. Lewontin. 1984. *Not in Our Genes*. New York: Penguin Books.

Rosenberger, A. L., & K. B. Strier. 1989. Adaptive radiation of the ateline primates. *J. Hum. Evol.*, 18, 717–750.

Rowell, T. E. 1979. How would we know if social organization were not adaptive? In: *Primate Ecology and Human Origins* (Ed. by I. S. Bernstein, & E. O. Smith), pp. 1–22. New York: Garland Press.

Rowell, T. E. 1988. Beyond the one-male group. *Behavior*, 104, 191–201.

Rowell, T. E. 1991. What can we say about social structure? In: *The Development and Integration of Behavior: Essays in Honour of Robert Hinde* (Ed. by P. Bateson), pp. 255–270. Cambridge, England: Cambridge University Press.

Rowell, T. E. 1993. Reification of social systems. *Evol. Anthropol.*, 3, 135–137.

Rowell, T. E. & D. K. Olson. 1983. Alternative mechanism of social organization in monkeys. *Behavior*, 86, 31–54.

Rutberg, A. T. 1983. The evolution of monogamy in primates. *J. Theor. Biol.*, 104, 93–112.

Seilacher, A. 1990. The Sand-Dollar syndrome: A polyphyletic constructional breakthrough. In: *Evolutionary Innovations* (Ed. by M. H. Nitecki), pp. 231–252. Chicago: University of Chicago Press.

Shively, C., S. Clarke, N. King, S. Schapiro, & G. Mitchell. 1982. Patterns of sexual behavior in male macaques. *Am. J. Primatol.*, 2, 373–384.

Sillén-Tullberg, B. 1988. Evolution of gregariousness in aposematic butterfly larvae: A phylogenetic analysis. *Evolution*, 42, 293–305.

Sillén-Tullberg, B., & A. P. Moller. 1993. The relationship between concealed ovulation and mating systems in anthropoid primates: A phylogenetic analysis. *Am. Nat.*, 141, 1–25.

Smuts, B. 1987. Sexual competition and mate choice. In: *Primate Societies* (Ed. by B. B. Smuts, D. L. Cheney, R. M. Seyfarth, R. W. Wrangham, & T. T. Struhsaker), pp. 385–399. Chicago: University of Chicago Press.

Smuts, B. B., D. L. Cheney, R. M. Seyfarth, R. W. Wrangham & T. T. Struhsaker, eds. 1987. *Primate Societies*. Chicago: University of Chicago Press.

Snowdon, C. T. 1990. Mechanisms maintaining monogamy in monkeys. In: *Contemporary Issues in Comparative Psychology* (Ed. by D. A. Dewsbury), pp. 225–251. Suderland, Massachusetts: Sinauer and Associates.

Sober, E. 1988. *Reconstructing the Past: Parsimony, Evolution and Inference*. Cambridge, Massachusetts: MIT Press.

Stearns, S. C. 1992. *The Evolution of Life Histories*. Oxford, England: Oxford University Press.

Stewart, C.-B. 1993. The powers and pitfalls of parsimony. *Nature*, 361, 603–607.

Strasser, E., & E. Delson. 1987. Cladistic analysis of cercopithecid relationships. *J. Hum. Evol.*, 16, 81–99.

Struhsaker, T. T. 1969. Correlates of ecology and social organization among African Cercopithecines. *Folia Primatol.*, 11, 80–118.

Struhsaker, T. T., & L. Leland. 1987. Colobines: infanticide by adult males. In: *Primate Societies* (Ed. by B. B. Smuts, D. L. Cheney, R. M. Seyfarth, R. W. Wrangham, & T. T. Struhsaker), pp. 83–97. Chicago: University of Chicago Press.

Swofford, D. L. 1991. *PAUP: Phylogenetic Analysis Using Parsimony, Version 3.1.* Champaign, Illinois: Illinois Natural History Survey.

Taylor, P. J. 1987. Historical versus selectionist explanations in evolutionary biology. *Cladistics*, 3, 1–13.

Terborgh, J. 1986. The social systems of New World primates: An adaptationist view. In: *Primate Ecology and Conservation* (Ed. by J. G. Else, & P. Lee), pp. 199–211. Cambridge, England: Cambridge University Press.

Terborgh, J. W. 1983. *Five New World Primates: A study in comparative Ecology.* Princeton, New Jersey: Princeton University Press.

Terborgh, J. W., & C. H. Janson. 1986. The socioecology of primate groups. *Annu. Rev. Ecol. Syst.*, 17, 111–35.

Thomas, R. D. K., & W.-E. Reif. 1993. The skeleton space: a finite set of organic designs. *Evolution*, 47, 341–360.

Van Schaik, C. 1983. Why are diurnal primates living in group? *Behavior*, 87, 120–144.

Van Schaik, C., & J. A. R. Van Hooff. 1983. On the ultimate causes of primate social systems. *Behavior*, 85, 91–117.

Van Schaik, C. P., & R. I. M. Dunbar. 1990. The evolution of monogamy in large primates: A new hypothesis and some crucial tests. *Behavior*, 115, 30–60.

Wake, D. B. 1991. Homoplasy: The result of natural selection, or evidence of design limitations? *Am. Nat.*, 138, 543–567.

Wanntorp, H.-E. 1983. Historical constraints in adaptation theory: traits and non-traits. *Oikos*, 41, 157–160.

Wanntorp, H.-E., D. Brooks, T. Nilsson, S. Nylin, F. Ronquist, S. Stearns, & N. Wedell. 1990. Phylogenetic approaches in ecology. *Oikos*, 57, 119–132.

Wenzel, J. W. 1992. Behavioral homology and phylogeny. *Annu. Rev. Ecol. Syst.*, 23, 361–381.

Wrangham, R. 1980. An ecological model of female-bonded primate groups. *Behavior*, 75, 262–300.

Wrangham, R. 1987. Evolution of social structure. In: *Primate Societies* (Ed. by B. B. Smuts, D. L. Cheney, R. M. Seyfarth, R. W. Wrangham, & T. T. Struhsaker), pp. 282–296. Chicago: University of Chicago Press.

CHAPTER 12

Using Cladistic Analysis of Comparative Data to Reconstruct the Evolution of Cognitive Development in Hominids

Sue Taylor Parker

> In discussing the mental development of the child we have both of these problems to solve: the two problems, i.e., whether the child's mental development recapitulates the stages of mental development in the animal's world, and second, whether it then goes on to show, or to recapitulate, the stages through which the human mind, after it arose in history, has passed in our race development. [Baldwin 1897, 1902, p. 189]

In his book *Mental Development in the Child and the Race* in 1894 (revised in 1906), James Mark Baldwin discusses the ontogeny and phylogeny of human mental development. Noting the analogies between development and evolution, qualified by recognition of departures from strict recapitulation, he lays out a grand design for future studies at the same time noting how far psychology is from fulfilling it:

> For it is clear that the stages of human life history may be built up from a wide series of observations of different children under varied conditions.... But such a science as comparative mental morphology — and even worse, that of mental embryology — is at present a chimera. How can we say anything about recapitulation when we

> know so little about mental ontogeny and less, perhaps about comparative mental physiology? [Baldwin 1906, p. 33]

At the turn of this century, American psychologists like James Mark Baldwin were fascinated by the notion that stages of human mental development recapitulated stages of human and primate evolution. Baldwin in particular suggested a blueprint for comparative evolutionary studies of development, though he realized that he lacked the tools to pursue it. The idea of recapitulation, which owes much to Haeckel and Spencer and Darwin, has its origins in theories about the evolution of animal development going back at least to embryologists in the 16th century if not to Aristotle (e.g., Gould 1977; Richards 1992). As Baldwin's statement implied, his desired research program lacked crucial data on the development of cognition in humans and other animals. Moreover, Baldwin's research program went against the tide of history.

By the twenties, American psychologists had renounced not only the concept of recapitulation but also the concept of psychological evolution and indeed of mentality. To this day, American psychologists and anthropologists react so negatively to the idea that ontogeny recapitulates the phylogeny of behavior that the mere mention of the word "recapitulation" elicits responses of avoidance or deflection. Undergraduates in psychology are taught to associate the term "recapitulation" with the primitive methodology of its best-known psychological proponent, G. Stanley Hall (1897, 1904) who described the development of fear and of play in children in terms of the imagined stages of their evolution in human ancestors. This attitude is transmitted, for example, in primary sources (Werner & Kaplan 1963) as well as many textbooks on psychology (Dacey & Travers 1994).

In psychology and anthropology, the opprobrium associated with the concept of recapitulation is only slightly greater than that associated with the concept of natural selection as it applies to human behavior (e.g. Fox 1971; Toobey & Cosmides 1992). Students soon catch the odor of contagion associated with these concepts without even knowing the sad history of fascist, racist, and sexist ideologies that they have been used to justify (e.g. Gould 1977; Richards 1992). For all these reasons, the validity of the idea of recapitulation in regard to any aspect of human evolution has rarely been addressed by modern anthropologists and psychologists.

On the other hand, the alternative model of human evolution through "neoteny" (i.e., juvenilization) seems to be widely accepted (e.g. Gould 1977; Montagu 1989). Ironically, considering that it is also an evolutionary concept that has been applied to human mentality, the notion that humans are juvenilized apes retaining the playful spirit of the young seems to be positively appealing to social scientists. Perhaps it is also the appeal of youth as compared to age. Perhaps it is the appeal of antithesis. Whatever the cause, the popular triumph of the idea that humans are juvenilized apes is reflected, for example, in discussions in textbooks: Following the lead of Gould (1977) and Montagu (1989), introductory textbooks in human development and human evolution show the famous photograph of the humanlike juvenile chimpanzee next to the apelike adult chimpanzee to illustrate the role of neoteny in human evolution (e.g. Stein & Rowe 1993; Relethford 1994).

The opposition between these two alternative models sets the stage for my discussion of the evolution of human mental development: In this paper, I assess the feasibility of Baldwin's research program in comparative developmental evolutionary psychology (CDEP) and his ideas about recapitulation. It is my contention that Baldwin's research program is now feasible and that sufficient progress has already been made in achieving the goal he laid out to justify a progress report.

At the time Baldwin wrote, little was known about human cognitive development and next to nothing was known about cognitive development in other species. More significant, theoretical frameworks in evolutionary biology necessary for comparative studies were in their infancy. Since that time, however, enormous strides have occurred in a number of areas of inquiry relevant to comparative developmental evolutionary psychology. As I explain in the following sections, these areas include not only developmental psychology and primatology, but also ethology, systematics, life history theory, and heterochrony.

Frames from evolutionary biology

Today, almost a century after Baldwin wrote, evolutionary biologists have begun to understand the mechanisms by which changes in ontogeny occur during evolution. Widespread interest in heterochrony, the evolution of ontogeny through changes in timing of development, was stimulated by Stephen J. Gould's *Ontogeny and Phylogeny* (1977), which provides both a history of ideas and a reinterpretation of heterochrony in terms of genetics and life history theory. Further interest has been stimulated by McKinney and McNamara's *Heterochrony* (1991), which provides a comprehensive review of mechanisms and outcomes of heterochrony. Heterochrony occurs through changes in the onset and/or offset of differentiation and growth and/or the rate of differentiation and growth. These mechanisms can be classified into two major categories: a) paedomorphosis, or juvenilization, or underdevelopment, which arises through progenesis, neoteny, and/or postdisplacement, and b) peramorphosis, or adultification, or overdevelopment, which arises through hypermorphosis, acceleration, and predisplacement. Neoteny involves slowing of the rate of development, while acceleration involves speeding the rate of development. Postdisplacement involves later onset of development while predisplacement involves earlier onset of development. Progenesis involves earlier offset of development (terminal deletion), while hypermorphosis involves later offset of development and addition of characters at the end of the life span (terminal addition). These mechanisms can operate globally on the entire organism or selectively on specific systems, resulting in global heterochrony or dissociated heterochrony, respectively (Fig 1).

Authors of both of these volumes have emphasized the relationship between heterochrony and life history theory, which treats the adaptive significance of species life cycle patterns (e.g., Stearns 1992). They have emphasized how mechanisms of heterochrony can produce evolutionary changes in such life history patterns as age at sexual maturity, which are advantageous under various conditions. They and others have discussed the kinds of small genetic changes that could result in rapid changes in development. Shea (1992), for example, has discussed the possible role of changes in responsiveness to growth hormones that could be involved in the evolution of small body stature.

Discussion of heterochrony, like any other evolutionary phenomena, must be placed in the context of systematics (e.g., Kluge & Strauss 1985). Like heterochrony, systematics has undergone a revolution since Baldwin's time. The Hennig (1966) school of taxonomy, cladism, develops classifications of species based solely on common ancestry,

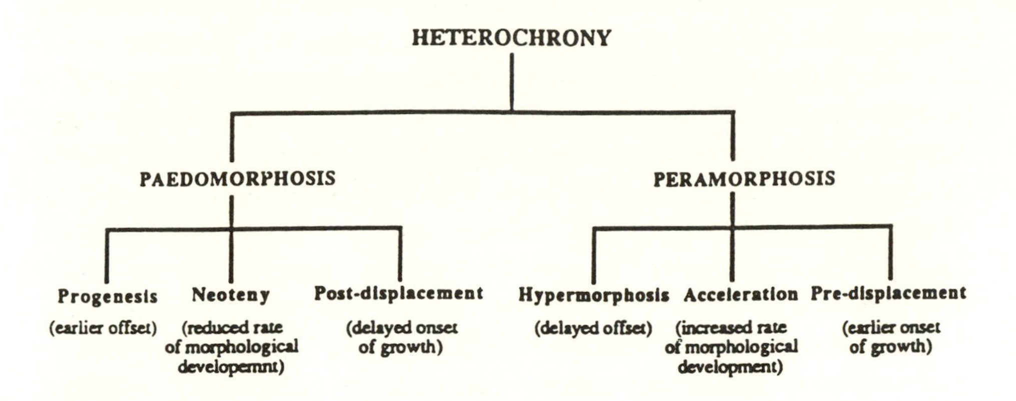

Figure 1. Diagram of processes of heterochrony from McKinney and McNamara (1991, p. 17).

giving no weight to degrees of divergence among sister species. All sister species of a monophyletic clade are included in the same taxonomic group. So, for example, humans must be classified with other great apes given that we share a more recent common ancestor with African apes than with Asian apes (Hennig 1966; Wiley 1981; Hull 1988). On the other hand, evolutionary taxonomists such as Mayr et al. (1953), Simpson (1961) and Hull (1988) allow some paraphyletic assemblages, so that species such as humans can be classified in a separate taxon owing to their marked divergence from their sister species.

Hennig (1966) argued that classification must be based on the distinction between a) homologous characters that are uniquely shared (shared derived character states) by a given group of sister species owing to their origin in a recent common ancestor and b) homologous characters that are broadly shared by a larger group of species (shared character states) owing to their ancient origin in a distant common ancestor (Wiley 1981). Cladists distinguish these two forms of homology by comparing character states among members of the "ingroup" with those of related "outgroups." So, for example, the homologous character state of brachiation, which occurs in all living apes but does not occur in the outgroup, Old World monkeys, is a shared derived character state among apes. On the other hand, the homologous character state of placental birth, which is shared by apes, is a shared (but not derived) character state because it is present in other primates and mammals. The contribution of the cladists was to argue that only shared derived character states should be considered indicative of common ancestry. Hennig's procedure of reciprocal illumination involves iterated mapping of character states in a group of related species onto logically possible branching trees until the most parsimonious tree is discovered. In addition to outgroup comparison, Hennig recognized the following additional criteria for distinguishing derived from shared characters: geological precedence, chorological (geographical) progression, correlation of character transformations, and ontogenetic character precedence (Hennig 1966, 1979; Wiley 1981; Hull 1988). "Geological precedence" refers to earlier appearance of one character as opposed to another in the fossil record of a monophyletic group; "chorological progression" refers to parallels between spatial distribution of species and their sequence of branching speciation; "correlation of character transformations" refers to parallel stages of transformation in characters in two or more species where the direction of transformation can be determined (morphoclines); and "ontogenetic character preference" refers to earlier appearance in ontogeny of one stage as compared to another assuming recapitulation (Hennig 1966, 1979).

Although it is more systematic, cladistic analysis is similar to ethological analysis of behavior in related species. Indeed ethology arose as a branch of systematics concerned with using behavioral characters for assessing phylogenetic relationships (Lorenz 1950; McLennan et al. 1988). The two disciplines offer much to each other: cladism offers a systematic approach to reconstructing phylogeny and ethology offers knowledge of characters which would seem to offer considerable insight into phylogenetic relationships. McLennan et al. (1988), for example, use ethological data on reproductive behavior in stickleback fish to resolve a taxonomic dispute in this taxon. Although it seems surprising that so little communication has occurred between ethologists and cladists, McLennan et al. argue that this is not a problem intrinsic to the data but a result of the relatively primitive treatment of ethological as compared to morphological data and lack of interest in behavior and life history by systematists.

It is important to note in this context that cladistic and ethological methods are useful not only for reconstructing branching relationships among species (phylogenies), but also for reconstructing the sequence of evolution of character states (Wiley 1981; McLennan et al. 1988; in this volume, Martins & Hansen, Chap. 2). This approach has been used, for example, to examine the relationship between the ontogeny and phylogeny of bird song (Irwin 1988) and the nest architecture of wasps (Wenzel 1993; see also, in this volume, Promislow, Chap. 10 and Chan, Chap. 11). Cladistic and ethological methods in comparative evolutionary developmental studies are useful for reconstructing the evolution of cognitive development in hominoids according to Baldwin's program. Detailed data on cognitive development in the great apes may contribute to resolution of the debate concerning the order of branching among the common ancestor(s) of gorillas, chimpanzees, and humans (e.g. Weiss 1987), but this remains to be seen.

Although heterochrony, cladistics, and ethology provide methodologies relevant to such reconstruction, important issues remain to be resolved. These issues include the identification of relevant characters and the determination of character polarities. Before addressing these issues, it is necessary to review other bodies of theory and data relevant to such an enterprise.

Frames from developmental psychology

As previously indicated (see above quotation) virtually nothing was known about cognitive development in human children in Baldwin's time. Ironically, a child born the year before Baldwin outlined his research program (1897) was destined to initiate a research program into child development that would provide a flood of data on this subject. Jean Piaget (1896–1980) was a Swiss-French psychologist who drew much of his inspiration from Baldwin's work (Kessen 1965; Case 1985).

Among the major concepts Piaget based on Baldwin's work were genetic epistemology, decalage (temporal displacement of stages relative to one another), epigenesis, the importance of action in adaptation, reflection on actions, role of assimilation (habit) and accommodation (imitation), stages of mental development, and circular reactions (repeated behavior patterns). Like Baldwin, Piaget (e.g., 1970) argued that cognitive development is epigenetic, that is, that each developmental stage was constructed from differentiation and coordination of the schemes of the preceding stage.

Piaget differed from Baldwin in the scope of his developmental studies of children, first his own children and later other children at the Jean-Jacques Rousseau Institute in Geneva, where he conducted research with Barbel Inhelder and many other colleagues for more than 50 years. Piaget identified a sequence of developmental periods, each one reflecting the highest achievements of children classified in that period: the sensorimotor period from birth to 2 years; the preoperations period from 2 to 6 years; the concrete operations period from 6 to 12 years; and the formal operations period from 12 years on (e.g. Inhelder & Piaget 1964; Piaget & Inhelder 1969).

Within each of these periods, Piaget distinguished stages of development across a variety of domains including objects, space, time, causality, and imitation; in postsensorimotor periods he also looked at the development of classification, number, geometry, and chance. He focused primarily on these nonsocial domains distinguishing logical-mathematical knowledge from physical knowledge. In later years he also worked on memory and perception. With the exception of his work on the development of imitation and symbolic play and on moral judgment, however, Piaget (1962, 1965) displayed little of Baldwin's interest in social intelligence and emotions (Table 12.1).

Piaget's (1962) stages of imitation in the sensorimotor period expanded on the stages of imitation outlined by Baldwin. Piaget's stages of sensorimotor intelligence elaborated on Baldwin's concept of circular reactions: "primary circular reactions" in stage 2 from 2 to 4 months are characterized by repetition of actions on the self elicited by sensations arising from contact between parts of the self, for example, thumb sucking; "secondary circular reactions" in stage 3 are characterized by repetition of actions which by chance produce contingent action in objects (e.g., shaking a rattle); while "tertiary circular reactions" in stage 5 are characterized by repetition of actions which produce interesting contingent variations in objects (e.g., different amounts of splashing from an object dropped from different heights into water). As in Baldwin's schemes, circular reactions underlie imitation (Piaget 1952). In addition to outlining the stages of development of imitation and circular reactions, Piaget also outlined parallel stages of development in four other sensorimotor period series: the object concept series, the causality, spatial, and temporal series (Piaget 1954). Throughout his work on the sensorimotor period, Piaget emphasized the interrelations among these six series with their six parallel stages (Table 12.2).

Piaget's work on the preoperations period is less elaborated and tightly connected than his work on the sensorimotor period or subsequent periods. He describes this period as an intermediate between sensorimotor and operative intelligence arising out of the representational abilities of the sixth sensorimotor stage and culminating in the true concepts and mentally reversible operations and dynamic imagery of concrete operations. Piaget traces the sensorimotor period domains through preoperational and into the operational period of development: during the preoperations period, the object concept develops toward conservation of identity and quantity; imitation develops into symbolic play and the other symbolic functions of drawing, mental imagery and language; the tertiary circular reaction gives rise to experiments with classes and series and arrays of objects; spatial representation develops into understanding of topological relationships (enclosure, proximity and separation), temporal representations develop into an understanding of temporal series, and causality develops into an understanding of directly mediated forces (e.g., Piaget & Inhelder 1969).

During the subsequent concrete operations period the child masters conservation of weight and volume, symbolic play transforms into games with rules, experimentation with classes develops into reversible hierarchical classification according to multiple criteria, seriation and number, topological concepts develop into projective and euclidean concepts, and understanding of directly mediated causality develops into understanding of invisible mediating forces. Finally, during formal operations the child transcends reliance on concrete relations and constructs hypothetical deductive systems and tests alternative hypotheses through systematic control of variables.

Although many of Piaget's ideas and claims have been disputed, his comprehensive model of cognitive development has largely set the agenda of 20th-century developmental psychology in both positive and negative senses (Flavell et al. 1993; Miller 1993). Much of American developmental research has been aimed at disproving various components of Piaget's developmental stages and processes (e.g.,

Meltzoff & Moore 1977; Gelman & Gallistel 1978; Baillargeon 1987; Meltzoff 1988), while other developmental research has been aimed at refining (e.g. Uzgiris & Hunt 1975) and elaborating his stages and theories (Langer 1980, 1986), especially through the introduction of information processing models (e.g., Case 1985; Siegler 1986). Other investigators have extended Piaget's work on moral judgment (e.g., Kohlberg 1969). Still others have extended his work on perspective taking to social perspective taking (Flavell et al. 1968). Finally, other bodies of developmental research have taken up the areas of social cognition that Piaget neglected.

Piaget's model of intellectual development not only arises from Baldwin's research program but provides a comprehensive model for human development. It also provides a consistent framework for comparative studies of cognitive development in other primate species. In other words, it provides a matrix of relationships within and across stages and periods of development and across domains (e.g. Antinucci 1989). This allows investigators to see what kinds of abilities and disabilities characterize a stage or period of development. It therefore provides possible characters for reconstructing the phylogeny of human cognition (Parker & Gibson 1979). Furthermore, it provides developmental stages that could be used to reconstruct the evolution of ontogeny.

Frames from primatology

At the time Baldwin wrote, little was known about the life histories and mentalities of nonhuman primates beyond a few anecdotal reports on pets (e.g., Romanes 1882) and a few laboratory studies, which Baldwin fails to cite. Since Baldwin's time, the study of nonhuman primate systematics, biogeography, morphology, physiology, genetics, paleontology and behavior has burgeoned into a large, worldwide interdisciplinary enterprise. As a result of such studies, the phylogeny of most primate taxa is known from molecular studies (e.g. Sarich & Cronin 1976; Cronin et al. 1980) (Fig. 2).

Table 12.1. Piaget's periods and domains of cognitive development.

	Cognitive domains		
	Physical knowledge	Logical-mathematical knowledge	Interpersonal knowledge
Sensorimotor (birth – 2 years)	Practical discovery of properties of objects, space, time, and causality	Practical discovery of object-object relations, creation of object sets (circular reactions as instruments of assimilation and accommodation)	Discovery of novel schemes though imitation as an instrument of assimilation
Preoperations (2–6 years)	Discovery of immediate causes of actions and reactions	Discovery of simple sets and classes (nonreversible schemes without compensations as instruments of assimilation & accommodation)	Discovery of new routines and social roles through symbolic play as an instrument of assimilation & accommodation (A&A)
Concrete Operations (6–12 years)	Discovery of indirect (mediated) causes of actions and reactions	Discovery of hierarchical classification, part-whole relations and numbers through reversible operations with compensations as instruments of A&A	Discovery of new routines and roles and rules through games with rules
Formal Operations (12–16 years)	Discovery of laws of physics through true measurement & control of variables	Discovery of logical necessity through hypothetical-deductive reasoning	Construction of universal rules and principles

Table 12.2. Piaget's sensorimotor period stages and series.

	Series					
	Sensorimotor	Causality	Space	Time	Object concept	Imitation
Stage 1 0– 2nd month			Reflexive Reactions			
Stage 2 2–3 months	Primary circular reactions	Diffuse sense of efficacy	Heterogenous perceptual spaces	Heterogenous perceptual durations	No search for hidden objects	Sporadic imitation of own schemes
Stage 3 3–8 months	Secondary circular reactions	Magicophenomenalistic causality	Coordination of subjective space	Subjective sense of the sequence of own schemes	Object permanence as exetnsion of movements	Systematic imitation of own schemes
Stage 4 8–12 months	Coordination of secondary schemes	Elementary externalization of causality	Transition to objective space; reversibility	Beginning notions of before and after	Active search for hidden object	Imitation of own schemes, invisible to self
Stage 5 12–18 months	Tertiary circular reactions (TCR)	Objectification of causality	Objectification through relative displacement of objects	Memory of sequential order of object displacements	Searching with understanding of sequential object displacement	Imitation of novel schemes through trial and error
Stage 6 18–24 months	Representation of TCR	Representation of causality	Representation of invisible displacements	Representation of temporal sequence of events	Representation of sequential invisible object displacement	Deferred imitation of novel schemes

Thanks to myriad field studies pursued since the 1960s, as well as demographic data from zoos and primate colonies, much is also known of the naturalistic behavior and life histories of many of the approximately 180 living species of primates (Harvey et al. 1986). Especially well studied are species of the great apes, lesser apes, baboons, macaques, guenons, and langurs, as well as the prosimian galagos in the Old World, and the marmosets, squirrel and cebus monkeys in the New World (e.g., Fedigan 1982; Richards 1985; Fleagle 1988). Living primates display an intriguing pattern of similarities and differences between and within families. As Schultz (1969) noted long ago, the prosimians, monkeys, lesser apes and great apes constitute a stepwise increase in such life history features as brain size, gestation period, life span, and age at sexual maturity (Table 12.3, Fig. 3).

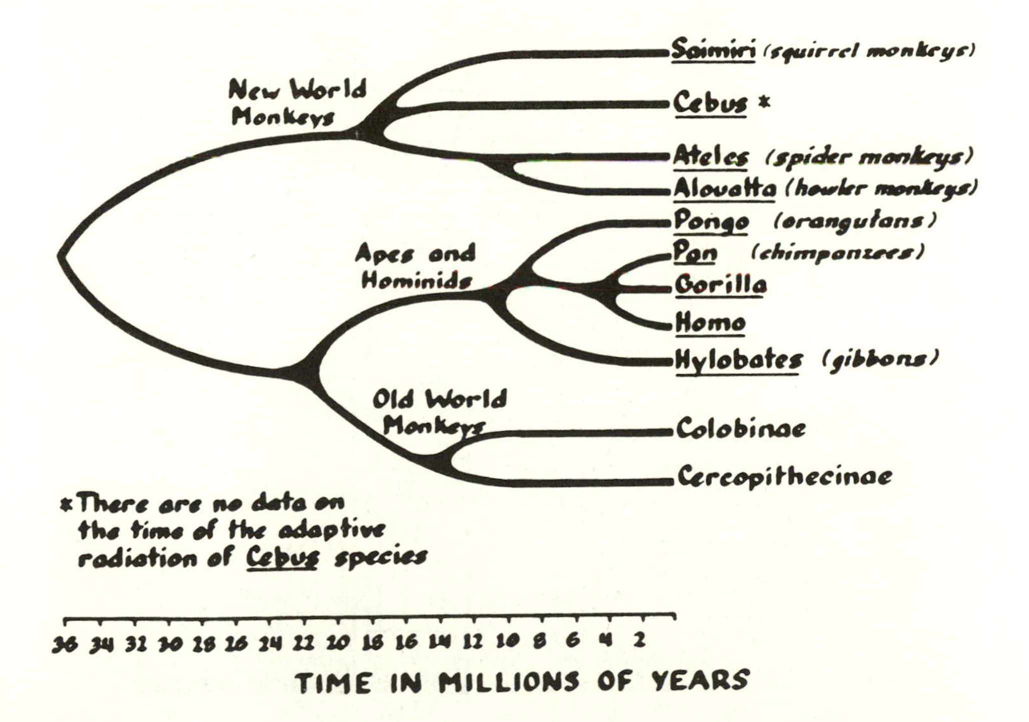

Figure 2. Primate phylogeny from Parker and Gibson (1977, p. 633).

On the other hand, and in contrast to their similarities in some life history features, taxa such as genera of Old World monkeys and

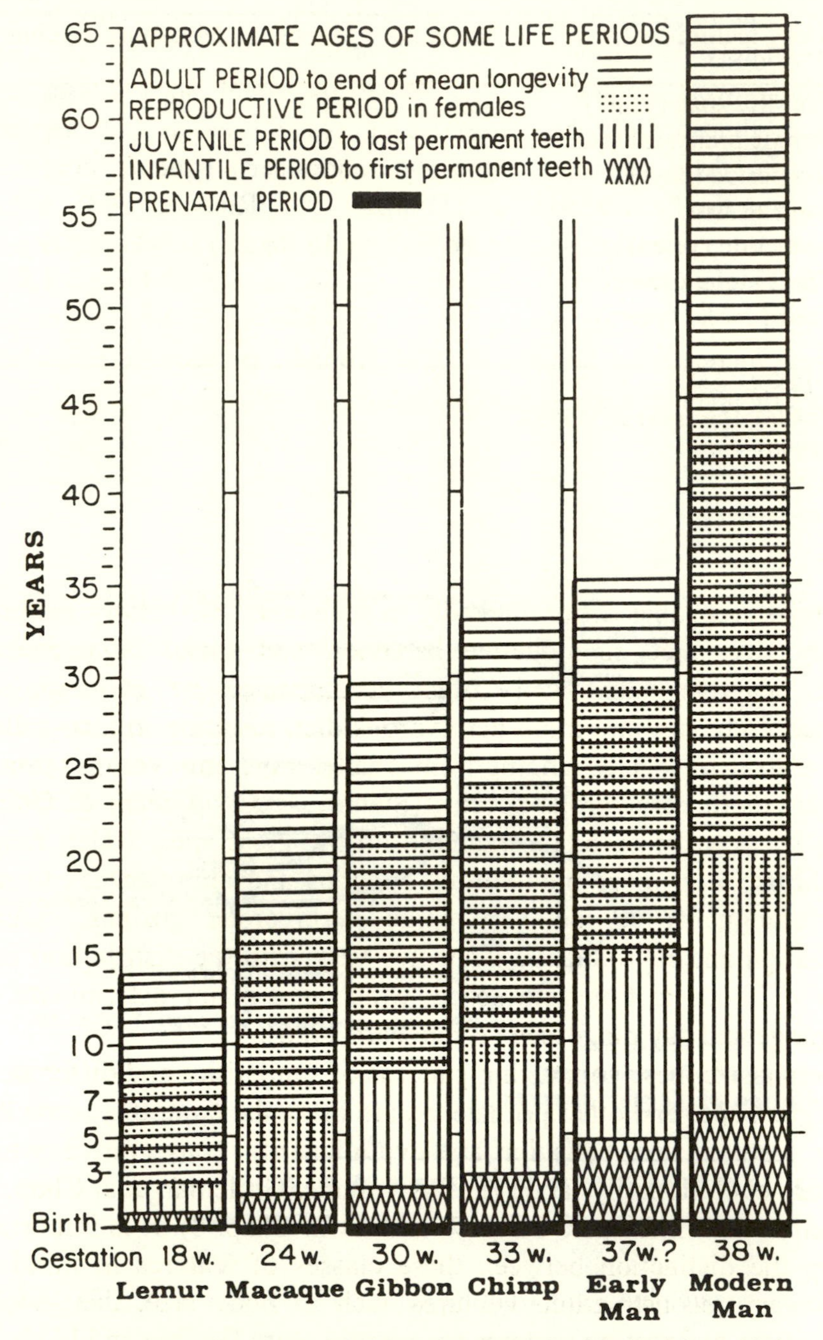

Figure 3. Diagram of primate life histories from Smith (1992, p. 136).

Table 12.3. Life history features of selected primate species.

Life history features	Humans	Apes	Macaques	Cebus
Maximum life span (years)[a]	95	45	30	45
Adult female brain size (cc)[b]	1250	350	100	71
Gestation period (days)[c]	277	230	170	155
Female age at first reprooduction (years)[c]	19[e]	10–14	3–4	6
Age of first molar (years)[d]	6	3	1.4	1.2
Age of third molar (years)[d]	20	11.5	5.5	–

[a]Cutler (1976)
[b]Tuttle (1986); Harvey et al. (1986)
[c]Watts (1990)
[d]Smith (1992)
[e]In hunter-gatherers

families of great apes differ markedly in such other life history features as body size, diet, reproductive behavior, and social organization. Indeed, adaptive radiation is often accompanied by evolution of increased body size and more herbivorous diet, resulting in a so-called adaptive array of species ranging from smaller more omnivorous species to larger, more herbivorous species (Stanley 1973; Pilbeam & Gould 1974). This can be seen, for example, in the great apes, which range from the smaller, more omnivorous bonobos and chimpanzees to the larger orangutans to the very large herbivorous gorillas. Social organization and reproductive strategies vary with the distribution and density of foods in time and space, which is a major determinant of home range and day range size (Wrangham 1986).

I emphasize these contrasting patterns of similarities and differences within great apes and other primate taxa to make the point that some life history features and behavior are evolutionarily conservative while others are evolutionarily labile (in this volume, Gittleman et al Chap. 6; de Queiroz & Wimberger, Chap. 7; and Irwin Chap. 8). It is important to keep the distinction between these classes of features in mind in evolutionary reconstruction. Features such as body size, diet, social organization, and mating systems are usually more flexible and likely to change during adaptive radiations than features such as gestation period, bone and tooth growth, life span, and brain size. Rapidly evolving features, like other highly variable features within and among species, are less likely to be taxonomically relevant characters (Clark 1964).

And what research has been done since Baldwin's time regarding the mentalities of nonhuman primates? Systematic studies of intelligence in captive primates go back at least to Kohler (1917, 1927) and Yerkes (1916). Until recently, however, most of these studies have focused on the same standardized tests of learning that were used on rats and pigeons. Beginning in the 1960s, however, a few investigators began to compare the linguistic abilities of chimpanzees to those of human children (e.g., Gardner & Gardner 1969, 1975). In the 1970's, ape language studies expanded to include other great ape species (e.g., Miles 1990; Patterson 1980). Also in the 1970s some investigators began using Piaget's sensorimotor period stages to compare the abilities of primate species (see Dore & Dumas 1987; Antinucci 1989; Parker 1990, for reviews).

During this period, comparative studies based on other frameworks in human cognitive development have also emerged, including work on imitation (e.g., Mitchell 1987; Russon & Galdikas 1993, 1994; Custance & Bard 1994), mirror self-recognition (e.g., Gallup 1982; Povinelli 1993; Parker et al. 1994) and theory of mind (e.g., Premack & Woodruff 1978; Premack 1988; Whiten 1991). Fortunately, many of these other domains can be connected via Piagetian stages and periods. As I show in the next section of this paper, these studies have revealed a pattern of similarities and differences among primate taxa that parallels the patterns in the more conservative life history features mentioned above.

Although they began earlier than field studies, studies of cognition in captive primates have been far fewer in number and have focused on a much smaller number of species. Owing to both historical and practical constraints, chimpanzees, macaques, and cebus monkeys have been the best-studied species (Parker 1990). Many more species of Old World and New World monkeys and lesser apes remain to be studied from this perspective. Despite these limitations, a significant step has been taken toward comparative developmental studies of cognition. In the following sections, I summarize the results of some of these studies and assess their implications for Baldwin's research program.

Progress toward Baldwin's research program of comparative developmental evolutionary psychology

Because a comparative developmental evolutionary approach to cognition immediately raises the specter of anthropomorphism, it is important to address this issue before going on to discuss comparative studies. The first point is that anthropomorphism may be the initial kick that stimulates investigators to conceive new studies of primate mentality. The second point is that anthropomorphism is the null hypothesis in comparative developmental studies, i.e., the hypothesis that there are no differences between humans and the other species. Any differences that emerge disconfirm the null hypothesis. The third point is that frameworks from human development are valuable in comparative research insofar as they generate useful data that reveal phenomena which had been inaccessible without them.

Frameworks from human development have several advantages over frameworks from learning theory. First, they provide greater discriminative power in that they are capable of identifying species differences, which are not revealed by most traditional learning tests (e.g., Rumbaugh & Pate 1984). Second, because they were devised on humans, they probe the highest levels of cognition species are capable of as opposed to traditional learning theory tests, which tend to test at lower levels of cognition. Third, because they are epigenetic, they provide a natural scale of increasing cognitive complexity which is useful for comparing the abilities of different species. Fourth, because they are developmental, they provide a built-in scale for comparing rates of development. Finally, because they are multifarious, they provide a range of domains for comparing rates and levels of development within and across species; that is, they provide a broad spectrum of data for reconstructing the evolution of ontogeny. (The validity of the stage concept is a contested issue. Although discussion of this issue is beyond the scope of a broad-stroke treatment such as this, there is considerable empirical support both for and against stages (Flavell et al. 1993; Fischer & Silvern 1985).

Comparative research on primate cognition using frameworks from developmental psychology has been of two varieties: studies of the terminal or highest-level abilities of subadult and adult animals and studies of the development of abilities in younger animals. Because the first variety of research has been more popular, there is considerable information on the highest-level abilities of the best-studied species (Dore & Dumas 1987). This information is valuable for comparing the terminal levels of cognitive development in various primates.

The general pattern of similarities and differences between human and nonhuman primates that has emerged from 25 years of comparative research based on various frameworks from human development can be summarized as follows: Adult great apes display a spectrum of cognitive abilities similar to those of human infants and children from 2 to 4 years of age, depending on the domain; adult macaque monkeys display a spectrum of cognitive abilities somewhat similar to those of children less than 1 year of age; adult cebus monkeys display an intermediate pattern somewhat similar to those of human children of approximately 2 years of age in certain limited domains.

More specifically, at least some adults of all of the great ape species have displayed the following complex of capacities similar to those of human children in late sensorimotor and early preoperations periods: a) intelligent tool use and practical understanding of physical causality (e.g., Kohler 1927; Lethmate 1977; McGrew 1992; Boysen 1995); b) imitation of novel behavior (e.g., Hayes & Hayes 1952; Russon & Galdikas 1993; Custance & Bard 1994); c) rudimentary symbolic capacities revealed in signing, drawing, and symbolic play (Patterson & Linden 1981; Gardner et al. 1989; Miles 1990; Jensvold & Fouts 1994); d) mirror-self- recognition, feature self-recognition, and self-labeling (e.g., Gallup 1982; Povinelli 1993; Parker et al. 1994); e) logical capacities for classification (e.g., Premack 1978; Spinozzi 1993) and number (e.g. Boysen & Capaldi 1993); f) rudimentary role playing and role reversal abilities and attribution of intentionality to others (e.g., Gomez 1990; Povinelli et al. 1992; Miles 1994). Macaques have displayed none of these characteristics (e.g., Chevalier- Skolnikoff 1977; Parker 1977; Antinucci 1989). While cebus monkeys have displayed intelligent tool use, they may or may not display most of these other abilities (e.g., Chevalier-Skolnikoff 1989; Fragaszy & Visalberghi 1990).

Given the similarities in their cognitive abilities and the broad stroke of this paper, I am lumping all the great apes together. For the sake of brevity, I am also lumping all the domains under late sensorimotor and early preoperational periods. I should note, however, that there are some species differences in the pace of development and in the frequencies and contexts of particular behavior. Not all species of great apes display all these abilities in the wild; gorillas, for example, apparently do not use tools in the wild; none of the great apes use mirrors in the wild. Despite these and other differences, I believe that it is reasonable to use these abilities as taxonomic characters. This is justifiable because behavior and abilities are more conservative evolutionarily than the contexts in which they originally evolved. Therefore, the behavior or ability itself rather than the behavior in a particular context is the proper unit for ethological and/or cladistic analysis (e.g., Lorenz 1950). In the case of tool use, for example, the character is tool use rather than tool use in the wild, since the ability to use tools is more conservative than the context in which it evolved (Parker in press).

If we consider the highest-level abilities in a taxon to be taxonomic characters, we can reconstruct the evolution of these abilities by mapping them onto an existing phylogeny of primate evolution. Cladistic mapping of late sensorimotor and early preoperational abilities demonstrates that they are shared derived character states which were present in the common ancestor of all the great apes. This fact provides a baseline for reconstructing the evolution of human cognition: the adult common ancestor of all hominids must have been displayed at least this level of cognitive abilities which is characteristic of 2-to-4-year-old children in the early preoperations period (Parker & Gibson 1979).

This analysis, in turn, suggests that all the subsequent periods of cognitive development seen in human children are the product of evolution in the hominid clade subsequent to branching from the common ancestor with chimpanzees and gorillas. It thus raises questions regarding the times at which and species in whom specific cognitive capacities emerged. To approach these questions, we need to examine paleo-anthropological and archeological evidence. This evidence suggests that the highest-level cognitive abilities emerged very late in human evolution.

Frames from Hominid paleontology and archeology

Another line of evidence key to reconstructing hominid evolution comes from hominid paleontology and archeology, fields which have grown immensely since Baldwin's time. The current state of knowledge in paleontology can be summarized briefly as follows: The probable common ancestor of all hominids was a small facultatively bipedal creature, *Australopithecus afarensis*, who lived in East Africa from about 3.5 to about 2.5 million years ago. This creature, who had an apelike brain and life history, probably was the common ancestor of two lineages, the Australopithecines (including an adaptive array of several other species of various sizes and dietary specializations, all of whom became extinct by about 1 million years ago) and beginning about 2.5 million years ago, of *Homo* species. The earliest members of this second lineage were *Homo habilis* and possibly a second species *Homo rudolfensis* which gave rise to *Homo erectus* about 1.6 million years ago in East Africa. Archaic *Homo sapiens* were probably descendants of this species, which was the first to migrate and settle out of Africa. Members of the *Homo* lineage beginning with *Homo habilis* had larger brains and manufactured stone tools (e.g., Lewin 1993; Schick & Toth 1993). By the time of *Homo erectus*, brain size was approximately two-thirds that of modern humans and life history was probably intermediate between that of great apes and modern humans (Smith 1992).

Little is known about the subsistence strategies and behavior of the earliest hominids, earlier notions of a hunting adaptation have been displaced by the lack of evidence of worked stone tools or butchery. Gathering has been proposed as an alternative (Lovejoy 1981), as has extractive foraging on embedded foods (Parker & Gibson 1979; Parker in press). Stanley (1992) has pointed out that these small bipedal creatures must have been highly vulnerable to predators. Based on anatomical evidence, including their long arm relative to leg length, he proposes that they not only climbed trees as a refuge, as others had recently proposed, but that their infants still clung to their mothers as great ape infants do. He proposes that full reliance on terrestrial bipedalism occurred at the time of *Homo erectus*. This, in turn, he argues, allowed brain size to increase as it freed infants from the need to cling.

Even after the emergence of worked stone tools, the manifestations of these artifacts remained essentially static for almost 2 million years. Although the complexity of these artifacts gradually increased during the middle Paleolithic from *Homo erectus* to archaic *Homo sapiens*, the tempo of change was extremely slow until the Upper Paleolithic, when a rapid pace of change set in coincident with the emergence of highly specialized regional tool cultures. This change of pace coincided with the appearance of art and intentional burials and other signs of full-blown culture, which apparently followed the emergence of fully modern *Homo sapiens* (e.g., Mellars & Stringer 1989; Stringer & Gamble 1993).

If we accept the implications of this trend, we have a marker for the emergence of formal operations, the highest level of modern cognitive abilities (Parker & Milbrath 1993). This marker brackets the top end of the spectrum of cognitive abilities. Combining this top-end marker with the bottom-end marker, which is implied by the shared derived cognitive characters of the great apes, we can identify the window of time and the range of species involved in the evolution of human intellectual development from late sensorimotor intelligence to formal operational intelligence: The earliest manifestations of these abilities evolved in the common ancestor of the great apes in the Miocene epoch about 14 million years ago, while the most advanced manifestations of these abilities originated in modern *Homo sapiens* in the Upper Pleistocene epoch about 0.3 million years ago.

The remaining challenge is to reconstruct the kinds of intellectual abilities that were typical of various hominid species. Although it is problematic to attribute culturally patterned phenomena to individual cognition, at least two attempts have been made to reconstruct the Piagetian intellectual stages of extinct hominids from archeological data (Parker & Gibson 1979; Wynn 1989). Both reconstructions attributed early concrete operational concepts to makers of biface Acheulean hand axes, *Homo erectus*.

Intimations of recapitulation in human cognitive development

Within cladistics, correlated transformation series in related species and ontogenetic characters are two lines of evidence for character polarity, i.e., the sequence of evolution of characters (e.g., Hennig 1966,1979; Wiley 1981). Cladistic analysis of cognitive abilities and their development among living anthropoid primates provides both kinds of evidence for the evolutionary polarity of cognitive abilities.

Evidence that the periods of cognitive development in anthropoid primates evolved in their current sequence and therefore form a transformation series can be found through cladistic mapping of the distribution of terminal-level cognitive abilities in macaques, great apes, and humans. This mapping reveals that human cognition evolved at least in part through a series of additions of new levels of cognitive abilities at the end of the development in a series of common ancestors: a) addition of late sensorimotor and early preoperational abilities in the common ancestor of great apes; b) addition of late preoperational and concrete operational abilities in two or more hominid ancestors; and finally, c) addition of formal operations abilities in modern humans (Parker & Gibson 1979). Since each new period arose in phylogenetic sequence, the developmental sequence forms a transformation series. Moreover, terminal addition of new stages at the end of development is a form of peramorphosis that results in ontogenetic recapitulation of stages of evolution (Gould 1977; McKinney & McNamara 1991) — just what Baldwin argued. This sort of analysis, however, is only a first step toward reconstructing the evolution of development.

In order to reconstruct the evolution of the timing of cognitive ontogeny in humans, we need not only comparative data on the direction and timing of cognitive development in humans, great apes, and monkeys but also a definition of ontogenetic characters appropriate for cladistic analysis of characters requisite to reconstructing the sequence of evolution of cognitive development. The traditional ontogenetic method in cladistics relies on analysis of ontogenetic transformations of instantaneous morphologies (i.e., those recognizable as a single morphological character; de Queiroz 1985). De Queiroz (1985), however, rejects the ontogenetic method on the grounds that it fails to distinguish between phylogenies of organisms and phylogenies of instantaneous morphologies (also see Kluge & Strauss 1985).

De Queiroz (1985) proposes an alternative definition of ontogenetic characters based on a view of organisms as life cycles. He argues that the ontogenetic transformations or lack of transformations themselves are the characters. In other words, he argues that instantaneous morphologies are incomplete characters which are actually parts of ontogenetic transformations which are themselves the characters (de Queiroz 1985, p. 298). According to de Queiroz's (1985) view, ontogenetic transformational characters have both ontogenetic and evolutionary polarities, but the ontogenetic polarities exist between instantaneous morphologies within the transformational characters, while phylogenetic polarities exist between ontogenetic transformational characters. He argues that only the phylogenetic polarities are relevant to evolutionary reconstruction, and therefore that only the out-group method (under which he subsumes the paleontogical method) is valid for determining evolutionary polarities or sequences.

In the previous analysis of terminal levels of cognitive ability, these terminal levels were instantaneous characters. To reconstruct the evolution of cognitive development, however, we need ontogenetic characters. According to de Queiroz, the definition of our ontogenetic characters should depend upon how large a segment of ontogenetic development is relevant to the problem at hand (de Queiroz 1985). In the case of intellectual development, I believe that the relevant segment should encompass the entire span of cognitive development from early sensorimotor stages through the terminal stage for each species. The most appropriate way to code such multistate (polymorphic) characters as ontogenetic sequences for phylogenetic analysis is in step matrices that reflect hypotheses regarding the order and distance among character states (Mabee & Humphries 1993) (Table 12.5).

Developmental data can be characterized according to achieved levels of cognitive development abbreviated as follows:

- Early sensorimotor period (ESM) (level 1)
- Late sensorimotor period (LSM) (level 2)
- Early preoperations period (EPO) (level 3)
- Later preoperations period (LPO) (level 4)
- Early concrete operations (ECO) (level 5)
- Late concrete operations (LCO) (level 6)
- Formal operations period (FO) (level 7)

Following de Queiroz's concept, ontogenetic characters for the relevant species then can be designated using the preceding abbreviations:

- Macaque monkeys, ESM
- Great apes, ESM — LSM — EPO
- Modern humans, ESM — LSM — EPO — LPO — ECO — LCO — FO

See Table 12.4 for listing of these ontogenetic characters, Table 12.5 for a step matrix of character state transformations which shows hypothetical steps and orders of transformation between ontogenies (Mabee & Humphries 1993). See Figure 4 for a mapping of ontogenetic charcters onto a phylogeny of primates.

Table 12.4. Ontogenetic characters coded for phylogenetic analysis

Taxon	Ontogeny	Code
A	ESM — LSM — EPO — LPO — ECO — LCO — FO	0
B	ESM — LSM — EPO	1
C	ESM — LSM — EPO	1
D	ESM — LSM — EPO	1
E	ESM	2

Key: A = modern humans, B = chimpanzees, C = gorillas, D = orangutans, E = macaque monkeys. EMS = early sensorimotor period - 1. LSM = late sensorimotor period - 2. EPO = early preoperations period - 3. LPO = later preoperations period - 4. ECO = early concrete operations - 5. LCO = late concrete operations - 6. FO = formal operations period - 7.

Table 12.5. Character state step matrix of observed ontogenetic characters (cognitive development) with pairwise distances indicating hypothetical distances and directions of transformation among ontogenies

123456	12345	1234	123	12	1	
123456	--	-1	-2	-3	-4	-5
12345	+1	--	-1	-2	-3	-4
1234	+2	+1	--	-1	-2	-3
123	+3	+2	+1	--	-1	-2
12	+4	+3	+2	+1	--	-1
1	+5	+4	+3	+2	+1	--

Key: ESM — LSM — EPO — LPO — ECO — LCO = 1 2 3 4 5 6
ESM — LSM — EPO — LPO — ECO = 1 2 3 4 5
ESM — LSM — EPO — LPO = 1 2 3 4
ESM — LSM — EPO = 1 2 3
EMS = 1

This breakdown provides a global definition of ontogenetic characters, but it leaves out the temporal dimension of onset and offset and rates of development (e.g., Kluge & Strauss 1985). A more useful designation would include species-typical ages in years as a part of each ontogenetic character as follows:

- Macaque monkeys, ESM (0–4 years)
- Great apes, ESM (0–2/3 years) — LSM (1–4 years) — EPO (4–8 years)
- Modern humans, ESM (0–2/3 years) — LSM (1–2 years) — EPO (2–4 years) — LPO (4–6 years)
- ECO (6–8 years) — LCO (8–10 years) — FO (12–16 years)

These data reveal that, while the pace of sensorimotor development is slower in humans and great apes than in macaques, the pace of late

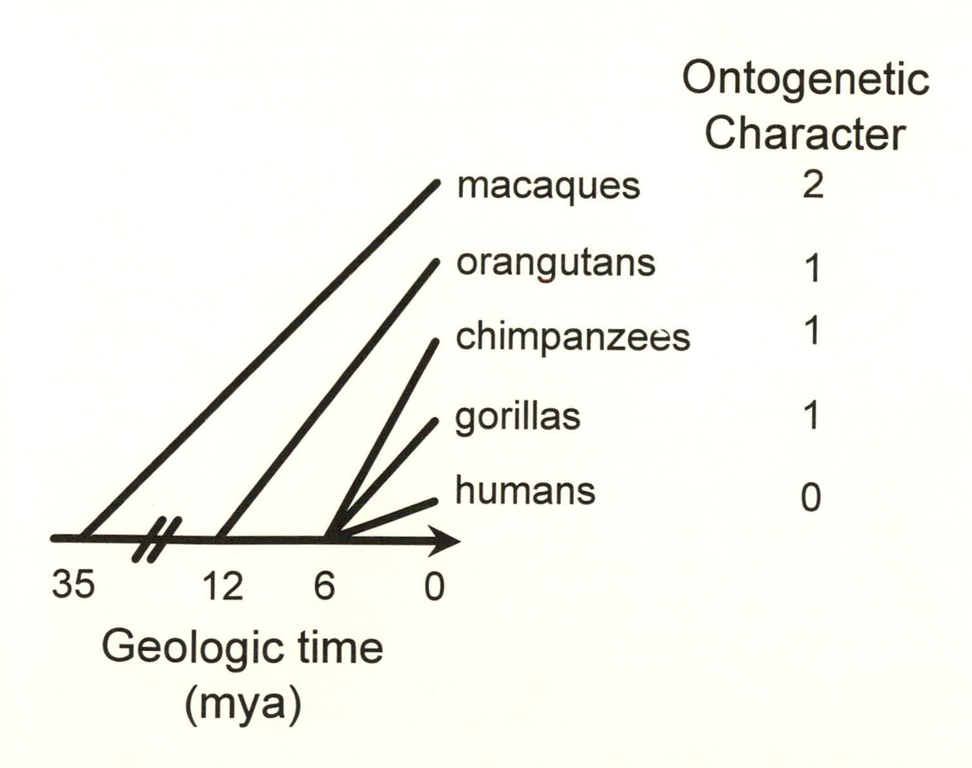

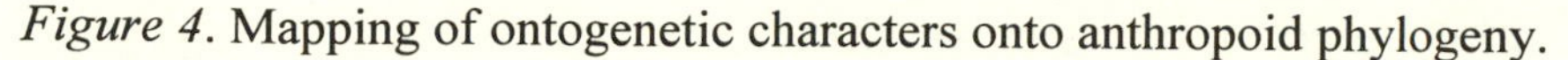
Figure 4. Mapping of ontogenetic characters onto anthropoid phylogeny.

sensorimotor development and early preoperational development is faster in humans as compared to great apes. The pace in humans increases at the end of the first year and during the second year of life and continues to accelerate in the preoperations period during the third and fourth years of life. This divergence in developmental rates in the latter part of the sensorimotor period and the early preoperations period contrasts with the similar rates of development in the two groups during the first two-thirds of the sensorimotor period. In humans, early preoperational abilities are superseded by late preoperational abilities by about 4 years, whereas in great apes early preoperational abilities probably continue to develop until about 7 or 8 years of age (Miles 1994; Povinelli et al. 1994; Russon, personal communication). In contrast to great apes, human children continue to traverse two additional periods of cognitive development — concrete and formal operations — after they complete early preoperations at about 4 years of age. Comparing the ages of offset/onset of each major Piagetian period in humans, great apes, and macaques (as an outgroup), reveals an increasing rate of development (as measured by relative onset and offset of periods) and an overall pattern of terminal addition in human phylogeny (Fig. 5).

Although this global model of developmental timing is an improvement over the terminal model, it leaves out a great deal of fine structure regarding temporal displacements (decalages) within and among series or domains which are highly significant for understanding the evolution of cognitive ontogeny. A finer scale analysis reveals that some series (e.g., the object concept series and the imitation series) within the sensorimotor period develop at different rates within macaques and great apes as compared to human children. These displacements imply that epigenesis operates within but not across series and domains.

During the sensorimotor period in nonhuman primates, for example, displacements occur between stages in the sensorimotor object concept and other series: macaque monkeys complete the fifth stage of the object concept series by 6 months while achieving only stage 3 in imitation and stage 4 in causality (Parker 1977; Antinucci 1989). Great apes apparently achieve all six stages in all six series in the sensorimotor period but achieve the sixth stage in various series at different times. While they achieve the first four stages in all series before the end of the first year (Hallock & Worobey 1984; Poti & Spinozzi 1994) and complete the sixth stage in the object concept series by 19 months, they apparently take up to 4 years to complete the sixth stage in the causality, sensorimotor intelligence, and imitation series (Chevalier-Skolnikoff 1977, 1983; Potí & Spinozzi 1994; Parker personal communication).

In addition to displaying a displacement between the early developing object concept and the later developing causality, imitation, and sensorimotor intelligence series, both macaques and great apes display displacements across modalities within some sensorimotor series: within the imitation series in great apes, imitation is limited primarily to the gestural and facial modalities (e.g., Chevalier-Skolnikoff 1977). In macaques, imitation remains at the third stage in these modalities. Within the sensorimotor intelligence series in great apes, the development of visually directed prehension and primitive means-ends precedes and outstrips the development of secondary and

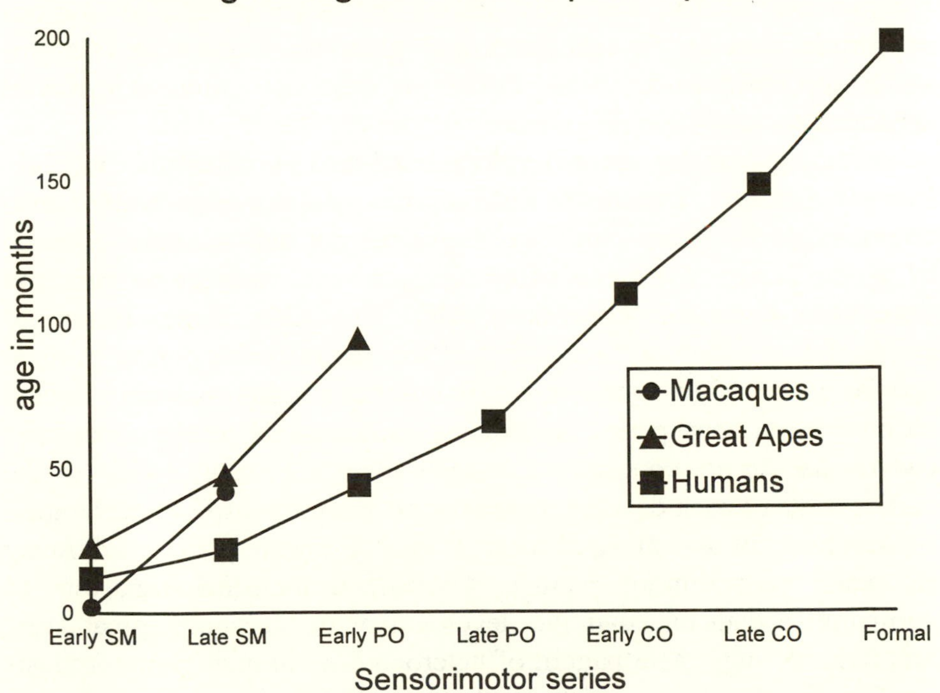

Figure 5. Comparison of trajectories of cognitive development in humans, great apes and macaques. SM = sensorimotor, PO = preoperations, CO = concrete operations.

tertiary circular reactions (Poti & Spinozzi 1994). Within the sensorimotor series in macaques, it is notable that the secondary and tertiary circular reactions and imitation of novel schemes never develop (Parker 1977; Antinucci 1989). In a broader frame, Antinucci (1989) and Langer (1989) argue that whereas, in human ontogeny, physical and logical-mathematical knowledge develop synchronously, in macaque and cebus ontogeny, the development of physical knowledge precedes the development of logical-mathematical knowledge. Langer (1993) attributes the emergence of representation and the subsequent elaboration of cognition in human development to the shift from asynchronous to synchronous development during human evolution.

Cladistic mapping of these and other sensorimotor series suggests that the object concept series is more primitive than the other sensorimotor series, and that visually directed prehension and simple means-ends schemes are more primitive than circular reactions are. This analysis also suggests that cognitive access to manual/gestural and facial schemes is more primitive than cognitive access to vocal schemes. Comparison of the displaced timing of development of various sensorimotor series in macaques and great apes also suggests this pattern. In other words, we see mosaic evolution of various domains of sensorimotor cognition (Figs. 6 and 7).

Although sketchy, these developmental data allow us to tentatively identify some mechanisms of heterochrony in the evolution of human cognitive development. Specifically, we see not only terminal addition of a new period at the end of ontogeny of each succeeding common ancestor (sequential hypermorphosis), but also three forms of peramorphosis. To wit, later offset of the sensorimotor period in great apes as compared to monkeys; earlier onset of the late sensorimotor and early preoperations period in humans as compared to the great apes (which has, as its concomitant, earlier offset). Finally, owing to the earlier onset of each developmental period, humans display acceleration of development as compared to great apes in all the periods following the early sensorimotor periods. Overall, the cladistic analysis of comparative data on cognitive development in macaques, great apes, and humans suggests a pattern of heterochrony through peramorphosis resulting in ontogenetic recapitulation of stages of evolution of adult forms. (see Ekstig 1994 for a similar conclusion.) Displacements can be seen as evidence for dissociated heterochrony both within and among sensorimotor series.

In addition to evidence for peramorphosis in cognitive development, there is evidence for retrospective elaboration of sensorimotor and preoperational cognition during human evolution. This elaboration takes the form of additional modalities, increased numbers of schemes, increased frequencies of schemes, and increased complexity of schemes, particularly of tertiary circular reactions and imitation in the sensorimotor period, and of symbolic schemes during the preoperations period. These elaborations might be considered a form of caenogenesis or introduction of new features into the ontogenetic sequence (McKinney & McNamara 1991).

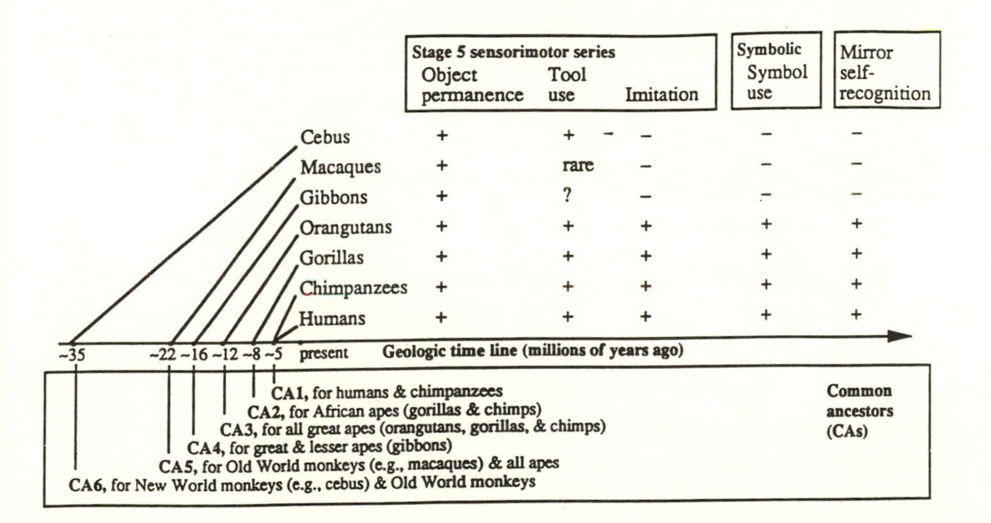

Figure 6. Cladistic comparison of late sensorimotor and early preoperational abilities in selected primates from Parker (in press). (Reprinted with permission from Parker, S. T. (in press). Apprenticeship in extractive foraging in great apes: The origins of imitation, teaching and self-awareness. In: *Reaching into Thought* (Ed. by A. Russon, K. Bard, & S. T. Parker). Cambridge, England: Cambridge University Press. Copyright Cambridge University Press.)

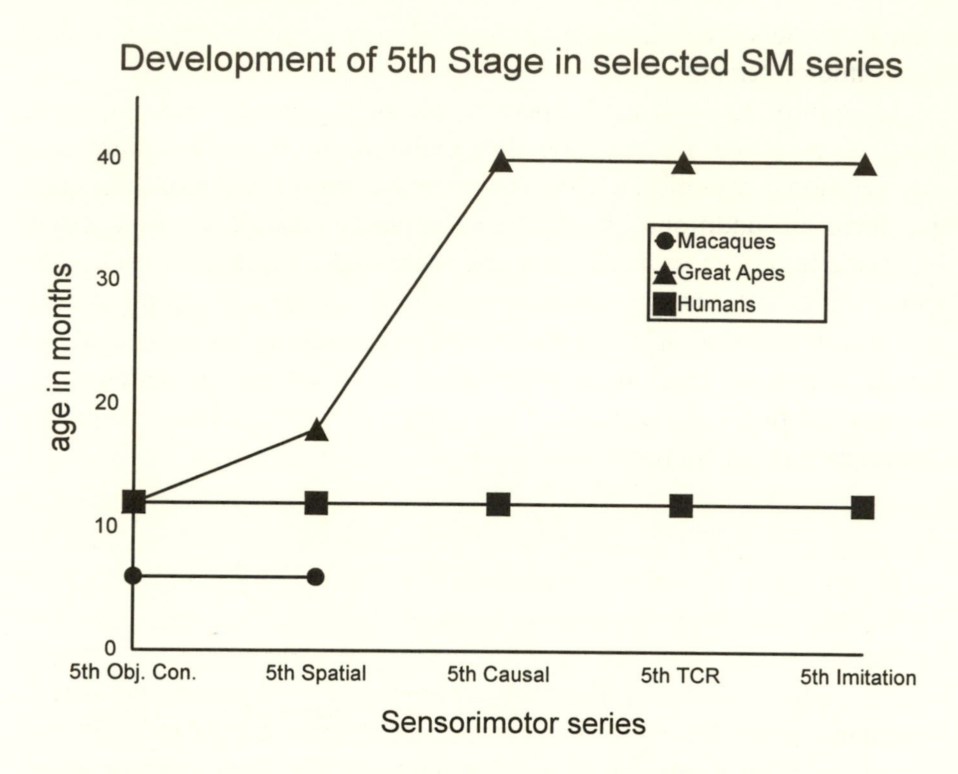

Figure 7. Comparison of fifth stage in selected sensorimotor series.

Although this reconstruction yields no evidence for paedomorphosis (juvenilization or "neoteny") in the evolution of human cognition, there are possible sources of errors in this model. It is, for example, possible that owing to sampling error the pattern that emerges from this analysis of the comparative data is not representative of the actual pattern of evolution (de Queiroz 1985). It is possible, for example, that extinct species of great apes showed different sequences of development and that the absence of these species from our sample obscures the fact that characters were lost through paedomorphosis. Certainly we need more data on cognitive development in the lesser apes and Old World monkeys for a more complete outgroup analysis. Two lines of evidence, however, militate against this possibility. The first is the paleontological evidence from fossil hominids. The second is the epigenetic nature of cognitive development.

As indicated earlier, Piagetian stages and periods of cognitive development — at least within each series or domain — are epigenetic, each higher stage and period being constructed from the schemes of the preceding stage or period (e.g., Piaget 1970). Therefore, the ontogenetic characters identified above are also epigenetic characters: "If a feature *x*, present in members of the taxon *T1*, is epigenetically dependent on feature *y*, present in members of taxon *T2* at any stage of their ontogeny, then the individual members of *T1*, must pass through an ontogenetic stage distinguished by *y*" (Løvtrop 1978, p. 350). Epigenetic characters act as developmental constraints biasing against the production of variant phenotypes (Smith et al. 1985). Developmental constraints and selection both operate in the evolution of ontogeny (McKinney & McNamara 1991). In this paper, my emphasis is on reconstructing the evolution of the patterns of cognitive development rather than on discovering the evolutionary mechanisms that have favored retention and or changes in life history and cognitive development.

If we can exclude simultaneous evolution of new periods of intellectual development and/or loss of later periods by ancestral primates, we can conclude that later periods could not have evolved before earlier periods within the same cognitive domain or series. Simultaneous evolution of late preoperational, concrete operational, and formal operational periods of development is difficult to imagine given the complexity of each period and the added brain size that each depends upon. Loss of later periods of intellectual development is unlikely to have occurred simultaneously in dozens of monkey taxa. If it has occurred in even one, it should be identifiable through comparative studies. In other words, the evidence that Piagetian stages form a phylogenetic series is based on an internal logic as well as outgroup analysis. Paleontological and archeological evidence for hominid evolution is also consistent with the terminal addition model in the sense that fossil hominids constitute a rough morphological series from smaller-brained to larger-brained creatures and that these creatures are associated with a rough artifactual series of simpler to more complex technology. Using data from fossil species to determine character polarity is, of course, another form of the outgroup method.

It is important to note in this context that terminal addition or sequential hypermorphosis with attendant recapitulation has also been identified in the evolution of human characteristics: human life history stages are longer in duration having a later offset than those of great apes and monkeys (McKinney & McNamara 1991); human brains grow larger than those of apes owing to later offset of fetal growth rates (McKinney & McNamara 1991); human leg bones (crucial elements in the bipedal complex) are hypermorphic owing to longer growth due to later offset as compared to the leg bones of chimpanzees (Shea 1988; McKinney & McNamara 1991).

Conclusions

I began with Baldwin's provocative idea that cognitive development in human children recapitulates the evolution of cognition, and proposed to reevaluate Baldwin's lost research program in comparative developmental evolutionary studies of cognition. This brief review of the current status of theory in evolutionary and developmental psychology and of data in human and nonhuman primate psychology demonstrates that the requisite conceptual tools are now available for the kind of comparative developmental evolutionary studies Baldwin envisioned at the turn of the century. Although the comparative data base for such studies is weaker and thinner than the theoretical structures for analyzing that data base, it is now possible — a Century later — to begin to do what Baldwin proposed to do. A preliminary analysis of the available data on cognitive development in macaques, great apes, and humans suggests that Baldwin was right: that in the domain of cognition, ontogeny recapitulates phylogeny. Cognitively, humans are overdeveloped rather than underdeveloped apes.

ACKNOWLEDGMENTS

I want to thank Dr. Emília Martins for inviting me to participate in the symposium at the Animal Behavior Society meetings in Seattle in 1994, and for her helpful comments on my manuscript. I also thank two anonymous reviewers for their helpful comments on the manuscript, and Dr. Michael McKinney for his encouragement in this project. Various weaknesses in the article must be attributed to the author.

References

Antinucci, F. (Ed.). 1989. *Cognitive Structure and Development in Nonhuman Primates*. Hillsdale, New Jersey: Erlbaum.

Baillargeon, R. 1987. Object permanence in 3 ½- to 4 ½ year old infants. *Dev. Psychol.*, 23, 655–664.

Baldwin, J. M. 1894/1906. *Mental Development in the Child and the Race*, 3rd ed. New York: Augustus Kelley (1968 reprint).

Baldwin, J. M. 1897/1902. *Social and Ethical Interpretations in Mental Development*. New York: Macmillian.

Boysen, S. T. In preparation. The development of tool use in gorillas. In: *The Mentality of Gorillas and Orangutans* (Ed. by Parker, S. T., H. L. Miles, & R. W. Mitchell).

Boysen, S. T., & E. J. Capaldi (Eds.). 1993. *The Development of Numerical Competence: Animal and Human Models*. Hillsdale, New Jersey: Erlbaum.

Case, R. 1985. *Intellectual Development from Infancy to Adulthood*. New York: Academic Press.

Chevalier-Skolnikoff, S. 1977. A Piagetian model for describing and comparing socialization in monkey, ape and human infants. In: *Primate Biosocial Development: Biological, Social and Ecological Determinants*. (Ed. by Chevalier-Skolnikoff, S., & F. Poirier), pp. 159–187. New York: Garland.

Chevalier-Skolnikoff, S. 1983. Sensorimotor development in orangutans and other primates. *J. Hum. Evol.*, 12, 545–563.

Chevalier-Skolnikoff, S. 1989. Spontaneous tool use and sensori-motor intelligence in *Cebus* compared with other monkeys and apes. *Behav. Brain Sci.*, 12, 561–588.

Clark, W. L. 1964. *The Fossil Evidence for Human Evolution*. Chicago: University of Chicago Press.

Cronin, J., R. Cann, & V. Sarich. 1980. Molecular evolution and systematics of the genus *Macaca*. In: *The Macaques: Studies in Ecology, Behavior and Evolution* (Ed. by D. G. Lindberg), pp. 31–50. New York: Van Nostrand Reinhold.

Custance, D., & K. Bard. 1994. The comparative and developmental study of self-recognition and imitation. In: *Self-Awareness in Animals and Humans*, pp. 207–227. New York: Cambridge University Press.

Dacey, J., & J. Travers. 1994. *Human Development Across the Life Span*, 2nd ed. New York: Brown & Benchmark.

Dassen, P. R., & A. de Ribaupierre. 1987. Neo-Piagetian theories: Cross-cultural and differential perspectives. *Int. J. Psych.*, 22, 793–832.

de Queiroz, K. 1985. The ontogenetic method for determining character polarity and its relevance to phylogenetic systematics. *Syst. Zool.*, 34, 280–299.

Dore, F. Y. & C. Dumas. 1987. Psychology of animal cognition: Piagetian studies. *Psych. Bull.*, 102, 219–233.

Ekstig, B. 1994. Condensation of developmental stages and evolution. *BioSci.*, 44, 158–164.

Fedigan, L. M. 1982. *Primate Paradigms: Sex Roles and Social Bonds*. Montreal: Eden Press.

Fischer, K. W., & L. Silvern. 1985. Stages and individual differences in cognitive development. *Annu. Rev. Psychol.*, 36, 613–648.

Flavell, J. H., P. Botkin, C. Fry, J. Wright, & P. Jarvis. 1968. *The Development of Role-taking and Communication Skills in Children*. New York: John Wiley and Sons.

Flavell, J., P. Miller & S. Miller. 1993. *Cognitive Development.* 3rd ed. Engelwood, New Jersey: Prentice-Hall.

Fleagle, J. 1988. *Primate Adaptation and Evolution*. New York: Academic Press.

Fox, R. 1971. The human animal. In: *Man and Beast: Comparative Social Behavior* (Ed. by Eisenberg, J., & W. Dillon), pp. 15–48. Washington, D. C.: Smithsonian Institution Press.

Fragaszy, D., & E. Visalberghi. 1990 Social processes affecting the appearance of innovative behaviors in Capuchin monkeys. In: *Adaptation and Adaptability of Capuchin Monkeys* (Ed. by Fragaszy, D., J. Robinson, & E. Visalberghi). *Folia Primatol.*, 54, 155–165.

Gallup, G. G., Jr. 1982. Self-awareness and the emergence of mind in primates. *Am. J. Primatol.*, 2, 237–248.

Gardner, R., & B. Gardner. 1969. Teaching sign language to a chimpanzee. *Science*, 165, 664–672.

Gardner, R., & B. Gardner. 1975. Early signs of language in child and chimpanzee. *Science*, 187,752–753.

Gardner, A., B. Gardner, & van Cantfort, T. 1989. *Teaching Sign Language to Chimpanzees*. New York: SUNY Press.

Gelman, R., & C. R. Gallistel. 1978. *The Child's Understanding of Number*. Cambridge, Massachusetts: Harvard University Press.

Gomez, J. C. 1990. The emergence of intentional communication as a problem-solving strategy in the gorilla. In: *"Language" and Intelligence in Monkeys and Apes*. (Ed. by S. T. Parker, & K. R. Gibson), pp. 333–355. New York: Cambridge University Press.

Gould, S. J. 1977. *Phylogeny and Ontogeny*. Cambridge, Massachusetts: Harvard University Press.

Hall, G. S. 1897. A study of fears. *Am. J. Psychol.*, 8, 147–249.

Hall, G. S. 1904. *Adolescence, Its Psychology and Its Relations to Physiology, Anthropology, Sociology, Sex, Crime, Religion and Education*, vols. 1 & 2. New York: Appleton.

Hallock, M., & J. Worobey. 1984. Cognitive development in chimpanzee infants. *J. Hum. Evol.*, 13, 441–447.

Harvey, P., R. D. Martin, & T. Clutton-Brock. 1986. Life histories in comparative perspective. In: *Primate Societies* (Ed. by Smuts, B., D. Cheney, R. Seyfarth, R. Wrangham, & T. Struhsaker), pp. 181–196. Chicago: University of Chicago Press.

Hayes, K., & C. Hayes. 1952. Imitation in a home-raised chimpanzee. *J. Comp. Physiol. Psychol.*, 45, 450–459.

Hennig, W. 1966, 1979. *Phylogenetic Systematics*. Champaign, Illinois: University of Illinois Press.

Hull, D. L. 1988. *Science as a Process*. Chicago: University of Chicago Press.

Inhelder, B., & J. Piaget. 1964. *The Early Growth of Logic in the Child.* New York: Norton.

Irwin, R. 1988. The evolutionary importance of behavioural development: The ontogeny and phylogeny of bird song. *Anim. Behav.*, 36, 814–824.

Jensvold, M. L. A., & R. S. Fouts. 1994. Imaginary play in chimpanzees (*Pan troglodytes*). *Hum. Evol.*, 8, 217–227.

Kessen, W. 1965. *The Child.* New York: John Wiley and Sons.

Kluge, A., & R. Strauss. 1985 Ontogeny and systematics. *Annu. Rev. Ecol. Syst.*, 16, 247–268.

Kohlberg, L. 1969. Stage and sequence: The cognitive-developmental approach to socialization. In: *Handbook of Socialization Theory and Research* (Ed. by A. A. Goslin). Skokie, Illinois: Rand McNally.

Kohler, W. 1917, 1927. *The Mentality of Apes*. New York: Vintage Press.

Langer, J. 1980. *The Origins of Logic: Six to Twelve Months*. New York: Academic Press.

Langer, J. 1986. *The Origins of Logic: One to Two Years*. New York: Academic Press.

Langer, J. 1989. Comparison with the human child. In: *Cognitive Structure and Development in Nonhuman Primates* (Ed. by F. Antinucci), pp. 229–243. Hillsdale, New Jersey: Erlbaum.

Langer, J. 1993. Comparative cognitive development. In: *Tools, Language and Cognition in Human Evolution* (Ed. by K. R. Gibson, & T. Ingold), pp. 300–313. New York: Cambridge University Press.

Lethmate, J. 1977. Problemlose Verhalten von Orang-utans (*Pongo pygmaeus*). *Fortschritte der Verhaltensforschung, Beihefte zur Zeitschrift fur Tierspsychologie*. No. 19. Berlin: Verlag Paul Parey.

Lewin, R. 1993. *Human Evolution: An Illustrated Introduction*, 3rd ed. New York: Blackwell Scientific Publications.

Lorenz, K. 1950. The comparative method in studying innate behavior patterns. *Symp. Soc. Exp. Biol.*, 4, 221–268.

Lovejoy, O. 1981. Human origins. *Science*, 211, 341–350.

Løvtrop, S. 1978. On von Baerian and Haeckelian recapitulation. *Syst. Zool.*, 27, 348–352.

Mabee, P., & J. Humphries. 1993. Coding polymorphic data: Examples from allozymes and ontogeny. *Syst. Biol.*, 42, 166–181.

Martin, R. E., & A. M. MacLarnon. 1990. Reproductive patterns in primates and other mammals: The dichotomy between altricial and precocial offspring. In: *Primate Life History and Evolution* (Ed. by C. J. DeRousseau), pp. 47–79. New York: Wiley-Liss.

Mayr, E., E. Linley, & R. Usinger. 1953. *Methods and Principles of Systematic Zoology*. New York: McGraw-Hill.

McGrew, W. 1992. *Chimpanzee Material Culture*. Cambridge, England: Cambridge University Press.

McKinney, M., & K. McNamara. 1991. *Heterochrony: The Evolution of Ontogeny*. New York: Plenum Press.

McLennan, D. A., D. R. Books, & J. D. McPhail. 1988 The benefits of communication between comparative ethology and phylogenetic systematics: A case study using gasterosteid fishes. *Can. J. Zool.*, 66, pp. 2177–2190.

Mellars, P., & C. Stringer (Eds.). 1989. *The Human Revolution*. Princeton, New Jersey: Princeton University Press.

Meltzoff, A. N. 1988. The human infant as *Homo imitans*. In: *Social Learning: Psychological and Biological Perspectives* (Ed. by T. R. Zentall, & B. G. Galef, Jr.) pp. 319–341. Hillsdale: Erlbaum.

Meltzoff, A., & M. Moore. 1977. Imitation of facial and manual gestures by neonates. *Science*, 198, 75–78.

Miller, P. H. 1993. *Theories of Developmental Psychology*, 3rd ed. New York: Freeman.

Miles, H. L. 1990. The cognitive foundations for reference in a signing orangutan. In: *"Language" and Intelligence in Monkeys and Apes* (Ed. by S. T. Parker, & K. R. Gibson), pp. 511–538. New York: Cambridge University Press.

Miles, H. L. 1994. "ME CHANTEK": The development of self-awareness in a signing orangutan. In: *Self-Awareness in Animals and Humans* (Ed. by S. T. Parker, R. W. Mitchell, & M. L. Boccia), pp. 254–272. New York: Cambridge University Press.

Mitchell, R. W. 1987. A comparative developmental approach to understanding imitation. In: *Perspectives in Ethology*, Vol. 7 (Ed. by P. Klopfer, & P. Bateson), pp. 183–215. New York: Plenum Press.

Montagu, A. 1989. *Growing Young*, 2nd ed. Granby, Massachusetts: Bergin & Garvery.

Parker, S. T. 1977. Piaget's sensorimotor series in an infant macaque: A model for comparing unstereotyped behavior and intelligence in human and nonhuman primates. In: *Primate Biosocial Behavior* (Ed. by S. Chevalier-Skolnikoff, and F. Poirier), pp. 43–113. New York: Garland.

Parker, S. T. 1990. The origins of comparative developmental evolutionary studies of primate mental abilities. In: *"Language" and Intelligence in Monkeys and Apes* (Ed. by S. T. Parker, & K. R. Gibson), pp. 3–63. New York: Cambridge University Press.

Parker, S. T. (in press). Apprenticeship in extractive foraging in great apes: The origins of imitation, teaching and self-awareness. In: *Reaching into Thought* (Ed. by A. Russon, K. Bard, & S. T. Parker). Cambridge, England: Cambridge University Press.

Parker, S. T., & K. R. Gibson. 1979. A developmental model for the evolution of language and intelligence in early hominids. *Behav. Brain Sci.*, 2, 367–408.

Parker, S. T., & C. Milbrath. 1993. Higher intelligence, propositional language, and culture as adaptations for planning. In: *Tools, Language, and Cognition in Human Evolution.* (Ed. by K. R. Gibson, & T. Ingold), pp. 314–344. Cambridge, England: Cambridge University Press.

Parker, S. T., R. W. Mitchell, & M. L. Boccia (Eds.). 1994. *Self-Awareness in Animals and Humans*. Cambridge, England: Cambridge University Press.

Patterson, F. 1980. Innovative uses of language by a gorilla: A case study. In: *Children's Language*, Vol. 2 (Ed. by K. E. Nelson), pp. 497–561. New York: Gardner Press.

Patterson, F., & E. Linden. 1981. *The Education of Koko*. New York: Holt, Rinehart & Winston.

Piaget, J. 1952. *The Origins of Intelligence in Children*. New York: Norton.

Piaget, J. 1954. *The Construction of Reality in the Child.* New York: Basic Books.
Piaget, J. 1962. *Play, Dreams and Imitation in Childhood.* New York: Norton.
Piaget, J. 1965. *The Moral Judgment of the Child.* New York: Free-Press.
Piaget, J. 1970. *Genetic Epistemology.* New York: Norton.
Piaget, J., & B. Inhelder. 1969. *The Psychology of the Child.* New York: Basic Books.
Pilbeam, D., & S. J. Gould. 1974. Size and scaling in human evolution. *Science*, 186, 892-901.
Potí, P., & G. Spinozzi. 1994. Early sensorimotor development in chimpanzees (*Pan troglodytes*). *J. Comp. Psych.*.
Povinelli, D., K. E. Nelson, & S. T. Boysen. 1992. Comprehension of social role reversal by chimpanzees: Evidence of empathy? *Anim. Behav.*, 44, 269–281.
Povinelli, D. 1993. Reconstructing the evolution of mind. *Am. Psychol.*, 48, 493–509.
Povinelli, D., A. Rulf, & D. Bierschwale. 1994. Absence of knowledge attribution and self-recognition in young chimpanzees (*Pan troglodytes*). *J. Comp. Psychol.*, 108, 74–80.
Premack, D. 1978. *Intelligence in Ape and Man.* Hillsdale, New Jersey: Erlbaum.
Premack, D., & G. Woodruff. 1978. Does the chimpanzee have a theory of mind? *Behav. Brain Sci.*, 1, 515–526.
Relethford, J. 1994. *The Human Species*, 2nd ed. New York: Mountain View, CA: Mayfield.
Richards, A. 1985. *Primates in Nature.* New York: Freeman.
Richards, R. J. 1992. *The Meaning of Evolution: The Morphological Construction and Ideological Reconstruction of Darwin's Theory.* Chicago: University of Chicago Press.
Romanes, G. 1882. *Animal Intelligence.* London: Kegan, Paul & French.
Rumbaugh, D. & J. Pate. 1984 The evolution a primate cognition: A comparative study of monkeys and apes. In: *Animal Cognition* (Ed. by T. Roitblatt, G. Bever, & H. S. Terrace), Hillsdale, New Jersey: Erlbaum.
Russon, A., & B. Galdikas. 1993. Imitation in free-ranging rehabilitant orangutans: Learning observationally about observational learning. *J. Comp. Psychol.*, 107, 147–160.
Russon, A., & B. Galdikas. 1994. Imitation in free-ranging rehabilitant orangutans: Model and action selectivity. *J. Comp. Psychol.*, 108.
Sarich, V., & J. Cronin. 1976. Molecular systematics of the primates. In: *Molecular Anthropology* (Ed. by M. Goodman, & R. Tashian). New York: Plenum Press.
Schick, K., & N. Toth. 1993. *Making Silent Stone Speak.* New York: Touchstone.
Schultz, A. 1969. *The Life of Primates.* New York: Universe Books.
Shea, B. 1988 Heterochrony in primates. In: *Heterochrony* (Ed. by M. McKinney). New York: W. L. Fink.
Shea, B. 1992. Developmental perspective on size change and allometry in evolution. *Evol. Anthropol.*, 1, 125–133.
Siegler, R. S. 1986. *Children's Thinking.* Englewood Cliffs, New Jersey: Prentice-Hall.
Simpson, G. G. 1961. *The Principles of Animal Taxonomy.* New York: Columbia University Press.

Smith, H. 1992. Life history and the evolution of human maturation. *Evol. Anthropol.*, 1, 134–142.

Smith, J. M., R. Burian, S. Kaufman, P. Alberch, J. Campbell, B. Goodwin, R. Lande, D. Raup, & L. Wolper. 1985. Developmental constraints and evolution. *Q. Rev. Biol.*, 60, 265–289.

Spinozzi, G. 1993. Development of spontaneous classificatory behavior in chimpanzees. *J. Comp. Psychol.*, 7, 193–200.

Stanley, S. M. 1973. An explanation for Cope's rule. *Evolution*, 27, 1–26.

Stanley, S. 1992. An ecological theory for the origin of *Homo*. *Paleobiol.*, 18, 237–257.

Stearns, S. 1992. *The Evolution of Life Histories*. Oxford, England: Oxford University Press.

Stein, P., & B. Rowe. 1993. *Physical Anthropology*, 5th ed. New York: McGraw Hill.

Stringer, C., & C. Gamble. 1993. *In Search of the Neanderthals*. New York: Thames and Hudson.

Toobey, J., & L. Cosmides. 1992. The psychological foundations of culture. In: *The Adapted Mind.* (Ed. by J. Barkow, L. Cosmides, & J. Toobey), pp. 19–136. Oxford, England: Oxford University Press.

Uzgiris, I. & J. McV. Hunt. 1975. *Assessment in Infancy: Ordinal Scales of Development*. Urbana, Illinois: University of Illinois Press.

Werner, H., & B. Kaplan. 1963. *Symbol Formation*. Hillsdale, New Jersey: Erlbaum.

Whiten, A. (Ed.). 1991. *Natural Theories of Mind*. London: Blackwell.

Weiss, M. 1987. Nucleic acid evidence bearing on hominoid relationships. *Yearbook Phys. Anthropol.*, 30, 41–74.

Wiley, E. O. 1981. *Phylogenetics: The Theory and Practice of Phylogenetic Systematics*. New York: John Wiley & Sons.

Wenzel, J. W. 1993. Application of the biogenetic law to behavioral ontogeny: A test using nest architecture in paper wasps. *J. Evol. Biol.*, 6, 229–247.

Wrangham, R. 1986. The evolution of social structure. In: *Primate Societies* (Ed. by B. Smuts, D. Cheney, R. Seyfarth, R. Wrangham, & T. Struhsaker), pp. 282–296. Chicago: University of Chicago Press.

Wynn, T. 1989. *The Evolution of Spatial Competence*. Champaign, Illinois: University of Illinois Press.

Yerkes, R. 1916. *The Mental Life of Monkeys and Apes*. New York: Holt.

CHAPTER 13

Why Phylogenies Are Necessary for Comparative Analysis

Sean Nee, Andrew F. Read and Paul H. Harvey

Constraints of time and space have meant that astronomy has developed largely as a nonexperimental science. Nevertheless, observations backed by appropriate theory have led to some remarkable predictions, such as the existence of the planet Neptune. Nowadays, evolutionary biologists face only one of the astronomer's constraints, that of time. It is frequently impractical to do experiments that will test ideas about why organisms have come to be as they are. But we can make observations on living organisms, and the patterns we find can often be used to test ideas about process. The comparative method, or at least that part of it which has been written about widely in recent years, uses information on character covariation among organisms to draw inferences about the reasons for character variation having arisen.

There are several reasons why the comparative method has become so fashionable. One is the rapid accumulation of data: methods are usually not developed until they are needed. When it became clear, for example, that a reasonably complete coverage of the ecological habits and the social behavior of almost all the world's living primate species had accumulated, it was reasonable to seek to develop methods that would allow primatologists to understand how variation in primate behavior can be explained by ecology. A second reason for the surge of interest in comparative studies is that phylogenetic relationships among extant organisms are being estimated with increasing precision, largely as a result of advances in molecular biology. But why is it that phylogenetic relationships are so central to pursuing the comparative program? That question can be answered in several ways, but the message is always the same: *improved phylogenetic information allows stronger inferences to be made from comparative studies.*

Our aim in this concluding chapter is to summarize the theme of this book by explaining why and, in the most general terms, how, phylogenetic information is incorporated into comparative analyses. There have been several reviews of comparative methodology in recent years and we do not intend to go over the same ground again. Instead, we shall use a somewhat different perspective to approach the problem of how inferences can be drawn from comparative data. Our argument will develop an experimental analogy which, we find, allows useful insights into what we can expect our comparative analyses to reveal and why phylogenetic information is so valuable. We emphasize two points throughout. First, phylogeny is a tool for inference, not a problem. The second point, which is more contentious, arises in the context of using comparative evidence to evaluate a hypothesis, such as something to do with the sex chromosomes is responsible for Haldane's rule. In this context, a particular model of evolution, although fully supported by the data, may not be useful for inference and one can develop a useful model for inference which is not, itself, a model of evolution.

Implicit in this second point is an important restriction on the scope of this chapter. We do not discuss, even indirectly, analyses which have as their purpose the inference of microevolutionary parameters and processes from comparative data. For example, one could assume that characters evolve under random genetic drift and perform an analysis based on Felsenstein's (1985) method to make an inference about the genetic variances and covariances of the characters studied. Or one could use comparative data to compare the performance of such a random drift model of microevolution with a model of stabilizing selection (Martins 1994). Such analyses, while important and becoming increasingly popular, fall outside our range and are discussed elsewhere in this volume (see, in this volume, Martins & Hansen, Chap. 2, and Gittleman et al. Chap. 6).

As will be evident throughout this chapter, detecting evidence for correlated evolutionary change between two characters or character states is a different issue from understanding the reasons for that correlated change when these correlations are sought in the context of hypotheses that one character, or ecological circumstance, directly influences the evolution of another. Even if two characters have changed in concert many times over evolutionary time in a manner consistent with the hypothesized causal argument, we can never be certain that the hypothesis is correct. In particular, it will always be possible that an undetected third variable has influenced change in each of the two being considered. However, sound statistical practice combined with plausibility arguments (often supported by analytical or numerical mechanical, engineering, or life history models), help to reduce the possibilities of incorrect inference. Also, one's hypothesis will make predictions about other variables which can be explored, which may or may not buttress the hypothesis in question.

Inference from non-experimental data

The fundamental problems facing comparative tests of the "*x* is responsible for *y*" type of hypotheses are often discussed in isolation from more general problems of inference. However, they are no different from the general problem of drawing inferences from evidence which has not been generated by a well designed experiment. In this more general context it becomes clear that phylogeny is not a "problem" for comparative analyses; it is, in fact, a basis of solutions to universal problems of inference from nonexperimental data (or data from a bad experiment), and we are fortunate to have it. In fact, there is no fully satisfactory solution to these problems except to design experiments, which we are usually unable to do.

So let us describe the problem faced by comparative analyses in a completely different context and see how an analog of phylogenetic hierarchical structuring of the data can be of help. Consider the hypothesis that the administration of laetrile improves the chances of recovery from cancer. The hypothetical evidence that we have pertaining to this hypothesis is that recovery rates are higher in clinics that use laetrile than those which do not and, all our commonly used statistical analyses tell us that this effect is "significant". So we make a statement of the form "either laetrile is connected to cancer recovery, or something highly improbable has happened." However, we should not be comfortable with this conclusion for the following reason. We know that factors other than laetrile might influence recovery, but we do not know what all of these are or how they are distributed amongst clinics with respect to laetrile prescription. Our discomfort would receive further support if someone informed us that all the clinics that use laetrile are in Mexico whereas those that do not are in the United States. We would no longer feel justified in drawing any inference about the efficacy of laetrile. The better recovery rates in Mexico could just as reasonably be owing to differences between Mexican and American doctors, or, on the basis of this evidence alone, even the cuisine. Any unstudied variable affecting recovery rates which differs between Mexico and the United States will result in the recovery rate observations being correlated. In this sense, the recovery rates for the clinics are not "independent", as our previous naive analyses had all assumed.

Another concern is that, indeed, laetrile may have an effect on recovery, but the effect is to reduce recovery chances, and our process of inference has fallen victim to Simpson's paradox (DeGroot 1989), usually discussed in the context of the dangers of aggregated data. DeGroot's example of the paradox is as follows. Forty subjects receive a new treatment for a disease and 40 receive the standard treatment. Of those receiving the new treatment, 50% are cured, whereas 60% of those receiving the standard treatment are cured. So the standard treatment looks better than the new one. But when we analyze the results for men and women separately, the opposite picture emerges. This is because women have a higher recovery rate than men regardless of the treatment, and that more men than women were in the group receiving the new treatment. So the recovery data were not "independent" because some uncontrolled variable associated with sex differed between the two groups, and, as a consequence, the subjects in the standard treatment group were more likely to recover, regardless of what treatment they received. (An example of this paradox in comparative biology can be found in Nee et al. (1991), where the relationship between abundance and body size within taxa is the opposite to that across species.) Such possibilities do not seem at all improbable to us when we look at our nonexperimental data on laetrile.

In this case something extraneous to the subject under investigation, the differences in law or social custom between two countries which may, for example, have resulted in more men than women being treated with laetrile, has produced what we would expect to be highly correlated systematic differences in factors which may indeed be relevant to the question. But the situation would not necessarily be improved if, in fact, laetrile use appeared random with respect to geographic location. The problem still remains that we know that there are other factors involved in cancer recovery rates, apart from laetrile use, but we do not know how they are distributed among the clinics. The sex difference that gave rise to Simpson's paradox in the above example might easily have arisen by chance.

So, although the variable we are looking at is laetrile use, we know that there are many other variables that may be involved in influencing recovery from cancer. Typically, these will be unknown or unmeasured. In such circumstances, in order to draw an inference about laetrile per se, we would like all these other variables to be held constant, so they are no longer variables. This is not the case with our data as it stands, hence our problem. In reality, of course, we would design an experiment to satisfy ourselves that the assumption that this situation prevails is not unreasonable, but let us continue to suppose that the experimental route is ruled out.

Statistical analysis of non-experimental data

If laetrile prescription is treated as a continuous variable (so clinics vary in the amount of laetrile they prescribe), and we observe a "significant" positive regression of recovery on the amount of laetrile prescribed, the same problem for inference remains, of course. We can easily imagine the following sort of situation. There is a geographical cultural gradient of conservatism which affects willingness to prescribe laetrile, and a geographical gradient in the values of any or all of the other relevant factors. This will produce a spurious significant effect of laetrile prescription on recovery rate.

We can rephrase the difficulty in a more statistical way, following Grafen (1989). If we intend to use a linear regression model to evaluate laetrile's efficacy, we need to make an assumption about the covariance structure of those "errors" arising not primarily from measurement error, but from all the unstudied variables which affect recovery rate. Usually, we design an experiment in such a way that we are happy to assume that the error terms are uncorrelated (i.e., that the observations, here recovery rates, are "independent"), and carry out the typical linear regression. But with nonexperimental data, we are entirely unwilling to do this.

Unable to construct an experiment designed to make the assumption of uncorrelated error terms reasonable, we are unwilling to use our simple regression model for inference. The data themselves cannot inform us of the adequacy of the statistical model as a basis for rational inference — by looking at the distribution of residuals, for example. If we simply had a list of *x*/*y* data, all our usual tests (for homoscedasticity, normally distributed residuals, and so on) might be passed with flying colors. We are simply fortunate enough to have been informed of the Mexico/United States distinction, or some continental gradient which put us on our guard. But even if such a distinction did not exist, we still cannot blithely assume uncorrelated error terms. However, suppose that we, somehow, could generate a statistical model which specified the variance-covariance matrix of the errors. Then we could perform a generalization of the usual regression analysis on our data, one which did not suppose that the matrix had zero covariance terms. This is one approach taken by Grafen (1989), and its analog in the present context is as follows.

Statistical inferences from hierarchically structured nonexperimental data

We are well aware that many variables display a hierarchical pattern of correlation, which is reflected in our hierarchical geographical categories: neighborhoods, cities, counties, states, and nations. (Modern mobility and communication is rapidly destroying this correlation, so the reader is asked to think back to an earlier time for us to continue with the analogy.) In all sorts of ways, institutions in Manhattan are more similar to each other than they are to institutions in the Bronx. Similarly, institutions in New York are more similar to each other than they are to institutions in San Francisco. And so on. The reasons for this hierarchical structure do not actually matter to us. We may now choose to exploit this structure by assuming that all the other factors which we do not know about, or do not have any data for, also display this structure. We are now in a position, in principle, to construct a model in which the covariance in the error of fitting recovery rate to degree of laetrile use between any two clinics is a function of where in the geographical hierarchy these two clinics co-occur. Then we simply perform a generalization of our usual regression analysis.

Another way to exploit geography is as follows. We observe that there are two clinics in many Mexican towns, and that in some cases laetrile is used in one clinic but not in another. So there are two clinics in Acapulco, two clinics in Chihuahua, two clinics in Tijuana, and so on, and, in each pair, one uses laetrile and one does not. We imagine that if we analyze, for each pair, the difference between laetrile use and the difference in recovery rate, then we may have an approximation to an ideal situation in which, for each pair, the multitude of other factors which may be involved are constant. We would then perform whatever paired comparison statistical test for an association between two variables that we felt appropriate. We know that the other factors are not constant, but this is the best we can do. If we got a "significant" result, we might then feel justified in making the following inference: either there is a relationship between laetrile use and recovery rate, or something improbable has happened.

However, we have still not actually solved the 'third variable' problem. It is still possible that some mystery variable, z, is determining the states of both x and y and that x and y are only related via z. So there is indeed a relationship between x and y, but it is indirect. For example, young and highly motivated doctors may tend to recognize the early signs of cancer when it is more effectively treated, and those same doctors may be the ones who are more likely to experiment with new drugs such as laetrile, as well as applying more conventional treatments. Constancy of such relevant variables as doctor's age and clinical training is what we would like, but we know it may not be possible. There is nothing we can do about this. Even if we set up an experiment, we can never be absolutely sure that we have achieved constancy of all possibly relevant variables, since there may always be a "mystery variable". One of the more interesting mystery variables that was discovered in medical experimentation is whether or not the doctor knows who is getting a drug and who is getting a placebo.

Leroi et al. (1994) describe quite correctly a suite of "third variable" concerns for comparative biology that could only be alleviated if one knew the genetic variance-covariance matrices for the characters. Determining such matrices before performing any comparative study is clearly an unrealistic general prescription, as the authors admit. Nevertheless, their paper does highlight the need to interpret data at the level at which it is available.

In summary, then, we feel greater confidence in drawing inferences about the relationship (possibly indirect) between two variables if (a) we can specify the covariances amongst the errors or (b) we can study contrasts. If we do not feel that we can do either, then we simply have to accept that the data are not useful. We are always aware of the third variable problem. In comparative biology, we are fortunate to have phylogenies. These allow us to specify paired comparisons (contrasts) or to construct phylogenetic models of the covariance matrices of the errors. Elaborations and combinations of these ideas are beyond our present scope.

Phylogenies as natural experiments

Another, closely related way of thinking about the role of phylogenies in comparative analysis is to think of them as providing clues as to the experimental protocol followed in a huge experiment, which was also, probably, very poorly designed. In that experiment, which started from a single species, species gave rise to species similar to themselves but, when environments changed, so did the characteristics of species. However, when environments changed, the current phenotype of the species also influenced its direction of phenotypic change. (Under an increased threat of predation, big species might get larger to better fight off predators, while small species might get smaller to better bolt down burrows into which predators cannot fit.) Some part of the similarity among species will, therefore, result from common ancestry endowing species with features which channel their subsequent evolution along similar paths, and some part of the similarity will result from living in similar environments. At one time it was fashionable to partition variance in traits among taxa to "phylogenetic" effects, considered to be inertial and constraining, versus "nonphylogenetic" effects, reflecting adaptive flexibility. It has increasingly become clear that such a distinction is not useful, in part because members of a taxon share a trait does not mean that the trait is in some sense non-adaptive and merely represents phylogenetic baggage. Returning to the "one huge experiment" analogy, what phylogeny does for us is to provide information on details of the experimental protocol (albeit something of a haphazard one in this case) about which we would otherwise be totally ignorant. We shall not pursue this particular analogy further because it is detailed elsewhere (Harvey & Pagel 1991, pp. 39–40).

Models of evolution and models of inference

It is not our intention here to review, yet again, the various models that have been suggested for comparative data analysis. Rather, we want to emphasize the point that a model of evolution is not necessarily useful as a model for the sort of inference we are discussing in this chapter. Consider Fig. 1, which shows the phylogenetic distribution of two discrete variables, solid/dotted (*s/d*) and thin/fat (*t/f*). We assume that this, true, phylogeny is a star phylogeny to simplify the discussion. One evolutionary model for the variable thin/fat is that each lineage has an independent and equal probability of evolving the state fat. We might then estimate this probability, say p_f. We could have a similar model for the variable solid/dotted and estimate the probability of evolving dotted, p_d. (In this case, since there are eight taxa, a reasonable estimate for p_d might be 1/8.) The illustrated data are perfectly compatible with this model.

Now suppose we have a hypothesis that the character whose states are fat/thin determines the state of the character solid/dotted, so the hypothesis is that "fat is responsible for dotted". Under the null hypothesis that the character fat/thin is unrelated to the character solid/dotted, the probability that the fat branch displays dotted while the others display solid is $p_d(1-p_d)^7$. If this probability is less than 0.05 (as it happens to be here), then either there is a relationship between the two variables, or something very improbable has happened. In other words, we reject the null hypothesis in favor of our alternative. Precisely this reasoning has been used to establish the "significance" of Haldane's rule

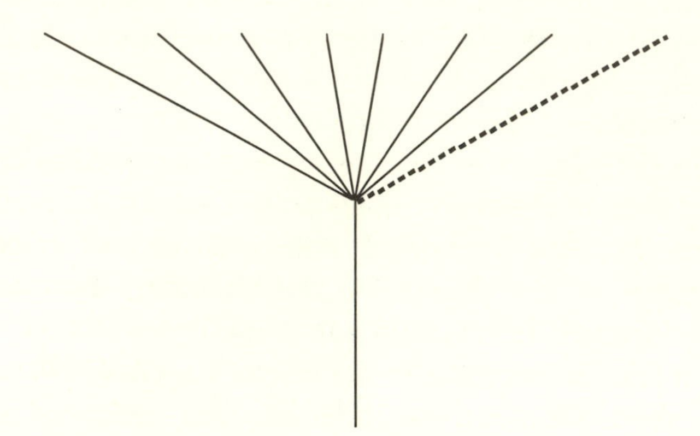

Figure 1. Eight taxa related by a star phylogeny, reflecting the instantaneous fission of one lineage into eight. The distribution of two characters is illustrated by the lines of the phylogeny. One character has two states, "fat" or "thin." The other has the states "solid" or "dotted." Seven lineages are thin and solid, while one lineage is fat and dotted.

(Brookfield 1993). But it is no more sensible than concluding that laetrile is effective if, in our list of 100 clinics, we had only one clinic that used laetrile and that clinic also had the highest recovery rate (although this might attract our attention and lead us to search for other sources of information). Indeed, the very fact that only one clinic uses laetrile suggests that that clinic probably differs in numerous other ways (perhaps it had the youngest, most motivated doctor). Any unique feature, in addition to fat, that the lineage possesses could account for dotted. Similarly, the other lineages, like the other clinics, might all share unstudied features, rendering the assumption that they provide independent information a dubious one as a basis for inference.

We could construct far more sophisticated tests: we could, for example, use a likelihood ratio test to compare the performances of two models, one assuming independent evolution of the characters and one assuming coevolution, and imbue these models with however much complexity we like. We could, for example, choose to model the evolution of the characters in the same way that gene sequence evolution has been modeled (Pagel 1994). We would decide that the coevolutionary model is a better model for the data. However, if we used this to make an inference about a causal relationship between the two traits we would have stepped beyond the bounds of credulity. So a model of evolution, no matter how well supported by the data, is not necessarily a reasonable model for inference.

The above models, and others like them, were designed to deal with the problem that species character values do not satisfy the assumption of independence, required by typical statistical tests of hypotheses. Viewing phylogeny as the reason for this problem, they explicitly incorporated phylogenetic information into the analysis. But they do this by assuming that branches or even, in the case of Pagel's (1994) method, infinitesimally short segments of branches are independent. The problem is that the assumption of independence has simply been swept up the phylogenetic tree (Read & Nee in press).

For this particular example we chose a phylogeny which is so simplified that there does not even appear to be a "problem of shared phylogeny" (unlike Fig. 2 below), and it illustrates starkly that phylogeny is a tool rather than a problem. Although there is now no "problem" of shared phylogeny to be dealt with (recall that this is the true phylogeny, revealed to us in some mysterious way), one cannot simply do a cross-species analysis treating the species as independent. There is no justification for supposing uncorrelated errors; this is still

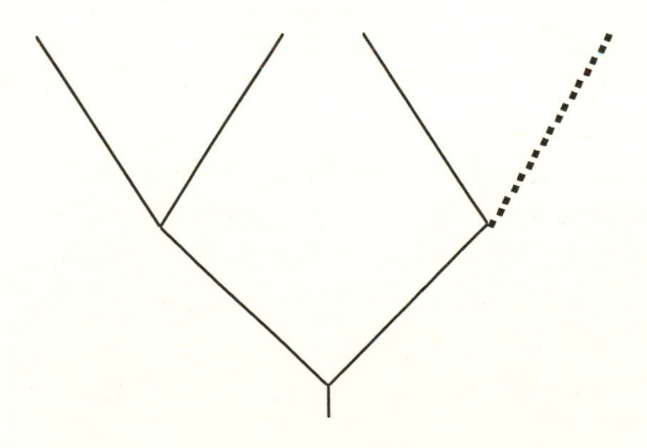

Figure 2. As Fig. 1, except that four taxa are related by a bifurcating phylogeny instead of a star. Each tip may represent a monophlyetic group containing many species, all of whose members display the characters in question.

nonexperimental data. Just as we might discover that the clinics which prescribe laetrile are all in Mexico, so too we might discover (or should fear the discovery) that half the animals in the phylogeny have evolved a typical mammallike morphology and lifestyle and the other half are more birdlike, for example. In the absence of phylogenetic structuring of the data, we are in no better position to make inferences about causal relationships than with any other non-experimental data. This does not necessarily mean that we cannot make any inferences — just that we do not have phylogeny to assist us. For example, we could perform elegant engineering analyses showing how wing size and shape in the flying organisms are well suited to the needs of the individual flying species. That is a valid scientific exercise in itself.

So a model of evolution does not necessarily make a good model for inference. Conversely, a model used for inference does not necessarily make many assumptions about how evolution has occurred. Consider the following model used for inference in the case of discrete variables (Read & Nee 1991), in the context of the bifurcating phylogeny of Fig. 2 (to simplify the discussion): We confine our attention to the node which exhibits variation in both traits, feeling that the lineages on either side are likely to be similar with respect to other variables which may be important. Similarly, in an experiment with humans, one might like to restrict the subjects to men of a certain age (to avoid problems like Simpson's paradox) or to the paired Mexican clinics. One does not suppose, then, that the subjects are identical, just as close as one can get given the shortage of identical twins. We argue that if the variable solid/dotted is unrelated to fat/thin, then the dotted state is as likely to occur on the fat as the thin branch. So the probability of the observation under this null hypothesis is 0.5. We require more nodes with variation in both traits on either side, to make any further progress. We are making rather few assumptions about evolution. We have restricted our attention to a single node not because we have some evolutionary model in mind but because of our general concerns about weak or fallacious inference.

It might seem that this restriction renders this approach inferior to methods which use all the data. While such an opinion seems reasonable when one is just looking at abstract lines on a page, it becomes untenable when one puts the issue into its proper context of inference from nonexperimental data. Similarly, with continuous characters, an investigator may reasonably decide simply to use contrasts between pairs of species and ignore information from higher nodes so long as the variance in character states among the pairs of species is high enough to give reasonable power to the statistical test used.

ACKNOWLEDGMENTS

We thank the BBSRC for postdoctoral fellowship and grant support (GR/H 53655) and Emília Martins for illuminating comments on a draft of this chapter

References

Brookfield, J. 1993. Haldane's rule is significant. *Evolution,* 47, 1885–1888.

DeGroot, M. H. 1989. *Probability and Statistics*. Tokyo: Addison-Wesley.

Grafen, A. 1989 The phylogenetic regression. *Phil. Trans. Roy. Soc. (Lond.) (B),* 326, 119–157.

Harvey, P. H., & M. D. Pagel. 1991. *The Comparative Method in Evolutionary Biology*. Oxford, England: Oxford University Press.

Leroi, A. M., M. R. Rose, & G. V. Lauder. 1994. What does the comparative method reveal about adaptation? *Am. Nat.,* 143, 381–402.

Nee, S., A. F. Read, & P. H. Harvey. 1991. The relationship between abundance and body size in British birds. *Nature,* 351, 312–313.

Pagel, M. D. 1994. Detecting correlated evolution on phylogenies: a general method for the comparative analysis of discrete characters. *Proc. R. Soc. Lond. B,* 255, 37–45.

Read, A. F. & S. Nee. 1991. Is Haldane's Rule significant? *Evolution,* 45, 1707–1709.

Read, A. F. & S. Nee. (in press). Inference from binary comparative data. *J. Theor. Biol.*

Subject Index